ENVIRONMENTAL BIOTECHNOLOGY

Biodegradation, Bioremediation, and Bioconversion
of Xenobiotics for Sustainable Development

ENVIRONMENTAL BIOTECHNOLOGY

Biodegradation, Bioremediation, and Bioconversion
of Xenobiotics for Sustainable Development

Edited by
Jeyabalan Sangeetha, PhD
Devarajan Thangadurai, PhD
Muniswamy David, PhD
Mohd Azmuddin Abdullah, PhD

APPLE
ACADEMIC
PRESS

Apple Academic Press Inc. | Apple Academic Press Inc.
3333 Mistwell Crescent | 9 Spinnaker Way
Oakville, ON L6L 0A2 | Waretown, NJ 08758
Canada | USA

ISBN-13: 978-1-77463-616-9 (pbk)
ISBN-13: 978-1-77188-362-7 (hbk)

Library and Archives Canada Cataloguing in Publication

Environmental biotechnology : biodegradation, bioremediation, and bioconversion of xenobiotics for sustainable development / edited by Jeyabalan Sangeetha, PhD, Devarajan Thangadurai, PhD, Muniswamy David, PhD, Mohd Azmuddin Abdullah, PhD.

Includes bibliographical references and index.
Issued in print and electronic formats.
ISBN 978-1-77188-362-7 (hardcover).--ISBN 978-1-77188-363-4 (pdf)
1. Bioremediation. 2. Biodegradation. 3. Xenobiotics. 4. Sustainable development.
I. Thangadurai, D., author, editor II. Sangeetha, Jeyabalan, author, editor III. David, Muniswamy, author, editor IV. Abdullah, Mohd zmuddin, author, editor

| TD192.5.E58 2016 | 628.5 | C2016-904275-8 | C2016-904276-6 |

Library of Congress Cataloging-in-Publication Data

Names: Sangeetha, Jeyabalan, editor. | Thangadurai, D. editor. | David, Muniswamy, editor. | Abdullah, Mohd Azmuddin, editor.
Title: Environmental biotechnology : biodegradation, bioremediation, and bioconversion of xenobiotics for sustainable development / editors, Jeyabalan Sangeetha, PhD, Devarajan Thangadurai, PhD, Muniswamy David, PhD, Mohd Azmuddin Abdullah, PhD.
Other titles: Environmental biotechnology (Apple Academic Press)
Description: 1st ed. | Toronto ; New Jersey : Apple Academic Press, 2015. |
Includes bibliographical references and index.
Identifiers: LCCN 2016027712 (print) | LCCN 2016028891 (ebook) | ISBN 9781771883627 (hardcover : acid-free paper) | ISBN 9781771883634 (eBook) | ISBN 9781771883634 ()
Subjects: LCSH: Bioremediation. | Biodegradation. | Green technology.
Classification: LCC TD192.5 .E59 2015 (print) | LCC TD192.5 (ebook) | DDC 628.5--dc23
LC record available at https://lccn.loc.gov/2016027712

Apple Academic Press also publishes its books in a variety of electronic formats. Some content that appears in print may not be available in electronic format. For information about Apple Academic Press products, visit our website at **www.appleacademicpress.com** and the CRC Press website at **www.crcpress.com**

ABOUT THE EDITORS

Jeyabalan Sangeetha, PhD, is an Assistant Professor in Central University of Kerala at Kasaragod, South India. She earned her BSc in Microbiology and PhD in Environmental Science from Bharathidasan University, Tiruchirappalli, Tamil Nadu, India. She holds also an MSc in Environmental Science from Bharathiar University, Coimbatore, Tamil Nadu, India. She is the recipient of the Tamil Nadu Government Scholarship and the Rajiv Gandhi National Fellowship of the University Grants Commission, Government of India, for her doctoral studies. She served as Dr. D.S. Kothari Postdoctoral Fellow and UGC Postdoctoral Fellow at Karnatak University, Dharwad, South India during 2012-2016 with funding from University Grants Commission, Government of India, New Delhi. Her main research interests are in environmental toxicology, environmental microbiology and environmental biotechnology and her scientific/community leadership have included serving as editor of an international journal, *Acta Biololgica Indica.*

Devarajan Thangadurai, PhD, is Senior Assistant Professor at Karnatak University in South India; President of the Society for Applied Biotechnology; and General Secretary for the Association for the Advancement of Biodiversity Science. In addition, Dr. Thangadurai is Editor-in-Chief of several journals, including *Biotechnology, Bioinformatics and Bioengineering; Acta Biologica Indica; Biodiversity Research International;* and the *Asian Journal of Microbiology.* He received his PhD in Botany from Sri Krishnadevaraya University in South India. During 2002–2004, he worked as CSIR Senior Research Fellow with funding from the Ministry of Science and Technology, Government of India. He served as a Postdoctoral Fellow at the University of Madeira, Portugal; University of Delhi, India; and ICAR National Research Centre for Banana, India. He is the recipient of the Best Young Scientist Award with a Gold Medal from Acharya Nagarjuna University and the VGST-SMYSR Young Scientist Award of the Government of Karnataka, Republic of India. He has edited/authored fifteen books including *Genetic Resources and Biotechnology* (3 vols.), *Genes, Genomes and Genomics* (2 vols.), and *Mycorrhizal Biotechnology* with publishers of national and international reputation.

Muniswamy David, PhD, joined Karnatak University, Dharwad, South India, in 1995 as Assistant Professor and advanced to the rank of full Professor of Zoology in 2012. He has vast teaching and research experience in the fields of ecotoxicology, molecular toxicology, microbial remediation of xenobiotics, and fishery biology. He has published more than 80 papers in national and internationally reputed journals. Dr. David obtained his PhD in Zoology from Sri Krishnadevaraya University, Anantapur, South India, and was awarded a postdoctoral fellowship by the University Grant Commission, New Delhi, India. He was also named Scientist of the Year in 2013 from the National Environmental Science Academy, New Delhi, India. He has successfully handled three major research projects funded by UGC, DST-SERB, and Hutti Gold Mines Pvt. Ltd., Karnataka.

Mohd Azmuddin Abdullah, PhD, is an Associate Professor of Bioprocess and Environmental Engineering at the Institute of Marine Biotechnology, Universiti Malaysia Terengganu. He was a visiting scientist (1997) at Kinki University, Japan, under the JSPS Fellowship and a postdoctoral scientist at the Massachusetts Institute of Technology, USA, under the Malaysia-MIT Biotechnology Partnership Program. He was a lecturer in the Department of Biotechnology, Universiti Putra Malaysia and a Senior Lecturer and Associate Professor in the Department of Chemical Engineering, Universiti Teknologi PETRONAS, Malaysia. He has published more than 120 articles in peer-reviewed journals and conference proceedings, and has published many book chapters and technical reports on cell culture engineering, secondary metabolites, natural products, biomaterials, bioenergy, environmental remediation, drug delivery, photocatalysts, and chemical sensors. Dr. Abdullah obtained an MEng degree in chemical engineering and biotechnology from the University of Manchester Institute of Science and Technology, United Kingdom, under a PETRONAS Scholarship, and a PhD in Bioprocess Engineering from Universiti Putra Malaysia under a UPM Scholarship.

CONTENTS

LIST OF CONTRIBUTORS

Mohd Azmuddin Abdullah
Institute of Marine Biotechnology, Universiti Malaysia Terengganu, Kuala Terengganu 21030, Terengganu, Malaysia

Ashfaq Ahmad
Department of Chemical Engineering, Universiti Teknologi PETRONAS, Bandar Seri Iskandar, Tronoh 31750, Perak, Malaysia

Al-Ashraf Amirul
School of Biological Sciences, Universiti Sains Malaysia, Pulau Pinang 11800, Malaysia

Kesaven Bhubalan
Malaysian Institute of Pharmaceuticals and Nutraceuticals, MOSTI, Bayan Lepas 11700, Pulau Pinang, Malaysia

Esiegbuya Ofeoritse Daniel
Plant Pathology Division, Nigerian Institute for Oil Palm Research (NIFOR), Benin City, Nigeria

Muniswamy David
Department of Zoology, Karnatak University, Dharwad 580003, Karnataka, India

Swapna Kishor Deshpande
Department of Botany, Karnatak University, Dharwad 580003, Karnataka, India

Prashant Kamalakar Dhakephalkar
Bioenergy, MACS Agharkar Research Institute, G.G. Agarkar Road, Pune 411004, India

Paraskevi A. Farazi
Mediterranean Center for Cancer Research, Department of Life and Health Sciences, University of Nicosia, 46 Makedonitissas Avenue, P.O. Box 24005, Nicosia 1700, Cyprus

Martin Ferenčík
Institute of Environmental and Chemical Engineering, Faculty of Chemical Technology, University of Pardubice, Studentská 573, Pardubice CZ-53210, Czech Republic

Sándor T. Forczek
Isotope Laboratory, Institute of Experimental Botany, Academy of Sciences of the Czech Republic, Vídeňská 1083, Prague CZ-14220, Czech Republic

Etigemane Ramappa Harish
Department of Zoology, Karnatak University, Dharwad 580003, Karnataka, India

Josef Holík
Isotope Laboratory, Institute of Experimental Botany, Academy of Sciences of the Czech Republic, Vídeňská 1083, Prague CZ-14220, Czech Republic

Okungbowa Francisca Iziegbe
Department of Plant Biology and Biotechnology, University of Benin, Benin City, Nigeria

Pradnya Pralhad Kanekar
Department of Biotechnology, Modern College of Arts, Science and Commerce, Shivajinagar, Pune 411005, India

Venkata Sai Badireenath Konkimalla
School of Biological Sciences, National Institute of Science Education and Research (NISER), Bhubaneswar, Odisha 751005, India

Paidi Murali Krishna
Department of Zoology, Karnatak University, Dharwad 580003, Karnataka, India

Devipriya Rabin Majumder
Department of Microbiology, Abeda Inamdar Senior College, 2390-KB Hidayatullah Road, Azam Campus, Camp, Pune 411001, Maharashtra, India

Kartheek Rajendra Malowade
Department of Zoology, Karnatak University, Dharwad 580003, Karnataka, India

Abhishek Channayya Mundaragi
Department of Botany, Karnatak University, Dharwad 580003, Karnataka, India

Subir Kumar Nandy
Technical University of Denmark, SøltoftsPlads, 2800 Kgs, Lyngby, Denmark

Ashish Vasantrao Polkade
Microbial Culture Collection, National Centre for Cell Science, First floor, Central Tower, Sai Trinity Building Garware Circle, Sutarwadi, Pashan, Pune 411021, India

Prathima Purushotham
Department of Botany, Karnatak University, Dharwad 580003, Karnataka, India

Kaliyaperumal Rani
Nanyang Technological University, 62 Nanyang Drive 637459, Singapore

Jagnyeswar Ratha
School of Life Sciences, Sambalpur University, Jyoti Vihar, Burla, Odisha 768019, India

Noor Fazielawanie Mohd Rashid
School of Marine and Environmental Sciences, Universiti Malaysia Terengganu, Kuala Terengganu 21030, Terengganu, Malaysia

Luděk Rederer
Povodí Labe, State Enterprise, Víta Nejedlého 951, Hradec Králové CZ-50003, Czech Republic

Barindra Sana
Agency for Science, Technology and Research (A*STAR), 1 Fusionopolis Way, #20-10 Connexis North Tower 138632, Singapore

Jeyabalan Sangeetha
Department of Environmental Science, Central University of Kerala, Kasaragod, Kerala 671316, India

Seema Shreepad Sarnaik
Microbial Sciences Division, MACS Agharkar Research Institute, G.G. Agarkar Road, Pune 411004, India

Raghunath Satpathy
School of Life Sciences, Sambalpur University, Jyoti Vihar, Burla, Odisha 768019, India

Prafulla Namdeo Shede
Department of Microbiology, MES' Abasaheb Garware College, Karve Road, Pune 411004, India

Jadhav Shrinivas
Department of Zoology, Karnatak University, Dharwad 580003, Karnataka, India

Ganapathi Sibi
Department of Biotechnology, Centre for Research and Post Graduate Studies, Indian Academy Degree College, Bangalore 560043, Karnataka, India

Devarajan Thangadurai
Department of Botany, Karnatak University, Dharwad 580003, Karnataka, India

Parag Avinash Vaishampayan
Biotechnology and Planetary Protection Group, Jet Propulsion Laboratory, California Institute of Technology, Pasadena, CA 91109, USA

Ain Farhana Mohd Yatim
School of Marine and Environmental Sciences, Universiti Malaysia Terengganu, Kuala Terengganu 21030, Terengganu, Malaysia

LIST OF ABBREVIATIONS

·OOH	Peroxide radicals
3CP	3-chloropropionic acid
3HB	3-hydroxybutyrate
3HB-*co*-3HV	Poly(3-hydroxybutyrate-*co*-hydroxyvaletare)
3HD	3-hydroxydecanoate
3HHx	3-hydroxyhexanoate
3HV	3-hydroxyvalerate
AgNPs	Silver nanoparticles
AM	Arbuscular mycorrhiza
AO7	Acid orange 7
AOX	Adsorbable organic halogens
APS	Ammonium persulfate
APX	Ascorbate peroxidase
ARDRA	Amplified ribosomal DNA restriction analysis
ASO	Ascorbate oxidase
BMP	Biochemical methane potential
BNF	Biological nitrogen fixation
BOD	Biological oxygen demand
C/N	Carbon/nitrogen
C_2Cl_4	Tetrachloroethylene
C_2HCl_3	Trichloroethylene
Ca	Calcium
CA	Clofibric acid
CCA	Chromated copper arsenic
*c*DCE	*Cis*-1,2-dichoroethene
CDM	Clean development mechanism
CER	Carbon emission reduction
CF	Concentration factor
CFCs	Chlorofluorocarbons
CFG	Cement flue gas
CFU	Colony forming units
$CHBr_2Cl$	Chlorodibromomethane
$CHCl_3$	Chloroform
CHIP	Chromatin immunoprecipitation

Cl	Chlorium
CLE	Council for Leather Export
CO_2	Carbon dioxide
COD	Chemical oxygen demand
Co-PCBs	Coplanar polychlorinated biphenyls
CSL	Corn steep liquor
Cu	Copper
CW	Constructed wetland
Da	Daltons
DAF	DNA amplification fingerprinting
DART	Dust at altitude recovery technology
DAS	Days after sowing
DCA	[UL-^{14}C]-3,4-dichloroaniline
DCP	Dichlorophenol
DCW	Dry cell weight
DDT	Dichloro-diphenyl-trichloroethane
DGGE	Denaturing gradient gel electrophoresis
DHA	Docosahexaenoic acid
DMSP	Dimethylsulphoniopropionate
DNAPLs	Dense non-aqueous phase liquids
DNT	Dinitrotoluene
DNX	Hexahydro-1,3-dinitroso-5-nitro-1,3,5-triazine
DOC	Dissolved organic carbon
DOE	Department of Energy
DSE	Dark septate endophyte
EDC	Endocrine disrupting chemical
EFB	Empty fruit bunches
EPA	Environmental Protection Agency
EST	Expressed sequence tag
ETC	Electron transport chain
eZVI	Emulsified ZVI
FAME	Fatty acid methyl ester
FBB	Fresh fruit bunch
FDA	Food and Drug Administration
FFA	Free fatty acids
FFB	Fresh fruit bunches
FGP	Fungal Genome Project
FIIRO	Federal Institute of Industrial Research Oshodi
FLT	Fluoranthene
FTIR	Fourier transform infrared spectroscopy

g m^{-2} d^{-1}	Gram per meter square per day
GC-MS	Gas chromatography-mass spectrometry
GHG	Greenhouse gases
GHS	Glutathione
GPX	Guaiacal peroxidase
GSTs	Glutathione S-transferases
H$_2$CO$_3$	Bicarbonate
HDTMA	Hexadecyltrimethylammonium
HE	Harvesting efficiency
HFCW	Horizontal flow constructed wetland
HFMD	Hand, foot and mouth disease
HMX	Octahydro-1,3,5,7-tetranitro-1,3,5,7-tetrazocine
HPLC	High performance liquid chromatography
HRT	Hydraulic retention time
IMG	Integrated microbial genomes
JGI	Joint Genome Institute
Kg	Kilogram
L	Liter
LAS	Linear alkylbenzene sulfonates
LCA	Life cycle assessment
LCFA	Long chain fatty acid
lcl-PHA	Long-chain-length PHA
LH-PCR	Length heterogeneity PCR
LiP	Lignin peroxidases
LMW	Low molecular weight
MBRT	Methylene blue dye reduction test
mcl-PHA	Medium-chain-length PHA
MDA	Malondialdehyde
MeP	Methyl parathion
MF	Membrane filtration
Mg	Magnesium
MIC	Minimum inhibitory concentration
MM	Minimal medium
MnP	Manganese peroxidase
MNX	Hexahydro-1-nitroso-3,5-dinitro-1,3,5-triazine
MPBR	Membrane photobioreactor
MPOC	Malaysian Palm Oil Council
MTs	Metallothioneins
MUFA	Monounsaturated fatty acids
NaCl	Sodium chloride

NH_4Cl	Ammonium chloride
NHGRI	National Human Genome Research Institute
NIFOR	Nigerian Institute for Oil Palm Research
nZVI	Nanoscale zero-valent iron
O_2^-	Superoxide ions
OH^-	Hydroxide ions
OLR	Organic loading rate
OMSW	Olive mill solid waste
P(3HB)	Poly(3-hydroxybutyrate)
PAHs	Polycyclic aromatic hydrocarbons
PBR	Photobioreactor
PCBs	Polychlorinated biphenyls
PCDDs	Polychlorinated dibenzo-ρ-dioxins
PCE	Perchloroethylene
PCP	Pentachlorophenol
PCR	Polymerase chain reaction
PCs	Phytochelatin synthase
PDB	Protein Data Bank
PEI	Polyethylenimine
PGPR	Plant growth promoting rhizobacteria
PHA	Polyhydroxyalkanoate
PKC	Palm kernel cake
PM_{10}	Particulate matter <10 mm
POME	Palm oil mill effluent
POMS	Palm oil mill sludge
PPi	Pyrophosphate
ppm	Parts per million
PUFA	Polyunsaturated fatty acids
PV	Peroxide value
RAPD	Random amplified polymorphic DNA
RAS	Rennin angiotensin system
RDX	Hexahydro-1,3,5-trinitro-1,3,5-triazine
RFLP	Restriction fragment length polymorphism
RISA	Ribosomal intergenic spacer analysis
RKF	Raw kapok fibers
RMX	Hexahydro-1,3,5-triaza-1,3,5-trinitrocyclohexane
RO	Reverse osmosis
scl-PHA	Short-chain-length PHA
SDS	Sodium dodecyl sulphate
SEM	Scanning electron microscopy

SFA	Saturated fatty acids
SHF	Separate hydrolysis and fermentation
SKF	Structurally modified kapok
SMKF	Surface-modified kapok fiber
SOD	Super oxide dismutase
SPC	Solid-phase cytometry
SPM	Suspended particulate matter
SRT	Solid retention time
SSCP	Single-strand conformation polymorphism
SSF	Simultaneous saccharification and fermentation
TAG	Triacyl glycerols
TBA	Terbuthylazine
TCA	Tricarboxylic acid
TCDD	2,3,7,8-tetrachloro dibenzo-ρ-dioxin
TCE	Trichloroethane
TDS	Total dissolved solids
THM	Trihalogenated methanes
TiO_2	Titanium dioxide
TMPA	Trimethylphenylammonium
TN	Total nitrogen
TNT	2,4,6-trinitrotoluene
TOC	Total organic carbon
TPH	Total petroleum hydrocarbon
T-RFLP	Terminal restriction fragment length polymorphism
UASB	Up-flow anaerobic sludge blanket reactor
UASFF	Upflow anaerobic sludge fixed film
UV-VIS	Ultraviolet-visible spectroscopy
VAM	Vesicular-arbuscular mycorrhiza
VBNC	Viable but non culturable
VFAs	Volatile fatty acids
VFCW	Vertical flow constructed wetland
VOCl	Volatile organochlorines
VP	Versatile peroxidase
WAS	Waste activated sludge
WHO	World Health Organization

PREFACE

Every fraction in our close proximities harbors inestimable living organisms with untold scientific mysteries. Given our widespread environment, this number would increase by leaps and bounds. This emphatically describes the plethora of information which could be derived if this surreptitious environment is premeditated. The ability to probe the environment at the molecular level to create a new awareness of fundamental biological processes therein has created an important new paradigm. This has been made possible by the successful integration of biotechnology into the many facets of our environment. Biotechnology is an emerging discipline that holds promise to offer long-range solutions to our developmental and environmental concerns. The future of our calling toward environment may in fact be in biotechnology. Modern biotechnology is considered to be one of the potential technologies of the current century to support sustainable development. That being the case, we can expect the merger to follow, giving rise to environmental biotechnology that involves specific application of biotechnology to the management of environment and environment related socio-economic developmental issues.

The present plight of the world as a victim to a surfeit of environmental setbacks ranging from global warming and ozone layer depletion to an alarming increase in world pollution levels is threatening the existence of the most intelligent species on earth. This has been enough for both environmentalists and laymen to wake up to the indisputable importance of environmental protection and ecofriendly alternatives. Environmental biotechnology is mainly the first alternative, with fields of application reaching out to every quarter. Environmental biotechnology utilizes the biochemical potential of microorganisms and plants for the preservation and restoration of the environment. The primary role of environmental biotechnology is to develop better approaches for sustainable development.

The purpose of this book is to present the reasonable manifestation of the realistic biological approaches currently employed in environmental biotechnology to address environmental problems. It covers the latest research by prominent scientists and researchers in this fast growing field. This most recent and updated book deals with various aspects of recent advances, current trends and practical applications on the understanding of microbial interactions, exploration of microbes in extreme environments, applications of nanomaterials for water remediation, impact of toxicants on microbial communities and soil bioprocesses, production of biofuel from microalgae, algal industrial waste treatment, reduction of tannery waste water pollution, phytoremediation of organic chemopollutants, bioconversion of palm oil and sugar industry wastes, environmental influence on postharvest deterioration of fruits, soil remediation, ecological restoration, heavy metal pollution, radioactive waste materials, dehalogenation, and bioremediation of xenobiotics using fungi.

In Chapter 1, S.K. Nandy gives an overview on the microbial evolution, ecological interactions, and the population dynamics with general methods and mathematical models. This is followed by the identification of microorganisms carried by Aeolian dust to other continents and their impact on public health by P.A. Farazi and in Chapter 3, S.T. Forczek, J. Holík, L. Rederer, and M. Ferenčík discuss the merits, risks, and uses of chlorinated compounds in natural and biotechnological processes. In Chapter 4, environmental impact of pesticide use on microbial communities and soil bioprocesses using molecular approaches is highlighted by J. Sangeetha and her colleagues. Discussion on the advances and applications of nanomaterials for water remediation is highlighted by K. Rani and B. Sana in Chapter 5, and fungal dehalogenation in Chapter 6 by R. Satpathy, V.B. Konkimalla, and J. Ratha. A comprehensive review on the prospect of microalgal utilization as renewable energy sources is made in Chapter 7 by G. Sibi, and on the integrated industrial waste treatment and bioenergy co-generation in Chapter 8 by M.A. Abdullah and his team have been presented with recent information. The use of green technology to reduce tannery waste water pollution with the help of plant-based bioambiant preservatives is elucidated by P.N. Shede, A.V. Polkade, P.P. Kanekar, P.K. Dhakephalkar, and S.S. Sarnaik in Chapter 9. P.P. Kanekar and her collaborators interpret on the phytoremediation of organic chemopollutants in Chapter 10, followed by the discussion on bioconversion of palm oil and sugar industrial wastes into polyhydroxyalkanoate by N.M.R. Fazielawanie, A.F.M. Yatim, K. Bhubalan, and A.A. Amirul. Environmental influence on the prevalence of postharvest deterioration of *Raphia* and Shea fruits is covered in Chapter 12 by O.F. Iziegbe and E.O. Daniel. Soil remediation and ecological restoration from heavy metal pollution and radioactive waste materials using fungi is discussed by J. Sangeetha and her group in Chapter 13. Finally, the discussion on the multifaceted bioremediation of xenobiotics using fungi is wrapped up in the last chapter by D.R. Majumder.

The mission of editing and publishing this book has received immense support of our foremost experts. We are deeply indebted to them for their devoted consideration, valuable guidance, constant assistance, support, endurance, and constructive suggestions for the successful completion of the book. We also wish to thank Sandy Jones Sickels, Vice President, and Ashish Kumar, Publisher and President, at Apple Academic Press, Inc., for quality production and humble attempts to publish this book. We owe our gratitude to the people who were directly or indirectly involved in preparing this book, for their co-operation and useful suggestions. We express thanks to our families for their unrelenting support and constant encouragement.

Jeyabalan Sangeetha, PhD
Devarajan Thangadurai, PhD
Muniswamy David, PhD
Mohd Azmuddin Abdullah, PhD

CHAPTER 1

THE ROLE OF GENERAL METHODS AND MATHEMATICAL MODELS ON MICROBIAL EVOLUTION, ECOLOGICAL INTERACTIONS, AND POPULATION DYNAMICS

SUBIR KUMAR NANDY

Department of Systems Biology, Technical University of Denmark, Lyngby, Denmark

CONTENTS

1.1   INTRODUCTION

Bacillus subtilis is a spore-forming bacterium that can persist for years in a resistant state. When faced with starvation, bacteria instead enter the complex developmental pathway of spore formation. González-Pastor et al. (2003) and other researchers (Engelberg-Kulka and Hazan, 2003) have found that bacteria *en route* to sporulation produce a toxin (sporulation-killing factor) similar to peptide antibiotics that lyses sibling cells not committed to sporulation. The killing operon also produces its own export pump and confers resistance to the killing peptide. The nutrient boost from the lysed cells allows the surviving cells to postpone sporulation, to escape its energetic costs, and to continue replication. A signaling factor (sporulation-delaying protein) mediates this escape from sporulation via a transcription factor that stimulates lipid oxidation and adenosine 5′-triphosphate production to restore energy reserves. The above studies of mixed culture require an easy, effective, and fast methodology to quantify the metabolic active cell count of different types of microorganisms. Both the survival strategy and the microbial interactions including stress response in pure and mixed culture are discussed in this chapter.

1.2   SURVIVAL STRATEGY

Microorganisms have evolved strategies to adapt in various stress environments such as change in temperature, pH, nutrient concentration, osmotic pressure, and media. Therefore, the obvious question that arises is that what are the mechanisms that underlie the evolution of these survival strategies? Several studies in molecular biology report the signaling pathways that establish the phenotypic response. However, in the ecosystem, organisms need to adapt in the presence of other microorganisms. The interactions of an organism with its environment are fundamental to the survival of that organism and also the functioning of the ecosystem as a whole. The interrelationship between various microorganisms under stress is relevant and has not been studied sufficiently. Different types of bacterial interactions exist in a mixed population. These are categorized as neutralism, mutualism, commensalism, ammensalism, antagonism, symbiosis, and competition.

Limited resources of nutrients or environment create competition in a mixed culture. The competition among the same species is termed as intraspecific competition, while that in different species is termed as interspecific competition. Cannibalism is an example for intraspecific competition.

Froelich et al. (1979) have shown in anaerobic marine sediments that a sulfate-reducing zone develops near sediment–water interface and is also underlined by a methanogenic zone. Studies have shown that this zone separation reflects the competition between sulfate reduces and methanogens for the same substrates (Lovely and Klug, 1983, 1986).

Under starvation condition, a majority of bacterial population die; however, some survive the nutrient-scarce environment by exploiting the nutrients made available from the dying cells. These surviving cells are highly dynamic in nature and the interaction is termed as cannibalism. The cannibalistic interaction has been studied by González-Pastor et al. (2003) and Nandy et al. (2007). It is also proposed that instead of surviving from the nutrients coming out of the dying cells, survivors become fitter mutants and simultaneously survival strategy becomes an evolutionary strategy. González-Pastor et al. (2003) have found that *B. subtilis* can produce killing factor under no nutrient medium to kill their own sister cells or siblings and survive on the released nutrients to delay sporulation. Table 1.1 shows the different existing ecological interactions.

TABLE 1.1 Different Ecological Interaction Processes are Summarized with Their System Name and Responses.

Interaction Process	Bacterial Response (Benefit: Y or N)	
	First	Second
Neutralism	N	N
Mutualism	Y	Y
Commensalism	Y	N
Ammensalism	–	N
Antagonism	Y	N

Y: yes; N: no.

1.3 CANNIBALISM

B. subtilis is a Gram-positive soil-living bacterium, which adapts to different phenotypes depending upon the nature of the surrounding environment. These phenotypes are regulated in both cell density and environment-dependent fashion. As bacteria move toward late log phase, they undergo sequentially competence development, cannibalism, and sporulation.

Cannibalism is a process in which bacteria produce a killing factor to kill their own sister cells to survive under no nutritional conditions. In *B. subtilis*, a regulatory protein Spo0A is responsible for cannibalism in the stress environment. Various interlinked signaling pathways regulate the activity of Spo0A and these signaling pathways are also known to be involved in the regulation of other physiological processes such as sporulation, synthesis of degradative enzymes, and secondary metabolites. Under the severity of nutrient limitation, cells start to secrete several digestive enzymes, such as proteases, which enable the bacteria to take food from several alternative nutrient sources. Cells also start to secrete several antibiotics, which kill the competitors, making the availability of nutrients for survival. If all these adaptive ways are closed to survive, *B. subtilis* maintains its integrity by undergoing sporulation.

skfA gene in skf operon is shown to be involved in the production of an extracellular killing factor. *Skf* operon also produces the factors skfE and skfF; these two factors confer resistance to killing factor. skfE behaves as a ATP-binding cassette and skfF acts as a transporter. Together, they offer resistance to mother cells by pumping out the killing factor. Sdp is another operon that is regulated by nutrient limitation through *spo0A*. *sdpC* synthesizes a 5 kDa protein, which regulates the expression of yvbA. Transcription factor yvbA upregulates the operon *atp* which is involved in ATP production and also operon yusLKG which synthesize the enzymes involved in lipid catabolism. Thus, higher expression of sdp operon generates energy under no nutrient condition to make cells metabolically active, which delays the sporulation in the cells expressing spo0A (González-Pastor et al., 2003; Westers et al., 2005).

González-Pastor and colleagues have shown that during the nutrient limitation, a subpopulation of bacteria become spo0A active cells, while remaining fraction of cells become spo0A inactive (Ellermeier et al., 2006). In spo0A active cells, spo0A protein upregulates expression of *skfA*, *skfE*, *skfF*, and *sdpC*. *skfE* and *skfF* together form killing factor transport protein complex, conferring resistance to the parent cells. Since some cells are spo0A inactive, *skf* operon remains inactive, making them sensitive to the killing factor. *sdp* operon is also switched on in spoA active cells, which synthesizes the signaling protein sdpC. sdpC upregulates yvbA which induce the expression of operon for ATP synthase (*atp*) and lipid catabolism enzymes (yusLKG), leading to energy production and delaying of sporulation. SdpC signaling molecule also induces the yvbA gene in spo0A-negative cells. yvbA represses the gene for sigma factor synthesis (Ellermeier et al., 2006).

It has been suggested that sigma factor confers resistance to the cells by making cells antibiotic resistant and also by detoxification of killing factor (González-Pastor et al., 2003; Ellermeier et al., 2006).

1.4   SPORULATION

Sporulation mechanism of Gram-positive bacterium involves a complex regulatory cascade which controls the expression of over 100 genes. This regulatory process is complex due to the temporal and spatial constraints. One of the best examples is *B. subtilis*, which forms spores in the environment under different stresses. At low nutrient condition, spores of *B. subtilis* are formed. These spores are metabolically dormant for years and are resistant to several factors such as heat and radiation. Metabolically and morphologically *B. subtilis* spores are different from the normal vegetative cells. Several genes are also identified to be involved in sporulation such as *spo0A, spo0B, spo0E, spo0F, spo0H, spo0J,* and *spo0K* (Westers et al., 2005). In the absence of either of these, the sporulation mechanism is blocked.

1.4.1   SPORULATION STAGES

This process involves a number of steps which can be monitored by light and electron microscopy and completes within 8 h (Driks, 1999). The whole process is divided into six parts and these are as follows. (A) During sporulation, morphology will change with sequential appearance of a series of transcription factors, called sigma factors binding to core polymerase. Initially, the mother cell was divided asymmetrically into two compartments. The mother cell which is the larger compartment will serve the spore until the complete development. This phenomenon will happen within 2 h. σ^E and σ^F become active in mother cell and forespore, respectively. (B) In the next level forespore engulfs into one membrane bound protoplast. σ^G becomes active in forespore whereas σ^K express gene in the mother cell. (C and D) Then one cortex forms in between forespore membranes, while the formation of the coat from synthesis of protein in mother cell that assembles around the forespore is visible by electron microscopy. (E) Finally the mother cell lyses and the mature spore will come out in the environment. (F) Under nutrient again spore can germinate and the nascent cell comes out in the environment (Driks, 1999). Our main objective is to explain the sporulation process only, and the whole process is described in Figure 1.1.

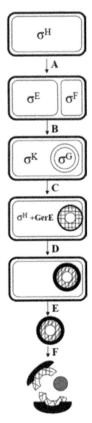

FIGURE 1.1 Sporulation steps: (A) When a cell commits sporulation, the cell divides into mother cell and forespore in asymmetrically, σ^H activity increases. (B) The forespore covers with a membrane where σ^G becomes active and σ^K directs a final phase of gene expression. (C) One cortex forms between the forespore membranes where final phase of gene expression has been taken place with the help of GerE. (D,E) The dark coat has been visible by microscope and after cell lysis spore comes out to the environment. (F) Spore can again germinate and cell can resume vegetative growth (Driks et al., 1999).

1.4.2 *METHODOLOGY TO STUDY SPORULATION*

1.4.2.1 *STAINING METHOD*

The spores are differentially stained by using dyes that penetrate the spore wall. An aqueous primary stain (malachite green) is applied and steamed to enhance the penetration of the impermeable spore coats. Once stained, the endospore do not readily decolorize and appear green. After fixing, the smears flood the malachite green over the slide and heat it for 5 min, adding more stain with time to time. Finally, wash the slides slowly with tap water and counter stain with safranin for 30 s. For example, the endospore of *B. subtilis* shows green color and vegetative cells are in red (Rao et al., 1965).

1.4.2.2 AGAR METHOD

Sporulation agar is employed to detect spore in the culture. The sample was boiled at 90–100°C for 10 min in a water bath and then spread on the sporulation agar to count the spores after incubation for 24 h (Driks, 1999).

1.4.3 REGULATION OF SPORULATION KILLING FACTOR

B. subtilis prefers cannibalism in the no nutrient condition. *B. subtilis* kills its own sister cells by producing and exports a peptide sporulation factor (skfA) where, skfA is a part of skf operon (skfA-H). This skf operon is responsible for immunity to SkfA and also for the production and export of SkfA. Allenby et al. (2006) reported that the transcription of skfA induced under phosphate starvation.

This phosphate starvation is responsible for the expression of genes in PhoP and SigB (σ^B) regulons of *B. subtilis* (Eymann et al., 1996; Hecker and Volker, 1998; Antelmann et al., 2000; Prágai and Harwood, 2002; Allenby et al., 2005). Spo0A is responsible for activation of PhoPregulon, and moreover activated Spo0A (Spo0A~P) is responsible for sporulation induction, and gene repression is induced by transition phase regulator AbrB. Thus, spo0A mutant should not show cannibalism. PhoPregulon is upregulated in spo0A null mutant and that is unable to initiate sporulation (Prágai et al., 2004). Researchers have also reported that Spo0A regulates skf operon which encodes sporulation killing factor (SkfA) (Fawcett et al., 2000; Molle et al., 2003). The siblings of *B. subtilis* cells are lysed by induced SkfA and they have not entered the sporulation pathway (i.e. Spo0A inactive), providing a source of nutrients to support this key differentiation process.

More than 125 genes are responsible for sporulation of *B. subtilis* that is governed by a program of gene transcription (Stragier and Losick, 1996) whereas in the early stages of development of DNA-binding Spo0A protein, 500 genes are expressed (Molle et al., 2003). Molle et al. (2003) demonstrated the direct control of Spo0A using chromatin immunoprecipitation in combination with gene microarray analysis to identify regions of the chromosomes at which an activated form of Spo0A binds *in vivo*. Therefore, Spo0A is a master regulator for sporulation, whereas many of its effects on the global pattern of gene transcription are likely to be mediated indirectly by regulatory genes under its control. Similarly, the gene expression during spore formation have very little information to describe properly.

Sporulation of *B. subtilis* is a developmental process that is most responsible for conversion of growing cell into a dormant cell type known as spore or endospore (Stragier and Losick, 1996). Normally sporulation is governed by DNA-binding protein, Spo0A acting like both activator and repressor of transcription. One of the members of response regulator family of proteins, Spo0A and its activity controlled through phosphorylation by a phosphorelay which consists of five histidine autokinases (KinA, KinB, KinC, KinD and KinE) and two phosphorelay proteins (Spo0F and Spo0B) (Jiang et al., 2000) that integrate environmental and physiological signals in the decision to sporulate (Burbulys et al., 1991). Spo0F is phosphorylated to Spo0F~P, while kinases feed phosphoryl group and in turn, the phosphoryl groups to Spo0B which in the final step of the relay, phosphorylates Spo0A to create Spo0A~P (Burbulys et al., 1991), target of phosphorylated form, involved in the formation of polar septum which divides cell into forespores and mother cell compartments (Levin and Losick, 1996). The transcription of genes and operons directed by Spo0A~P activates the compartment-specific regulatory proteins σ^F and σ^E which direct transcription in forespore and mother cell, respectively (Stragier and Losick, 1996). Recently it is also shown that Spo0A~P continues to function after formation of the polar septum when it accumulates to high levels and directs transcription in the mother cell (Fujita and Losick, 2003). Spo0A also activates transcription of certain genes indirectly by repressing the gene for the "transition-state regulator" AbrB (Strauch et al., 1990). Among the targets of AbrB, stationary phase sigma factor σ^H is the gene influenced by Spo0A, where asymmetric cell divides. Genome-wide analysis of σ^H regulon (Britton et al., 2002) has confirmed that both mutant and Spo0A mutant have similar transcriptional profile. Recently Stragier and Losick (1996) showed that σ^F and σ^E are replaced by forespore σ^G and mother cell σ^K, respectively. Whereas computational searches for binding sites for each regulator in sequences upstream of σ^H- and Spo0A-regulated genes indicated that the list of presumed direct targets of each regulatory protein is largely different (Fawcett et al., 2000; Britton et al., 2002). Therefore, researchers used a recently developed procedure for carrying out chromatin immunoprecipitation (CHIP) in conjunction with DNA microarray analysis (chip) to identify chromosome regions where Spo0A binds *in vivo*

(Ren et al., 2000; Iyer et al., 2001; Molle et al., 2003). In addition, some analysis also enabled to discover many additional genes that are likely to be part of the sporulation program (Fawcett et al., 2000).

1.5   MATHEMATICAL MODELS RELATED TO ECOLOGICAL INTERACTIONS

1.5.1   GENERAL METHODS

In this chapter, competition between different species is discussed. Competition refers to dependence of two factors on several factors such as nutrient, light, and temperature. It is interesting to check the natural advantage of one species over another. One of the best examples of competition is cannibalism. Several models for the dynamics of a cannibalistic population are derived under the assumption that cannibals attack only weaker and smaller victims. Vito Volterra has developed a model for the growth of two competing species, which is an attractive model in mathematical ecology (Lotka, 1925; Volterra, 1926). The following equation represents a simple model for predator–prey interaction (Bailey and Ollis, 1986):

$$\frac{dH}{dt} = rH - aHP$$

and

$$\frac{dP}{dt} = bHP - mP$$

It has two variables (P, H) and several parameters: H is the density of prey, P is the density of predators, r is the intrinsic rate of prey population increase, a is the predation rate coefficient, b is the reproduction rate of predators per one prey eaten, and m is the predator mortality rate.

Many mathematical modeling and theoretical analyses have been conducted for the decision-making of an individual microorganism so that to maximize the currency of the individual's fitness. Several models have been developed on the population dynamics of cannibalistic population (Cushing, 1991; Hansen, 1993). Structured and discrete models are available to explain cannibalism for both continuous models (Diekmann, 1986) and discrete models (Cushing, 1991). Cells are shown to adapt to the fluctuating environments by bringing diversity within the population stochastic phenotype switching mechanism. The idea of randomization of phenotype has been analyzed theoretically and computationally (Diekmann, 1986).

1.5.2 POPULATION DYNAMICS

On the one hand, cannibalism can be an effective mechanism for the regulation and equilibration of population density. On the other, it can result in population oscillations too. Thus, cannibalism can promote either equilibration or oscillations depending upon the exact circumstances under which it is practiced. Again in some cases cannibalism can be a crucial mechanism for population survival. The so-called lifeboat strategy asserts that the resources obtained from cannibalism can permit a population to survive during periods of non-cannibalistic resource scarcity or in other circumstances under which it could not otherwise survive (Cushing, 1991). For example, the cannibalism of the young by adults may provide access to resources available to young individuals that are otherwise unavailable to adults. Therefore, the interplay between the negative and positive feedbacks loops in regulatory network which are inherent in cannibalism can result in multiple steady states and hysteresis. This can lead to catastrophic crashes to lower equilibrium levels, as population parameters are changed below critical values that cannot be reversed by increasing the parameters back above their critical values. This fact has been used by researchers, for example, to explain the collapse of certain harvested fish populations and their failure to return to pre-harvested levels after harvesting is ceased (although other explanations are possible, such as evolutionary effects as studied). Only a handful of dynamical models for populations practicing cannibalism appear in the literature. There have been some studies of the extreme case of egg cannibalism that address the destabilization effect of cannibalism. The possibility of oscillations in a discrete age-structured cannibalism model was studied by Cushing (1991). The effects of cannibalism can be developed using a simple discrete age-structured model, one that is both analytically and numerically very tractable. The simple model suggests that the phenomena are very likely to be common in model cannibalistic populations and consequently might also be expected to occur in more sophisticated models. Cannibalism, like interspecies predation, is most commonly practiced by larger individuals on smaller individuals. If we assume that age correlates with body size, then a simple model could be built by distinguishing just two age classes, a juvenile class and an adult class, as is done in the simple model for intraspecific competition introduced by Cushing (1991).

1.6   CANNIBALISM *VS* PREDATION

There are very few organisms that exhibit the rare phenomena called cannibalism and predation in our ecosystem. In our previous study, both the properties have been shown in *B. subtilis* and *Escherichia coli* in a mixed culture system (Nandy et al., 2007, 2008; Nandy and Venkatesh, 2008, 2014).

B. subtilis resorts to cannibalism to delay sporulation under nutritional stress. However, in ecosystem *B. subtilis* responds in presence of other microorganisms. Methylene blue dye reduction test (MBRT) (Bapat et al., 2006) was developed to quantify viable cell count of different microorganisms in pure and mixed cultures. MBRT was also described to check the bacterial contamination in the industrial reactor to save batch time. This method was further used to study the survivability of *B. subtilis* and *E. coli* as a mixed culture in phosphate buffered saline (PBS)-lacking nutrients. Pure culture of *B. subtilis* demonstrated cannibalism under stress condition or limiting nutrient condition. A regulatory protein, SpoOA, present in fraction of the culture is responsible for delaying sporulation that produces a killing factor by activating skf operon and an associated pump to export the factor. Other cells not containing *spoOA* are lysed. Therefore, there is a competition started in the culture to survive in the nutritional limitation condition which demonstrates cannibalism. Our extended study has also shown the predatory behavior of *B. subtilis* in the presence of *E. coli* under severe nutritional limitation (Nandy et al., 2007).

Further, the model developed in the previous study was empirical and phenomenological, and structured model based on the molecular understanding of the process of SpoOA regulation can be carried out. Environmental changes are known to have an effect on the cannibalistic behavior of species. Abiotic factors like temperature alter the interaction between species in ecology and affect the population dynamics. It is important to study the effect of temperature on the cannibalistic behavior of *B. subtilis*. It has been demonstrated that the cannibalistic property of *B. subtilis* under extreme nutrient deficiency is dependent on the medium temperature. The data were analyzed by proposing a model using a delay differential equation. The dynamics of growth and death in medium lacking nutrition result in an oscillatory behavior of cell count. This was captured by simulating the dynamics using a delay differential equation. The key parameters that determined the oscillatory behavior were the kinetic constants for growth and death. Since these parameters were dependent on temperature, the dynamics of cannibalism were also strongly related to the medium temperature. At high temperatures (beyond

40°C) due to high kinetic constant of death and the absence of killing factor, the dynamics demonstrated a first-order death kinetics. At low temperatures, the growth rate was low which resulted in a very low amplitude oscillation. The presence of the killing factor which indicates the process of cannibalism was essential for observing the oscillating behavior. The presence of the killing factor was not observed at low initial cell counts (less than 1000 cells) and at high temperatures, resulting in the absence of cannibalism (Nandy et al., 2008). The study is also extended to predatory behavior and the dynamics show very satisfactory result.

Our previous results on the effect of external carbon and nitrogen on the cannibalistic behavior of *B. subtilis* have already been described (Nandy et al., 2008). It has been found that when carbon was introduced into a medium in the absence of any other nutrients, the cannibalistic tendency was delayed. This delay increased with the increase in the amount of glucose which is a major source of carbon. Hence, the cells could not totally consume glucose. Further, the cannibalism occurred at a later stage after the maximum utilization of carbon in the medium. Thus, the cannibalism was observed to be very sensitive to the amount of carbon present in the medium. However, when the medium contained only nitrogen and was devoid of carbon, the effect on cannibalism was minimal. Therefore, cannibalism was more sensitive to carbon than to nitrogen, indicating that the phenomenon of cannibalism was more energy dependent than nitrogen assimilation (Nandy et al., 2008). Further, predatory behavior, another interesting property associated with ecosystem, was also rare to observe in the microbial world. In the extended work, *B. subtilis* maintains its viable cells under different combinations of carbon and nitrogen effects with *E. coli* cells. In this work, as the concentration of carbon sources increases, a number of *B. subtilis* cells maintain more than normal predation, whereas there is no effect in the presence of nitrogen. Cannibalistic tendency was delayed with the increase in the concentration of carbon source. For higher concentration of carbon, maximum glucose was consumed, with some of the carbon sources remaining unused. Therefore, predation was also sensitive to carbon and then nitrogen, which proves the dependency on carbon in both cannibalism and predation.

1.7 CONCLUSION AND FUTURE PERSPECTIVES

Microorganisms under stress usually form endospores for their survival. The endospores transform into vegetative cells under suitable conditions for growth. Suitable conditions may be biotic in nature, such as media constituents or abiotic, such as temperature and pH. For example, *B. subtilis*, a Gram-positive soil microorganism, can form endospores under nutritional stress (González-Pastor et al., 2003). However, the formation of endospore is energy intensive and the organism resorts to spore formation only under sustained stress conditions. Competence is a distinct phase from vegetative growth and sporulation that takes place during the onset of late log phase and under specific nutritional conditions. During competence, cells acquire extracellular DNA and get transformed. Since spore formation is a last re-sort for survival under extreme condition, cells have evolved strategies to delay spore formation. The organism uses strategies such as cannibalism and competence to delay spore formation. In the nature, microorganisms have to deal with competition from other microorganisms present in the en-vironment. Thus, the response to stress has to be analyzed in the presence of other microorganisms. This requires study of mixed culture under condition of stress. Mixed culture can respond in various ways in an ecosystem. The response varies from competition, symbiosis, predation, cannibalism, neu-tralism, commensalism, and ammensalism.

In this chapter, a review of our previous study on cannibalistic and preda-tory behavior of *B. subtilis* is reported. The organism is metabolically ac-tive as demonstrated in previous studies. An extension of this study involves quantifying metabolic state during cannibalism using molar balance or metabolic flux analysis. Further, the model developed in this study was em-pirical and phenomenological, and structured model based on the molecular understanding of the process of Spo0A regulation can be carried out. These studies can also be extended to other species and also be used to study other interactions such as competition and mutualism.

KEYWORDS

- ***Bacillus subtilis***
- **Cannibalism**
- **Ecological interactions**
- **Endospore**
- ***Escherichia coli***
- **Forespore**
- **Gram-positive bacterium**

- **Killing factor**
- **Mathematical models**
- **Metabolites**
- **Methylene blue dye reduction test**
- **Microbial interaction**
- **Mixed culture**
- **Phosphorylation**
- **Population dynamics**
- **Predation**
- **Spore**
- **Sporulation**
- **Survival strategy**

REFERENCES

Allenby, N. E. E.; O'Connor, N.; Pragai, Z.; Ward, A. C.; Wipat, A.; Harwood, C. R. Genome-wide transcriptional analysis of the phosphate starvation stimulation of *Bacillus subtilis*. *J Bacteriol* 2005, 187, 8063–8080.

Allenby, N. E. E.; Watts, C. A.; Homuth, G.; Pragai, Z.; Wipat, A.; Ward, A. C.; Harwood, C. R. Phosphate starvation induces the sporulating killing factor of *Bacillus subtilis*. *J Bacteriol* 2006, 188, 5299–5303.

Antelmann, H.; Scharf, C.; Hecker, M. Phosphate starvation-inducible proteins of *Bacillus subtilis* proteomics and transcriptional analysis. *J Bacteriol* 2006, 182, 4478–4490.

Bailey, J. E.; Ollis, D. F. *Biochemical engineering fundamentals*. McGraw Hill: New York, 1986.

Bapat, P.; Nandy, S. K.; Wangikar, P.; Venkatesh, K. V. Quantification of metabolically active biomass using methylene blue dye reduction test (MBRT): Measurement of CFU in about 200 s. *J Microbiol Met* 2006, 65, 107–116.

Britton, R. A.; Eichenberger, P.; Gonzalez-Pastor, J. E.; Fawcett, P.; Monson, R.; Losick, R.; Grossman, A. D. Genome-wide analysis of the stationary-phase sigma factor (sigma-H) regulon of *Bacillus subtilis*. *J Bacteriol* 2002, 184, 4881– 4890.

Burbulys, D.; Trach, K. A.; Hoch, J. A. Initiation of sporulation in *Bacillus subtilis* is controlled by a multicomponent phosphorelay. *Cell* 1991, 64, 545–552.

Cushing, J. M. A simple model of cannibalism. *Mat Bios* 1991, 107, 47–71.

Diekmann, O.; Nisbet, R. M.; Gurney, W. S. C.; Van den Bosch, F. Simple mathematical models for cannibalism: A critique and a new approach. *Mat Bios* 1986, 78, 21–46.

Driks, A. *Bacillus subtilis* spore coat. *Microbiol Mol Biol Rev* 1999, 63, 1–20.

Ellermeier, C. D.; Hobbs, E. C.; Gonzalez-Pastor, J. E.; Losick, R. A. Three-protein signaling pathway governing immunity to a bacterial cannibalism toxin. *Cell* 2006, 124, 549–559.

Engelberg-Kulka, H.; Hazan, R. Cannibals defy starvation and avoid sporulation. *Science* 2003, 301, 467–468.

Eymann, C.; Mach, H.; Harwood, C. R.; Hecker, M. Phosphate-starvation-inducible proteins in *Bacillus subtilis*: A two-dimensional gel electrophoresis study. *Microbiol* 1996, 142, 3163–3170.

Fawcett, P.; Eichenberger, P.; Losick, R.; Youngman, P. The transcriptional profile of early to middle sporulation in *Bacillus subtilis*. *Proc Natl Acad Sci USA* 2000, 97, 8063–8068.

Froelich, P. N.; Klinkhammer, G. P.; Bender, M. L.; Luedtke, N. A.; Heath, G. R.; Cullen, D.; Dauphin, P.; Hammond, D.; Hartman, B.; Maynard, V. Early diagnosis of organic matter in pelagic sediments of the eastern equatorial Atlantic: Suboxicdiagenesis. *Geochim Cosmochim Acta* 1979, 43, 1075–1090.

Fujita, M.; Losick, R. The master regulator for entry into sporulation in *Bacillus subtilis* becomes a cell specific transcription factor after asymmetric division. *Genes Dev* 2003, 17, 1166–1174.

González-Pastor, J. E.; Hobbs, E. C.; Losick, R. Cannibalism by sporulating bacteria. *Science* 2003, 301, 510–513.

Hansen, J. N. Antibiotics synthesized by post translational modification. *Ann Rev Microbiol* 1993, 47, 535–564.

Hecker, M.; Volker, U. Non-specific, general and multiple stress resistance of growth-restricted *Bacillus subtilis* cells by the expression of the σ^B regulon. *Mol Microbiol* 1998, 29, 1129–1136.

Iyer, V. R.; Horak, C. E.; Scafe, C. S.; Botstein, D.; Snyder, M.; Brown, P. O. Genomic binding sites of the yeast cell-cycle transcription factors SBF and MBF. *Nature* 2001, 409, 533–538.

Jiang, M.; Shao, W.; Perego, M.; Hoch, J. A. Multiple histidine kinases regulate entry into stationary phase and sporulation in *Bacillus subtilis*. *Mol Microbiol* 2000, 38, 535–542.

Levin, P. A.; Losick, R. Transcription factor Spo0A switches the localization of the cell division protein FtsZ from a medial to bipolar pattern in *Bacillus subtilis*. *Gen Dev* 1996, 10, 478–488.

Lotka, A. J. *Elements of physical biology*. Williams and Wilkins Co.: Baltimore, 1925.

Lovely, D. R.; Klug, M. J. Sulfate reducers can outcompete methanogens at freshwater sulfate concentrations. *Appl Environ Microbiol* 1983, 45, 187–192.

Lovely, D. R.; Klug, M. J. Model for the distribution of sulfate reduction and methanogenesis in freshwater sediments. *Geochim Cosmochim Acta* 1986, 50, 11–18.

Molle, V.; Fujita, M.; Jensen, S. T.; Eichenberger, P.; Gonzalez-Pastor, J. E.; Liu, J. S.; Losick, R. The Spo0A regulon of *Bacillus subtilis*. *Mol Microbiol* 2003, 50, 1683–1701.

Nandy, S. K.; Bapat, P.; Venkatesh, K.V. Sporulating bacteria prefers predation to cannibalism in mixed cultures. *FEBS Lett* 2007, 581, 151–156.

Nandy, S. K.; Prasad, V.; Venkatesh, K. V. Effect of temperature on the cannibalistic behavior of *Bacillus subtilis*. *Appl Environ Microbiol* 2008a, 74, 7427–7430.

Nandy, S. K.; Venkatesh, K. V. Effect of carbon and nitrogen on the cannibalistic behavior of *Bacillus subtilis*. *Appl Biochem Biotechnol* 2008b, 151, 424–432.

Nandy, S. K.; Venkatesh, K. V. Study of CFU for individual microorganisms in mixed cultures with a known ratio using MBRT. *AMB Exp* 2014, 4, 38.

Prágai, Z.; Harwood, C. R. Regulatory interactions between the Pho and σ^B-dependent general stress regulons of *Bacillus subtilis*. *Microbiol* 2002, 148, 1593–1602.

Prágai, Z.; Allenby, N. E.; O'Connor, N.; Dubrac, S.; Rapoport, G.; Msadek, T.; Harwood, C.R. Transcriptional regulation of the *phoPR* operon in *Bacillus subtilis*. *J Bacteriol* 2004, 186, 1182–1190.

Rao, V. G.; Desai, M.; Kulkarni, N. B. An account of some physiological studies in two species of *Phytophthora*. *Mycopathol* 1965, 30, 121–128.

Ren, B.; Robert, F.; Wyrick, J. J.; Aparicio, O.; Jennings, E. G.; Simon, I. Genome-wide location and function of DNA binding proteins. *Science* 2000, 290, 2306–2309.

Stragier, P.; Losick, R. Molecular genetics of sporulation in *Bacillus subtilis*. *Ann Rev Gen* 1996, 30, 297–341.

Strauch, M.; Webb, V.; Spiegelman, G.; Hoch, J. A. The Spo0A protein of *Bacillus subtilis* is a repressor of the *abrB* gene. *Proc Nat Acad Sci USA* 1990, 87, 1801–1805.

Volterra, V. Variazioni e fluttuazioni del numero individui in specie animali conviventi. *Memorie della R. Acc. dei Lincei* 1926, 2, 31–113.

Westers, H.; Braun, P. G.; Westers, L.; Antelmann, H.; Hecker, M.; Jongbloed, J. D. H.; Yoshikawa, H.; Tanaka, T.; Dijl, J. M. V.; Quax, W. J. Genes involved in SkfA killing factor production protect a *Bacillus subtilis* lipase against proteolysis. *Appl Environ Microbiol* 2005, 71, 1899–1908.

CHAPTER 2

IDENTIFICATION OF MICROORGANISMS CARRIED BY AEOLIAN DUST TO OTHER CONTINENTS AND THEIR IMPACT ON PUBLIC HEALTH

PARASKEVI A. FARAZI

Mediterranean Center for Cancer Research, Department of Life and Health Sciences, University of Nicosia, 46 Makedonitissas Avenue, P.O. Box 24005, Nicosia 1700, Cyprus

CONTENTS

2.1 AEOLIAN DUST

2.1.1 *CHARACTERISTICS, SOURCES, AND TRAVEL OF AEOLIAN DUST*

Aeolian dust originates from wind erosion of the regolith - the loose rock and dust layer of bedrock - and consists of soil particles found in deserts or arid regions. The term "Aeolian" comes from the Greek word "Aeolus," who was the God of winds in ancient Greece. Aeolian dust events arise from deserts of different continents such as the Sahara and Sahel deserts in Africa, Australian deserts (the Simpson and Strzelecki deserts) as well as Lake Eyre Basin and the western sector of the Murray–Darling Basin in Australia, the Taklamakan, Gobi and Badain Jaran deserts as well as the Loess plateau in Asia. The desert dust can actually be transported over long distances by air currents that are able to carry it even to other continents (Revel-Rolland et al., 2006; Yamaguchi et al., 2012). For example, more than one million tons of Asian dust particles, which travel a distance of 3000–5000 km, reach Japan every year (http://www.nies.go.jp/index-e.html). Furthermore, dust from North Africa has been reported in many European countries including Greece (Crete), Spain, Italy, UK, France (Alps), and Scandinavia (Stevenson, 1969; Ricq de Bouard and Thomas, 1972; Bergametti et al., 1989; Nihlén and Mattsson, 1989; Rodá et al., 1993; Franzén et al., 1994). Similarly, dust from Africa reaches the United States and Caribbean (Prospero, 1999; Prospero and Lamb, 2003). At the global level, estimates have shown that 0.5–5 billion tons of desert dust migrates by air annually (Perkins, 2001). The majority of the dust (50–75%) comes from the North African deserts, even though in the last few decades there has been an increased migration of dust from Asia, due to changes in the climate and desertification (Moulin et al., 1997; Goudie and Middleton, 2001; Prospero and Lamb, 2003; Zhang et al., 2003).

In order for dust to enter the atmosphere, there needs to be a large supply of it as well as strong surface winds, such as the low-level jet described in the Bodele region in Africa (Washington and Todd, 2005). In addition, the velocity of the wind matters on whether the dust will be raised in the atmosphere and winds in the range of 6.5–13 m/s have been reported to raise dust in the air in Western Sahara region (Helgren and Prospero, 1987). Strong vertical transport is needed to carry the dust into the troposphere, which occurs under conditions of unstable thermal stratification as well as large-scale synoptic upward motions (Estoque et al., 1986). A large supply of dust is usually found in the most arid settings where there is no vegetation (New et al., 2002). Atmospheric humidity has also been found to be extremely important in the generation and transport of dust. Interestingly, dust concentration increases with relative humidity (Csavina et al., 2014).

Aeolian dust is rich in many mineral and pollutant elements such as zinc (Zn), chlorium (Cl), copper (Cu), lead (Pb), and sulfur (S) (Zhang et al., 2010). In addition, the dust has been shown to carry persistent organic pollutants such as pesticides, polychlorinated biphenyls, and polycyclic aromatic hydrocarbons (Garrison et al., 2006). Nitrate and sulfate ions have also been shown to be transported along with Aeolian dust (Moria et al., 2003). Finally, quartz, crystalline aluminosilicates, iron oxides, and hydroxides have been identified on travelling dust particles (Tondera et al., 2007).

2.1.2 AEOLIAN DUST AS A CARRIER OF MICROORGANISMS

Bacteria have been shown to attach on Aeolian dust with the potential to migrate globally. Bacteria species such as actinobacteria, bacilli, and sphingobacteria were found to be the predominant species in a study of Asian dust (Yamaguchi et al., 2012). Furthermore, bacilli and sphingobacteria were actually able to grow on media suggesting that these species are not merely transported by the dust but are also physiologically active (Yamaguchi et al., 2012). Analysis of African dust reaching the Eastern Mediterranean has revealed several bacteria such as firmicutes, actinobacteria, gammaproteobacteria, betaproteobacteria, and cyanobacteria being carried by the dust. Within spore-forming bacteria (such as firmicutes), bacilli and clostridia have been identified. About 24% of sequenced clones were actually closely related to human, plant, and animal pathogens, which have been linked to diseases such as pneumonia, meningitis, bacteremia, and pathologic reactions like endocarditis (Polymenakou et al., 2008). Bacteria carried by dust have been identified in the Himalayas at a 6000 m altitude and interestingly these bacteria had good viability and growth potential. The most abundant bacteria phyla were acidobacteria, proteobacteria, verrucomicrobia, and actinobacteria (Stres et al., 2013). Similarly, viable bacteria were identified in tropospheric samples from the Gulf of Mexico (DeLeon-Rodriguez et al., 2013), Asian dust traveling to Japan (Hara and Zhang, 2012; Yamaguchi et al., 2012) and North America (Smith et al., 2012). These findings have important implications as bacteria carried by the dust can potentially reach their destinations alive and thus affect the host ecosystem. Bacteria genera such as *Microbacterium*, *Sphingomonas*, *Bacillus*, and *Streptomyces*, as well as the opportunistic pathogen *Pseudomonas aeruginosa* were isolated in the Caribbean during African dust events (Griffin et al., 2001; Griffin et al.,

2003). In addition, bacteria such as *Bacillus*, *Gordonia*, and *Staphylococcus* have been found in African dust reaching the mid-Atlantic (Griffin et al., 2006).

Fungi have also been shown to be carried by Aeolian dust. In California, an outbreak of coccidioidomycosis caused by the fungus *Coccidioides immitis* was reported after a severe dust storm (Williams et al., 1979). More recently, the events of yellow fever associated with coccidioidomycosis were studied in Phoenix Arizona, and associations between yellow fever events and dust storm days were found (Sprigg et al., 2014). Many other types of fungi such as *Acremonium*, *Alternaria*, *Arthrinium*, *Aspergillus*, *Cladosporium*, *Curvularia*, *Emericella*, *Fusarium*, *Nigrospora*, *Paecilomyces*, *Pithomyces*, *Phoma*, *Penicillium*, *Torula*, *Trichophyton*, and *Ulocladium* have been identified to be carried by dust (Griffin, 2007). These fungi are pathogenic and therefore could have important implications in human health and disease outbreaks following dust storms.

Viruses have not been investigated as extensively in terms of their ability to be carried by dust during storms, however, there have been some timing associations between disease outbreaks (such as foot and mouth disease in livestock populations) and dust storms. These studies failed to show the presence of the relevant viruses on dust, however, the occurrence of these disease outbreaks right after dust storms is strongly suggestive of the ability of dust to carry viruses (Ozawa et al., 2001; Joo et al., 2002; Sakamoto and Yoshida, 2002). A more recent study has shown higher concentrations of ambient influenza A virus during Asian dust storms, suggesting that the dust is able to carry viruses during its travel (Chen et al., 2010). Another study has shown the ability of dust particles to induce inflammatory response of macrophages, further suggesting that dust could contain microorganisms such as viruses and bacteria (Higashisaka et al., 2014). In the United States, bioaerosols (which can include bacteria, fungi, viruses, pollen, cell debris, and bio-films), ranging in size between 10 nm and 100 μm have been identified to be transported via dust storms, further implicating dust acting as a carrier of different types of microorganisms, including viruses (Hallar et al., 2011).

2.1.3 THE IMPACT OF AEOLIAN DUST ON PUBLIC HEALTH

Considering that Aeolian dust carries all the aforementioned elements, pollutants, and microorganisms, it comes as no surprise that the dust has a huge impact on public health throughout its travel. For example, in Cyprus, an island in Southeastern Mediterranean, increased hospitalization, as well as increased mortality linked to cardiovascular causes was reported during sand storm days (Middleton et al., 2008; Neophytou et al., 2013). More specifically, a 2.43% increase in cardiovascular mortality was reported for each 10 $\mu g/m^3$ increase in PM_{10} concentration (Neophytou et al., 2013). Ambient concentrations of particulate matter consisting of particles smaller than 10 μm (PM_{10}) have been shown to increase during Aeolian dust events (Pey et al., 2012). Interestingly, in New Zealand all-cause mortality increased by 7% per 10 $\mu g/m^3$ increase in PM_{10} exposure (Hales et al., 2012). Lung cancer and respiratory-associated deaths showed higher associations. One could assume that this effect would also apply to the increased concentrations of PM_{10} associated with Aeolian dust. In fact, a study in Spain revealed an increase of 8.4% in mortality per 10 $\mu g/m^3$ increase in PM_{10} exposure during Sahara dust storm days and only 1.4% increase in mortality during non-Saharan dust days (Perez et al., 2008). A similar effect was reported in a study in Italy, where a 22% increase in respiratory mortality of elderly people over the whole year was associated with Saharan dust events and 33.9% increase in respiratory mortality during the hot season (Sajani et al., 2011). In Athens, Greece, a 0.71% increase in overall mortality was reported to be associated with 10 $\mu g/m^3$ increase in PM_{10} exposure during dust storm days, with greater effects of people older than 75 years of age than younger age groups (Samoli et al., 2011). In Japan, increased hospitalization of children due to asthma was reported during days of Aeolian dust events (Kanatani et al., 2010). Similar results were obtained in a study in the United States (Texas), where increased number of hospital admissions due to asthma and acute bronchitis were observed after dust events (Grineski et al., 2011). In a two-year study in the Caribbean, scientists failed to show an association between asthma-related hospital admissions and dust storm days, however, it is unclear whether this has to do with the environmental setting of the study (i.e., lower pollution levels in Barbados to start with) (Prospero et al., 2008). Finally, allergic skin reactions in Japan during Asian dust events were suggested to be linked to metals bound to and carried by the dust during its travel (Otani et al., 2012).

2.2   METHODS IN AEOLIAN DUST RESEARCH

To reach the conclusion that the Aeolian dust is a carrier of microorganisms and has an impact in public health, several studies have been conducted in different regions of the world. Although these studies have the same overall goal, their methodology differs especially in the following three aspects: (1)

the choice of method of dust collection, (2) the geographic location of dust collection, and (3) the method of microorganism characterization and identification. The next sections summarize these different ways of studying the problem of microorganism transfer by the Aeolian dust to other destinations.

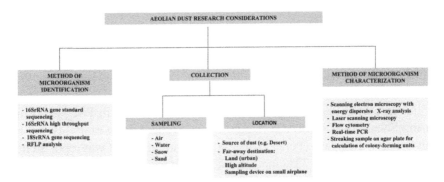

FIGURE 2.1 The important issues a scientist has to consider in the study of the microorganism contents of Aeolian dust. The figure summarizes the considerations a scientist has to make in regards to methods of dust sampling as well as the options for sampling location. In addition, the methods used thus far in order to characterize and identify microorganisms carried by the dust are shown.

2.2.1 METHODS OF AEOLIAN DUST COLLECTION

2.2.1.1 SAMPLING

An interesting method of dust sampling involves use of a fabricated dust sampler with wet beads, which are able to absorb dust particles. Such a method was used by Yamaguchi et al. (2012) who studied the dispersion of bacterial cells on Asian dust (Yamaguchi et al., 2012). The wet beads were packed inside a stainless can and the air was collected through a sterilized Teflon inlet tube. The sampling device can easily be attached on a small airplane to collect air and dust at a higher altitude, as was done in the aforementioned study. The beads are suspended in a certain volume of particle-free water and filtered onto a sterilized polycarbonate membrane filter of small pore size (such as 0.4 µm) to collect the dust particles. The dust is thus trapped on these filters, which can be used for DNA extraction or microorganism growth studies by placing the filters on agar media.

Air samples can also be collected using multi-stage impactors. These impactors are attached to high-volume air samplers and can fractionate particulate matter into different fractions that differ in size and then they are captured on filters. They can be set to collect air at a constant flow rate. Such a sampler was used in the Greek study and was set to collect air particles at constant flow rate of 740 L/min. The dust was trapped on glass fiber filters (Polymenakou et al., 2008). Such impactors are easy to attach on air samplers and use in different land-settings. They are usually put on a platform to avoid contamination from nearby surfaces. For example, Polymenakou et al. (2008) put the impactor on a 5 m high platform.

Other types of collectors can be used for the collection of air samples containing dust. For example, in-house designed portable sampling systems have been used, which consist of a vacuum pump attached to a manifold (PVC-pipe) secured inside a carrying case. Cellulose nitrate membrane filters are placed on the manifold in order to capture dust particles from the air drawn. Such a sampling system was used in a study in Mali in West Africa (Kellogg et al., 2004).

When sampling is performed at higher altitudes (e.g., mountains), where the level of air pollution is minimal compared to urban settings, it can be done in much simpler ways. For example, even a simple collection of air for a longer period of time (e.g., weeks to a month) in tubes might work in a clean environment setting. Such a study was performed to study microorganism diversity in the Himalayas (Stres et al., 2013). In this study, the method of collection involved setting 50 ml sterile falcon tubes prefilled with a layer of 2 cm tall 100% polyester wool (pre-treated with a solution of 10% formaldehyde in 70% glycerol to prevent *in situ* microbial growth). The tubes were left open for 30 days at an altitude of 5000–6000 m every 200 m intervals. The tubes were then rinsed with sterile water and the water was passed through a filter (0.22 μm size) to capture any microorganisms and subsequently isolate DNA for additional analyses (DNA and microorganism count).

At high altitudes in mountainous areas, another form of sample is snow. It is easy to collect and transport for further analysis. For example, Stres et al. (2013) collected snow samples in 50 ml sterile falcon tubes immediately after snowfall to ensure it was clean and fresh, then allowed the snow to thaw and fixed it using formaldehyde to avoid microbial growth other than what was in the snow sample initially. The thawed and fixed snow sample was then filtered on a membrane and processed for further analysis as described earlier (Stres et al., 2013). An alternative method to collecting snow in a sterile way is to remove the top layer of snow and collect snow from below. This was actually done in a study investigating the microbial diversity carried by Aeolian dust to the Alps (Chucochina et al., 2011). In this study, to ensure sterile collection of snow, 2–5 cm of snow in contact with air was carefully removed with a clean shovel and clean snow was collected underneath using sterile plastic containers. To further ensure sterility, the researchers were wearing single-use Tyvek coveralls and vinyl gloves to avoid any bacterial contamination other than the bacteria contained in the snow. In this case, the snow was allowed to thaw and was concentrated using centrifugal filter devices which allow concentration and desalting of solutions. Concentration is achieved by ultrafiltering the sample solution through an anisotropic membrane. The centrifugal force essentially directs solvents and low molecular weight solutes through the membrane into the filtrate vial, whereas macrosolutes are retained above the membrane inside the sample reservoir.

An interesting new approach to collect dust samples for investigation of potential microorganisms carried by the dust involves the use of an air-sampling device that is placed under the wing of an F-104 Starfighter jet and can capture particles directly from clouds at a maximum altitude of 25000 ft (http://uk.mobile.reuters.com/article/environmentNews/idUK-BRE9B510F20131206). The device (called DART - Dust at Altitude Recovery Technology) is currently being used by researchers at the University of Florida and the results should come out in the coming years (http://www.insidescience.org/content/tracking-dust-around-world/1643; http://phys.org/news/2013-12-high-altitude-device-airborne-pathogens. html). The fact that collection of microorganisms can be performed right before the entry of the dust into the destination route allows the identification of microorganisms carried by the dust, which would have not mixed with local microorganisms, thus making the results much clearer.

All the methods of Aeolian dust collection described thus far involve collection of the dust from air or snow samples. Dust can also be collected from water samples, for example water from lakes. A study like this was conducted in Spain, where the method of collection involved taking water samples from the surface of lakes in the Pyrenees ranging in altitude from 1620–2240 m above sea level. The water samples were filtered twice through 0.2 µm polycarbonate membranes to exclude local contaminating microbes (Hervàs et al., 2009). The filtered water sample can then be processed further through a filter as described above to capture microorganisms or directly as a source of the microorganisms for further characterization and identification studies.

Aeolian dust originates from the deserts and thus a sample of it can be collected directly from the source that is the desert sand. In fact, many studies have collected desert sand samples in search of microorganisms. Typically, a certain amount of sand is collected and used for further analysis. For example, in a study aimed at identifying gamma radiation resistant bacteria in the Sahara desert, sand samples were collected, subjected to a dose of 15 kGy gamma radiation, and subsequently plated on agar plates to identify colony-forming bacteria. Such bacteria would be good candidates for travelling to far-away destinations and surviving harsh conditions (de Groot et al., 2005). Another study used sand samples from the Republic of Chad in Africa, which were collected aseptically in tubes and used for microbial characterization and DNA isolation analyses (Giongo et al., 2013).

2.2.1.2   GEOGRAPHIC LOCATION

Aeolian dust can be collected in different environmental settings. For example, it can be collected at the source (i.e., the desert sand itself or air samples from the desert of dust origin) (Kellogg et al., 2004; de Groot et al., 2005; Giongo et al., 2013). Since the impact of Aeolian dust on destinations far from the source is huge, it is also important to collect dust samples from far away destinations of Aeolian dust travel. The latter can be achieved in different ways. Some investigators choose to collect air samples on land (just a few feet from the ground) (Polymenakou et al., 2008). For example, in a Greek study sampling was done on a 5 m high platform (located at the University of Crete Campus) with the aim to reduce possible effects from near-surface sources (Polymenakou et al., 2008). The problem with such studies is that their urban setting increases the probability that microorganisms of local origin (due to local air pollution) will be collected along with the Aeolian dust sample. Thus, in such a setting it would be important to compare dust collection on days with Aeolian dust events and days with pollution from other sources. Of course, this increases the expenses as more sequencing will need to be undertaken.

Other investigators try to avoid the local contamination problem by setting up air collection at high altitudes (e.g., in the Alps or the Himalayas) (Chucochina et al., 2011; Stres et al., 2013). In the Himalayas for example, air sampling has been performed at an altitude of 5000–6000 m (Stres et al., 2013). At high altitudes, sample collection can also be achieved by collecting snow samples (Chucochina et al., 2011; Stres et al., 2013). For example, in the Alps in France snow was sampled at an altitude of 4250 m from days with Saharan dust events (during which the snow had a reddish-brown color) and clean days (during which the snow had a clean color) (Chucochina et al., 2011). Even at high altitudes investigators still choose to sample at other points for comparison. For example, in the study of dust in the Alps samples from a nearby city of Grenoble (at 200 m altitude) during rain deposition were collected as well as samples from the Sahara desert for comparisons of microorganism contents.

An interesting way to collect air samples containing Aeolian dust is by collecting air near the destination route at a high altitude using an airplane. For example, a study in Japan was conducted by taking samples at an altitude of 900 m about 10 km away from the coasts, which ensures that the dust collected is not mixed with local pollution and that the microorganisms identified were carried by the dust (Yamaguchi et al., 2012). Similarly, an impactor device was mounted within a collector housing located on the underside of a Lockheed Martin ER-2 to collect air samples at 20000 m in two different locations in the United States (Griffin, 2004). Samples collected at such high altitudes minimize the chance of picking up local microorganisms, thus making the identification of microorganisms carried by the Aeolian dust easier. Of course they require a more refined set up, however, once the set up is in place the study of microorganism identification is more straight-forward.

2.2.2 METHODS OF CHARACTERIZATION OF MICROORGANISMS CARRIED BY AEOLIAN DUST

Once dust samples are collected from various sources as described above, it is important to characterize their contents. Techniques such as scanning electron microscopy (SEM) with energy dispersive X-ray analysis have been used previously to study the chemical composition of the dust in elements such as hydrogen, metals, and so forth as well as the size of the dust particles (Matsuyama et al., 2008; Yamaguchi et al., 2012). SEM focuses an electron beam across the surface and detects secondary or backscattered electron signals that allow the researcher to obtain detailed high-resolution images of the sample. An energy dispersive X-ray analyzer is used along with the SEM and it provides identification of elements and information on the quantitative composition of the sample. This technique was used to study dust in Japan for example (Yamaguchi et al., 2012).

To visualize bacteria on the surface of dust particles (in order to quantify bacterial abundance) scientists can use a laser-scanning microscope attached to a microspectrophotometer. This technique was used in the aforementioned study of dust in Japan (Yamaguchi et al., 2012). Nucleic acids were stained with a fluorescent dye to distinguish bacteria from dust. Essentially, dust particles are first suspended in water, fixed, and filtered on polycarbonate membrane filters for staining. The technique can be quite tricky because of high fluorescence background from soil itself. Efforts to reduce sample high auto-fluorescence have been successful by using SYBR Green-I (which stains the nucleic acids) and treating the specimens with hydrofluoric acid (Morono et al., 2009). Bacterial cell concentrations in dust samples can also be measured using flow cytometry as was done in a study of dust in the Alps (Chucochina et al., 2011). In this technique, samples are fixed (e.g., in glutaraldehyde), mixed with a dye (e.g., SYBR Green I) and put through a flow cytometer. Bacterial abundance has also been successfully determined using 16S rRNA gene quantification by real-time PCR (Yamaguchi et al., 2012). The colony-forming units of bacteria present per gram (g) of sand are calculated by collecting the sand sample and then cultivating bacteria on agar by streaking a specific amount of sand (suspended in saline) on agar (Giongo et al., 2013).

2.2.3   METHODS OF IDENTIFICATION OF MICROORGANISMS CARRIED BY AEOLIAN DUST

A popular method for identifying microorganisms involves sequencing of the 16S rRNA gene which is shared among all bacteria but bears some sequence differences since it has nine variable regions (Weisburg et al., 1991; Cox et al., 2013). Sequencing of this gene allows the identification of bacteria on the dust. The 16S rRNA has been described as a molecular chronometer (Woese, 1987). It is highly conserved due to its critical function in the cell as a component of the ribosomes, which play an essential role in protein translation. The 16S rRNA gene is about 1550 bp in length and is composed of both conserved and variable regions (Cox et al., 2013). Universal primers, which are complementary to the conserved regions, are used to amplify the gene in any bacterium and subsequently sequencing allows the identification of the slight sequence differences between bacteria. Genbank, which contains the sequences of all sequenced bacteria thus far, and other databases such as the ribosomal database project (http://rdp.cme.msu.edu/), Greengenes (http://greengenes.lbl.gov/cgi-bin/nph-index.cgi), and SILVA (http://www.arb-silva.de/) are used to make comparisons of the sequences retrieved from experimental samples with the database sequences so as to enable matching and identification of bacteria (Clarridge, 2004). Various software products have been developed to assist in these comparisons such as QIIME (http://qiime.org/) and Mothur (http://www.mothur.org/).

Typically the sample DNA is amplified by PCR using universal primers for 16S rRNA, cloned into a vector and then a certain number of clones are sequenced using the same primers for bacteria identification (Kellogg et al., 2004; Polymenakou et al., 2008; Chucochina et al., 2011). The same technique can be used for 18S rRNA in order to identify fungi, which can also be identified by microscopy (Kellogg et al., 2004). When DNA is isolated from bacterial colonies on agar plates then cloning is not necessary and bacterial DNA is PCR-amplified and sequenced immediately (de Groot et al., 2005). Sequencing can also be coupled to restriction fragment length polymorphism (RFLP) analysis, which allows classification of bacteria into operational taxonomic units based on their restriction pattern (Wei et al., 2007; Polymenakou et al., 2008; Chucochina et al., 2011). In addition to standard sequencing, high throughput sequencing can be used for bacteria identification, which does not require cloning of the 16S rRNA PCR products. In fact, pyrosequencing has been successfully used for this type of studies (Stres et al., 2013). Pyrosequencing is a sequencing method based on the generation of light every time a nucleotide is added in the sequencing reaction. Essentially pyrophosphate (PPi) is released after the addition of a nucleotide to the growing chain. The PPi is then converted into adenosine triphosphate (ATP) by the enzyme ATP sulfurylase in the presence of ammonium persulfate (APS). ATP then drives the conversion of luciferin to oxyluciferin with the help of the enzyme luciferase and ultimately light is generated and detected to allow for reading the sequence.

A study of sand samples from the desert attempted to use three different molecular approaches to identify bacteria in order to assess their efficiency. To this end, the scientists: (1) amplified the 16S rRNA gene by PCR, cloned the PCR product and then sequenced, (2) amplified the 16S rRNA gene and then performed high throughput sequencing, and (3) performed high throughput sequencing directly on the DNA sample without amplifying it. Cloning the PCR products reduced the number of bacteria the scientists were able to identify to 13 genera. High throughput sequencing was more efficient. A total of 270 genera were identified using high throughput sequencing of amplified DNA and 260 genera were identified using high throughput sequencing of non-amplified DNA. Of note, high throughput sequencing of non-amplified DNA allowed the identification of Archaea, which was not possible with the other two methods (Giongo et al., 2013). This development is important, as sequencing a DNA sample directly without having to amplify it first is more efficient.

2.2.4 METHODS OF IDENTIFICATION OF LIVE MICROORGANISMS CARRIED BY AEOLIAN DUST

Identifying the microorganisms carried by the dust from the deserts to far-away destinations is important, however, it is even more important to determine which microorganisms actually make it to the destination alive. The aforementioned methods allow for the identification of bacteria, however, this could be meaningless if the bacteria are not alive and therefore would have no impact on health. Along these lines several researchers tried to investigate the growth potential of the bacterial cells on the dust particles by incubating dust particles in different types of media (nutrient-rich and nutrient-poor liquid media as well as on agar media). To this end, Hervàs et al. (2009) tried to identify microbes that are actually viable in lakes in the Pyrenees in Spain ranging in altitude from 1620–2240 m above sea level (after a major dust storm event) as well as sandy soil from Mauritania in the Sahel region. The samples were filtered twice to exclude local contaminating microbes and then supplemented with different nutrients to enrich for any bacteria carried by the dust that are actually viable. The supplements included: (1) organic carbon (sodium acetate 1 mM, final concentration), (2) nitrogen (casamino acids 0.04% w/v, final concentration), and (3) phosphorous (KH_2PO_4 10 mM, final concentration) (Hervàs et al., 2009). Identification of the viable bacteria was then carried out by 16S rRNA gene sequencing as already described.

Several studies have attempted to obtain counts of viable microbes from air samples of small volumes (< 200 L) and have shown an order of hundreds of bacteria and fungi that are capable of surviving dust transport over long distances. Live microorganisms identified in the Virgin Islands included plant pathogens (25%) and opportunistic human pathogens (10%) (Griffin et al., 2001). In this study, filters onto which dust from air was collected were plated and grown on R2A agar, which is a low-nutrient medium suitable for culturing stressed microbes (Reasoner and Geldreich, 1985). Similar investigations in Africa have shown live animal pathogens (10% of bacteria), plant pathogens (5% of bacteria), and opportunistic human pathogens (27% of bacteria) (Kellogg et al., 2004). An in-house designed portable sampling system containing a pump and a manifold was used to collect air samples.

The samples were collected on a cellulose nitrate membrane filter of 0.2 μm size. The filters were plated on R2A agar and grown as in the previous study.

2.2.5 METHODS OF INVESTIGATION OF THE IMPACT OF AEOLIAN DUST ON PUBLIC HEALTH

To investigate the impact of microorganisms carried by the Aeolian dust on human health, scientists try to make associations of dust episodes, identification of microorganisms, and increases in the incidence of particular diseases, infections, and so forth. One example is the investigation of the link between Asian dust in Japan and hand, foot, and mouth disease (HFMD). To achieve this, the scientists collected bioaerosol samples (after dust events) from rooftops in an attempt to investigate whether the enterovirus was present in these air samples. At the same time, they analyzed hospital records and studied the hospitalization rate for HFMD during dust days as a function of suspended particulate matter (SPM, which increases following dust events). Indeed, hospitalization rates increased in association with SPM, however, scientists failed to identify the enterovirus (http://ehp.niehs.nih.gov/isee/p-2-02-13/). Studies of asthma and allergies in children have identified dust as a predictor for increased incidence of these conditions using logistic regression (Bener et al., 1996). In addition, another hint that dust has implications in respiratory diseases and public health comes from the fact that areas that are severely affected by desert dust such as the Aral Sea and the Caribbean have the highest asthma rates in the world (Howitt, 2000). In Barbados for example, asthma rates increased 17 times in just over two decades (highest rates in the 18–23% range) (Howitt et al., 1998). In Trinidad, a retrospective study has shown increased hospital visits due to pediatric asthma during African dust events (Gyan et al., 2005). It has been found that fungal and bacterial spores as well as molecules of the microbes such as endotoxins and mycotoxins can trigger respiratory stress. Thus, one could infer from these observations that a potential explanation of higher asthma rates after dust storms involves the contents of dust in microorganisms and toxins (Braun-Fahrlander et al., 2002; Griffin and Kellogg, 2004).

Another way to study the impact of the dust on human health is to study its effect in various biological processes. Dust for example, has been shown to cause lipid peroxidation and DNA damage (Athar et al., 1998). In addition, study of the elements and other chemicals carried by dust can reveal ways in which the dust can have an impact on public health. The finding that dust carries radioactive elements, such as ^{90}Sr and ^{137}Cs (which can be toxic and carcinogenic), suggests that Aeolian dust can have significant long-term effects in public health, including an increase in the burden of cancer (Igarashi et al., 2011). In addition, chromium which is carcinogenic in some forms has been identified as one of the largest elemental components of Aeolian dust, raising the possibility of an impact of the dust on cancer (Liu et al., 2011; Aydin et al., 2012). The dust has also been shown to carry persistent organic pollutants such as pesticides, polychlorinated biphenyls, and polycyclic aromatic hydrocarbons (Garrison et al., 2006). There has not been a demonstrated association between Aeolian dust and cancer yet nor has there been direct evidence that dust has carcinogenic properties at the molecular level. This represents an area open for investigation.

2.3 FUTURE PERSPECTIVES IN AEOLIAN DUST RESEARCH METHODS

2.3.1 ALTERNATIVE METHODS TO CAPTURE AND CHARACTERIZE MICROORGANISMS IN AEOLIAN DUST

2.3.1.1 MICROORGANISM COLLECTION

Bacteria and fungi have also been successfully captured and identified from the stratosphere at heights up to 41 km. The collection of air sample in this case was achieved using the deployment of balloon-borne cryosamplers as reported earlier (Lal et al., 1996; Brugger et al., 2012). The probes in these cryosamplers were carefully sterilized to ensure that the microorganisms captured would be those from the stratosphere and no other contaminating microorganisms. The microorganisms in this study were identified and studied by SEM. Even though this method has not been used on Aeolian dust samples, it can easily be adopted for this sort of study. Of course, it needs to be adjusted to the altitude that dust can be lifted to during dust storms. Aeolian dust might only be able to reach lower levels of the stratosphere (Griffin, 2004).

2.3.1.2 MICROORGANISM QUANTIFICATION AND CHARACTERIZATION

Quantifying bacteria on dust samples can be tricky as mentioned in a previous section due to high fluorescence background of the soil. However, if bacteria are captured on agar plates, then quantifying them would be easier. Techniques such as automated counting of bacterial colony-forming units on agar plates have been developed which make use of an illuminator, a camera, and colony-counter software algorithms. The application of such techniques can help quantify the bacterial abundance in dust samples across different conditions (Wainwright et al., 2003). Such techniques can be applied to quantify bacterial number on Aeolian dust samples.

Another interesting technique used to characterize bacteria and other microorganisms such as fungi is solid-phase cytometry (SPC) which combines epifluorescence microscopy with flow cytometry. In this technique the sample containing suspect microorganisms is filtered, fluorescently labeled, scanned through a laser which excites the fluorophore and analyzed through computer software. The advantage of using SPC is that it is fast and works well with low numbers of microorganisms. One limitation of the technique is the limited availability of compatible stains and the requirement for filterable samples that are clear, aqueous solutions (Williamson et al., 2003).

2.3.1.3   ISOLATION OF VIRUSES FROM DUST

Bacteria are relatively easy to isolate from dust samples, however, viruses present many challenges. Work on the extraction of viruses from soil has shown that the success of the extraction depends on the elution buffer used as well as the enumeration technique used. For the latter, plaque assay, epifluorescence microscopy, and transmission electron microscopy were compared. It was shown that the constitution of each soil affected the success of each of these methods, thus requiring extensive testing to determine the most efficient method of isolating viruses in a particular soil setting (Lies et al., 2010). When it comes to dust investigations, dust from different deserts or dust from the source versus the destination should be experimentally tested to identify the best way to isolate viruses. This is essential in order to ensure complete characterization of viruses in these samples. Viruses represent a major public health concern, thus their efficient identification and characterization would be essential. These findings should be taken into consideration in studies of Aeolian dust aimed at identifying viruses.

2.3.2   ALTERNATIVE METHODS TO CAPTURE LIVE MICROORGANISMS FROM AEOLIAN DUST

There are various ways to capture live microorganisms from air samples and grow them, which can be utilized in studies of Aeolian dust. The simplest way is sedimentation, which basically involves exposing agar plates to the air. The disadvantage of this method is that larger particles will settle on the plate first and thus prohibit smaller particles from settling. In addition, agar can dry out, which reduces the chances of microbial growth. Finally, the method is very crude as there is no way to measure the volume of air sampled and thus makes reproducing the sampling impossible (http://www.foodquality.com/details/article/878155/Air_Sampling_101. html?tzcheck=1). Another way to capture live microorganisms involves the use of an impactor, which basically is a jet that draws air in the sampler by vacuum and directs the air to the surface of a petri dish with agar. The dish is on a turntable, thus allowing the dish to turn and collect different microorganisms in different areas of the plate. One of the drawbacks of this method is that it can potentially miss smaller particles. Media dehydration in this case as well can limit the growth of microorganisms (http://www. appliedphysicsusa.com/MicrobialAirSamplers.html; http://electroiq.com/ blog/1997/06/examining-ways-to-capture-airborne-microorganisms/).

Sieve samplers represent another way to sample air. The setup of these samplers involves a perforated plate which is in front of an agar plate and allows air to pass through. A sieve sampler can also be set up in a stacked way, where several perforated plates and agar plates are set one after the other with the size of the perforation reducing as the air goes through the arrangement of plates. This allows for the capture of larger particles in the first plates and smaller particles in the last plates. One of the problems with the sieve samplers is that if the sampling time is not short or if there is not high humidity the area around the perforation on the agar plate can dry out consequently inhibiting the growth of certain microorganisms. In addition, sieve samplers are not very good at capturing small particles, unless the air flow rate is fast, which can affect the viability of the microorganisms.

Another way to obtain air samples is through centrifugal samplers. This type of sampler basically uses an impeller inside an open shallow drum, which draws air in by creating a vortex. A centrifugal force accelerates the air into the drum, which is lined with a thin layer of agar, thus allowing capture and growth of microorganisms. Once again, these samplers fail to capture very small particles, for example, less than 10% of particles smaller than 2 µm actually get deposited.

To overcome the difficulty of capture of small particles, glass impingers can be used. These basically draw air through a curved suction tube and accelerate the air toward the bottom through a jet. The particles are captured in a liquid at the bottom and then processed for microorganism growth studies. As already mentioned, these impingers actually allow the capture of particles of all sizes including very small particles. Their major disadvantage is that because they are made of glass they are not disposable and actually need to be prepared every time for sampling. However, for the purpose of collecting air samples during a dust event these impingers might work quite well. Cooling of the sample here is essential to avoid evaporation of the liquid and thus loss of microorganisms from the sample.

A very efficient sampling technology is gelatin membrane filtration. The flow rate can be programmed and microbes are captured on a gelatin membrane filter that is about 300 μm thick. The filter is porous allowing microorganisms to be captured inside as well, not just at the surface and actually, it allows for the capture of even the smallest microorganisms, such as viruses. In addition, since the filter is made up of 50% water it does not present problems of drying out and thus the viability of microorganisms is unaffected, contrary to all other air collection methods described. The filter is easy to remove and place on a standard agar petri dish for microorganism growth. In addition, the filter dissolves very easily in the agar. If microorganisms are to be captured on filters instead of agar plates, then the choice of filters when collecting air samples is crucial in ensuring the viability of captured microorganisms. It was found that HEPA filters for example support the viability of captured microorganisms for even up to 210 days for some types (Mittal et al., 2011).

Finally, to overcome the low capture efficiency and low viability problems encountered when sampling microorganisms on agar plates due to drying of the agar, a new instrument was recently developed, which rotates the agar plates, thus allowing for continuous more efficient collection of microorganisms and water retention on the agar (essentially rotation prevents drying of the agar surface and improves microorganism viability) (http://www.mbv.ch/documents/paper_continuos_monitoring_2012.pdf).

2.4   CONCLUSION

The study of the microorganism contents of Aeolian dust is an exciting area of research that is expected to benefit public health tremendously. If scientists manage to put together a detailed catalogue of all the microorganisms the dust is able to carry with it alive, this will pave the way for better understanding the dust's impact on human disease. In addition, by knowing what microorganisms to expect with dust episodes, health officials can more efficiently screen dust particles for these microorganisms, identify them quickly and eradicate them in an attempt to prevent infectious disease outbreaks following Aeolian dust episodes. The methods that are necessary to collect dust particles and identify microorganisms carried by the dust are in place as described in earlier sections (Fig. 2.1). Up to now research studies have been performed on an individual group basis. It is now time to think about creating a network of scientists that work together to study the microorganism contents of the dust in different continents using standard protocols and collection methods. This will help better understand the global impact of Aeolian dust and manage its negative health effects.

KEYWORDS

- **16S rRNA sequencing**
- **Aeolian dust**
- **Air sampling device**
- **Asthma**
- **Microorganisms**
- **Multi-stage impactors**
- **Public health**
- **Respiratory diseases**
- **Scanning electron microscopy**

REFERENCES

Athar, M.; Iqbal, M.; Beg, M. U.; Al-Ajmi, D.; Al-Muzaini, S. Airborne dust collected from Kuwait in 1991–1992 augments peroxidation of cellular membrane lipids and enhances DNA damage. *Environ Int* 1998, 24, 205–212.

Aydin, F.; Aydin, I.; Erdogan, S.; Akba, O.; Isik, B.; Hamamci, C. Chemical characteristics of settled particles during a dust-storm. *Pol J Environ Stud* 2012, 21, 533–537.

Bener, A.; Abdulrazzaq, Y. M.; Al-Mutawwa, J.; Debuse, P. Genetic and environmental factors associated with asthma. *Hum Biol* 1996, 68, 405–414.

Bergametti, G.; Gomes, L.; Remoudaki, E.; Desbois, M.; Martin, D.; Buat-Menard, P. Present transport and deposition patterns of African dusts to the north-western Mediterranean. In *Palaeoclimatology and palaeometeorology: Modern and past patterns of global atmospheric transport*, Leinen, M., Sarnthein, M., Eds.; Kluwer Academic Publishers: Dordrecht, 1989, pp 227–252.

Braun-Fahrlander, C.; Riedler, J.; Herz, U.; Eder, W.; Waser, M.; Grize, L.; Maisch, S.; Carr, D.; Gerlach, F.; Bufe, A.; Lauener, R. P.; Schierl, R.; Renz, H.; Nowak, D.; von Mutius, E. Environmental exposure to endotoxin and its relation to asthma in school-age children. *N Engl J Med* 2002, 347, 869–877.

Brugger, S.; Baumberger, C.; Jost, M.; Jenni, W.; Brugger, U.; Mühlemann, K. Automated counting of bacterial colony forming units on agar plates. *PLoS One* 2012, 7(3), e33695.

Chen, P. S.; Tsai, F.; Lin, C.; Yan, C. U.; Chan, C. C.; Young, C. Y.; Lee, C. H. Ambient influenza and avian influenza virus during dust storm days and background days. *Environ Health Perspect* 2010, 118(9), 1211–1216.

Chucochina, M.; Marie, D.; Chevaillier, S.; Petit, J.; Normand, P.; Alekhina, I.; Bulati, S. Community variability of bacteria in Alpine snow (Mont Blanc) containing Saharan dust deposition and their snow colonisation potential. *Microbes Environ* 2011, 26(3), 237–247.

Clarridge, J. Impact of 16S rRNA gene sequence analysis for identification of bacteria on clinical microbiology and infectious diseases. *Clin Microbiol Rev* 2004, 17(4), 840–862.

Cox, M.; Cookson, W.; Moffatt, M. Sequencing the human microbiome in health and disease. *Hum Mol Genet* 2013, 22(1), 88–94.

Csavina, J.; Field, J.; Félix, O.; Corral-Avitia, A.; Sáez, E.; Betterton, E. Effect of wind speed and relative humidity on atmospheric dust concentrations in semi-arid climates. *Sci Total Environ* 2014, 487, 82–90.

De Groot, A.; Chapon, V.; Servant, P.; Christen, R.; Saux, M.; Sommer, S.; Heulin, T. *Deinococcus deserti* sp. nov., a gamma-radiation tolerant bacterium isolated from the Sahara desert. *Int J Syst Evol Microbiol* 2005, 55, 2441–2446.

DeLeon-Rodriguez, N.; Lathem, T. L.; Rodriguez-R, L. M.; Barazesh, J. M.; Anderson, B. E.; Beyersdorf, A.; Ziemba, L.; Bergin, M.; Nenes, A.; Konstandinidis, K. Microbiome of the upper troposphere: Species composition and prevalence, effects of tropical storms, and atmospheric implications. *Proc Natl Acad Sci USA* 2013, 110(7), 2575–2580.

Estoque, M.; Fernandez-Partagas, J.; Helgren, D. M.; Prospero, J. M. *Genesis of major dust storms in West Africa during the summer of 1974.* U.S. Army Research Office: Florida, 1986.

Franzén, L. G.; Hjelmroos, M.; Kållberg, P.; Brorström-Lundén, E.; Juntto, S.; Savolainen, A. L. The yellow snow episode of northern Fennoscandia, March 1991 - a case study of long-distance transport of soil, pollen and stable organic compounds. *Atmos Environ* 1994, 28(22), 3587–3604.

Garrison, V. H.; Foreman, W. T.; Genualdi, S.; Griffin, D. W.; Kellogg, C. A.; Majewski, M. S.; Mohammed, A.; Ramsubhag, A.; Shinn, E. A.; Simonich, S. L.; Smith, G. W. Saharan dust - a carrier of persistent organic pollutants, metals and microbes to the Caribbean? *Rev Biol Trop* 2006, 54(S3), 9–21.

Giongo, A.; Favet, J.; Lapanje, A.; Gano, K.; Kennedy, S.; Davis-Richardson, A.; Brown, C.; Beck, A.; Farmerie, W.; Cattaneo, A.; Crabb, D.; Aung, Y.; Kort, R.; Brumsack, H.; Schnetger, B.; Broughton, W.; Gorbushina, A.; Triplett, E. Microbial hitchhikers on intercontinental dust: High-throughput sequencing to catalogue microbes in small sand samples. *Aerobiologia* 2013, 29(1), 71–84.

Goudie, A. S.; Middleton, N. J. Saharan dust storms: Nature and consequences. *Earth Sci Rev* 2001, 56, 179–204.

Griffin, D.; Kellogg, C. Dust storms and their impact on ocean and human health: Dust in earth's atmosphere. *Eco Health* 2004, 1, 284–295.

Griffin, D. Atmospheric movement of microorganisms in clouds of desert dust and implications for human health. *Clin Microbiol Rev* 2007, 20(3), 459–477.

Griffin, D. Terrestrial microorganisms at an altitude of 20,000 m in Earth's atmosphere. *Aerobiologia* 2004, 20(2), 135–140.

Griffin, D. W.; Garrison, V. H.; Herman, J. R.; Shinn, E. A. African desert dust in the Caribbean atmosphere: Microbiology and public health. *Aerobiologia* 2001, 17, 203–213.

Griffin, D. W.; Kellogg, C. A.; Garrison, V. H.; Lisle, J. T.; Borden, T. C.; Shinn, E. A. African dust in the Caribbean atmosphere. *Aerobiologia* 2003, 19, 143–157.

Griffin, D. W.; Westphal, D. L.; Gray, M. A. Airborne microorganisms in the African desert dust corridor over the mid-Atlantic ridge, Ocean Drilling Program, Leg 209. *Aerobiologia* 2006, 22, 211–226.

Grineski, S.; Staniswalis, J.; Bulathsinhala, P.; Peng, Y.; Gill, T. Hospital admissions for asthma and acute bronchitis in El Paso, Texas: Do age, sex, and insurance status modify the effects of dust and low wind events? *Environ Res* 2011, 111(8), 1148–1155.

Gyan, K.; Henry, W.; Lacaille, S.; Laioo, A.; Lamesee-Eubanks, C.; McKay, S.; Antoine, R.; Monteil, M. African dust clouds are associated with increased paediatric asthma accident and emergency admissions on the Caribbean island of Trinidad. *Int J Biometeorol* 2005, 49, 371–376.

Hales, S.; Blakely, T.; Woodward, A. Air pollution and mortality in New Zealand: Cohort study. *J Epidemiol Community Health* 2012, 66, 468–473.

Hallar, G.; Chirokova, G.; McCubbin, I.; Painter, T.; Wiedinmyer, C.; Dodson, C. Atmospheric bioaerosols transported via dust storms in the western United States. *Geophys Res Lett* 2011, 38, L7801.

Hara, K.; Zhang, D. Bacterial abundance and viability in long-range transported dust. *Atmos Environ* 2012, 47, 20–25.

Helgren, D. M.; Prospero, J. M. Wind velocities associated with dust deflation events in the Western Sahara. *J Appl Meteorol* 1987, 26, 1147–1151.

Hervàs, A.; Camarero, L.; Reche, I.; Casamayor, E. Viability and potential for immigration of airborne bacteria from Africa that reach high mountain lakes in Europe. *Environ Microbiol* 2009, 11(6), 1612–1623.

Higashisaka, K.; Fujimura, M.; Taira, M.; Yoshida, T.; Tsunoda, S.; Baba, T.; Yamaguchi, N.; Nabeshi, H.; Yoshikawa, T.; Nasu, M.; Yoshioka, Y.; Tsutsumi, Y. Asian dust particles induce macrophage inflammatory responses via mitogen-activated protein kinase activation and reactive oxygen species production. *J Immunol Res* 2014, 856154, http://dx.doi.org/10.1155/2014/856154.

Howitt, M. E. Asthma management in the Caribbean - an update. *Postgraduate Doctor – Caribbean* 2000, 16, 86–104.

Howitt, M. E.; Naibu, R.; Roach, T. C. The prevalence of childhood asthma and allergy in Barbados: The Barbados national asthma and allergy study. *Am J Respir Crit Care Med* 1998, 157, A624.

Igarashi, Y.; Fujiwara, H.; Jugder, D. Change of the Asian dust source region deduced from the composition of anthropogenic radionuclides in surface soil in Mongolia. *Atmos Chem Phys* 2011, 11, 7069–7080.

Joo, Y. S.; An, S. H.; Kim, O. K.; Lubroth, J.; Sur, J. H. Foot-and mouth disease eradication efforts in the Republic of Korea. *Can J Vet Res* 2002, 66, 122–124.

Kanatani, K.; Ito, I.; Al-Delaimy, W.; Adachi, Y.; Mathews, W.; Ramsdell, J. The Toyama Asian desert dust and asthma study team. Desert dust exposure is associated with increased risk of asthma hospitalization in children. *Am J Respir Crit Care Med* 2010, 182, 1475–1481.

Kellogg, C.; Griffin, D.; Garrison, C.; Peak, K.; Royall, N.; Smith, R. R.; Shinn, E. A.. Characterization of aerosolized bacteria and fungi from desert dust events in Mali, West Africa. *Aerobiologia* 2004, 20, 99–110.

Lal, S.; Archarya, Y. B.; Patra, P. K.; Rajaratnam, P.; Subbarya, B. H.; Venkataramani, S. Balloon-borne cryogenic air sampler experiment for the study of atmospheric trace gases. *Ind J Radio Space Phys* 1996, 25, 1–7.

Lies, M. E.; D'Haese, E.; Cools, I.; Nelis, H. J.; Coenye, T. Detection and quantification of bacteria and fungi using solid-phase cytometry. In *Detection of bacteria, viruses, parasites and fungi, NATO science for peace and security, Series A: chemistry and biology*, Magni, M. V., Ed.; NATO, Springer, Netherlands, 2010, pp 25–41.

Liu, D.; Abuduwaili, J.; Lei, J.; Wu, G. Deposition rate and chemical composition of the aeolian dust from a bare saline playa, Ebinur Lake, Xinjiang, China. *Water Air Soil Pollut* 2011, 218, 175–184.

Matsuyama, S.; Ishii, K.; Yamazaki, H.; Kikuchi, Y.; Kawamura, Y.; Oyama, R.; Yamanaka, K.; Yamamoto, T.; Watanabe, M.; Tsuboi, S.; Arao, K. Microbeam analysis of yellow sand dust particles. *X-Ray Spectrometry* 2008, 37, 151–155.

Middleton, N.; Yiallouros, P.; Kleanthous, S.; Kolokotroni, O.; Schwartz, J.; Dockery, D.; Demokritou, P.; Koutrakis, P. A 10-year time-series analysis of respiratory and cardiovascular morbidity in Nicosia, Cyprus: The effect of short-term changes in air pollution and dust storms. *Environ Health* 2008, 7, 39.

Mittal, H.; Parks, S.; Pottage, T.; Walker, J.; Bennett, A. Survival of microorganisms on HEPA filters. *Appl Biosaf* 2011, 16(3), 163–166.

Moria, I.; Nishikawa, A.; Tanimura, T.; Quan, H. Change in size distribution and chemical composition of kosa (Asian dust) aerosol during long-range transport. *Atmos Environ* 2003, 37, 4253–4263.

Morono, Y.; Terada, T.; Masui, N.; Inagaki, F. Discriminative detection and enumeration of microbial life in marine subsurface sediments. *ISME J* 2009, 3, 503–511.

Moulin, C.; Lambert, C. E.; Dulac, F. L.; Dayan, U. Control of atmospheric export of dust from North Africa by the North Atlantic oscillation. *Nature* 1997, 387, 691–694.

Neophytou, A.; Yiallouros, P.; Coull, B.; Kleanthous, S.; Pavlou, P.; Pashardis, S.; Dockery, D.; Koutrakis, P.; Laden, F. Particulate matter concentrations during desert dust outbreaks and daily mortality in Nicosia, Cyprus. *J Expo Sci Environ Epidemiol* 2013, 23, 275–280.

New, M. G.; Lister, D.; Hulme, M.; Makin, I. A high-resolution data set of surface climate for terrestrial land areas. *Clim Res* 2002, 21, 1–25.

Nihlén, T.; Mattsson, J. O. Studies on eolian dust in Greece. *Geogr Ann* 1989, 71, 269–274.

Otani, S.; Onishi, K.; Mu, H.; Yokoyama, Y.; Hosoda, T.; Okamoto, M.; Kurozawa, Y. The relationship between skin symptoms and allergic reactions to Asian dust. *Int J Environ Res Public Health* 2012, 9, 4606–4614.

Ozawa, Y.; Ong, B. L.; An, S. H. Traceback systems used during recent epizootics in Asia. *Rev Sci Tech* 2001, 20, 605–613.

Perez, L.; Tobias, A.; Querol, X.; Künzli, N.; Pey, J.; Alastuey, A.; Viana, M.; Valero, N.; González-Cabré, M.; Sunyera, J. Coarse particles from Saharan dust and daily mortality. *Epidemiology* 2008, 19(6), 800–807.

Perkins, S. Dust, the thermostat. *Sci News* 2001, 160, 200–201.

Pey, J.; Querol, X.; Alastuey, A.; Forastiere, F.; Stafoggia, M. African dust outbreaks over the Mediterranean basin during 2001–2011: PM10 concentrations, phenomenology and trends, and its relation with synoptic and mesoscale meteorology. *Atmos Chem Phys Discuss* 2012, 12, 28195–28235.

Polymenakou, P.; Mandalakis, M.; Stephanou, E.; Tselepides, A. Particle size distribution of airborne microorganisms and pathogens during an intense African dust event in the Eastern Mediterranean. *Environ Health Perspect* 2008, 116(3), 292–296.

Prospero, J. M. Long-term measurements of the transport of African mineral dust to the southeastern United States: Implications for regional air quality. *J Geophys Res* 1999, 104 (15), 917–927.

Prospero, J. M.; Lamb, P. J. African droughts and dust transport to the Caribbean: Climate change implications. *Science* 2003, 302, 1024–1027.

Prospero, J.; Blades, E.; Naidu, R.; Mathison, G.; Thani, H.; Lavoie, M. Relationship between African dust carried in the Atlantic trade winds and surges in pediatric asthma attendances in the Caribbean. *Int J Biometeorol* 2008, 52(8), 823–832.

Reasoner, D. J.; Geldreich, E. E. A new medium for the enumeration and subculture of bacteria from portable water. *Appl Environ Microbiol* 1985, 49(1), 1–7.

Revel-Rolland, M.; De Deckker, P.; Delmonte, B.; Hesse, P.; Magee, J.; Basile-Doelsch, I.; Grousset, F.; Bosch, D. Eastern Australia: A possible source of dust in east Antarctica interglacial ice. *Earth Planet Sci Lett* 2006, 249, 1–13.

Ricq de Bouard, M.; Thomas, A. *La Météorologie (Paris)* 1972, 24, 65–83.

Rodá, F.; Bellot, J.; Avila, A.; Escarré, A.; Piñol, J.; Terradas, J. Saharan dust and the atmospheric inputs of elements and alkalinity to Mediterranean ecosystems. *Water Air Soil Pollut* 1993, 66(3), 277–288.

Sajani, S.; Miglio, R.; Bonasoni, P.; Cristofanelli, P.; Marinoni, A.; Sartini, C.; Goldoni, C.; De Girolamo, G.; Lauriola, P. Saharan dust and daily mortality in Emilia-Romagna (Italy). *Occup Environ Med* 2011, 68, 446–451.

Sakamoto, K.; Yoshida, K. Recent outbreaks of foot and mouth disease in countries of East Asia. *Rev Sci Tech* 2002, 21, 459–463.

Samoli, E.; Kougea, E.; Kassomenos, P.; Analitis, A.; Katsouyanni, K. Does the presence of desert dust modify the effect of PM_{10} on mortality in Athens, Greece? *Sci Total Environ* 2011, 409(11), 2049–2054.

Smith, D. J.; Jaffe, D. A.; Birmele, M. N.; Griffin, D. W.; Schuerger, A. C, Hee, J.; Roberts, M. Free tropospheric transport of microorganisms from Asia to North America. *Microb Ecol* 2012, 64, 973–985.

Sprigg, W. A.; Nickovic, S.; Galgiani, J.; Pejanovic, G.; Petkovic, S.; Vujadinovic, M.; Vukovic, A.; Dacic, M.; DiBiase, S.; Prasad, A.; El-Askary, H. Regional dust storm modelling for health services: The case of valley fever. *Aeolian Res* 2014, http://dx.doi.org/10.1016/j.aeolia.2014.03.001

Stevenson, C. M. The dust fall and severe storms of July 1, 1968. *Weather* 1969, 24, 126–132.

Stres, B.; Sul, W.; Murovec, B.; Tiedje, J. Recently deglaciated high-altitude soils of the Himalaya: Diverse environments, heterogenous bacterial communities and long-range dust inputs from the upper troposphere. *PLoS One* 2013, 8(9), e76440.

Tondera, A.; Jabłońska, M.; Janeczek, J. Mineral composition of atmospheric dust in Biebrza national park, Poland. *Polish J Environ Stud* 2007, 16(3), 453–458.

Wainwright, M.;Wickramasinghe, N. C.; Narlikar, J. V.; Rajaratnam, P. Fungi and bacteria at 41 km over the earth's surface: Microorganisms cultured from stratospheric air samples obtained at 41 km. *FEMS Microbiol Lett* 2003, 218, 161–165.

Washington, R.; Todd, M. Atmospheric controls on mineral dust emission from the Bodélé Depression, Chad: The role of the low level jet. *Geophys Res Lett* 2005, 32, L17701, doi: 10.1029/2005GL023597

Wei, W.; Davis, R.; Lee, I. -M.; Zhao, Y. Computer-simulated RFLP analysis of 16S rRNA genes: Identification of ten new phytoplasma groups. *Int J Syst Evol Microbiol* 2007, 57, 1855–1867.

Weisburg, W.; Barns, S.; Pelletier, D.; Lane, D. 16S ribosomal DNA amplification for phylogenetic study. *J Bacteriol* 1991, 173(2), 697–703.

Williams, P. L.; Sable, D. L.; Mendez, P.; Smyth, L. T. Symptomatic coccidioidomycosis following a severe natural dust storm. An outbreak at the Naval Air Station, Lemoore, California. *Chest* 1979, 76, 566–570.

Williamson, K. E.; Wommack, E.; Radosevich, M. Sampling natural viral communities from soil for culture-independent analyses. *Appl Environ Microbiol* 2003, 69(11), 6628–6633.

Woese, C. R. Bacterial evolution. *Microbiol Rev* 1987, 51, 221–271.

Yamaguchi, N.; Ichijo, T.; Sakotani, A.; Baba, T.; Nasu, M. Global dispersion of bacterial cells on Asian dust. *Sci Rep* 2012, 2, 525.

Zhang, R.; Shen, Z.; Cheng, T.; Zhang, M.; Liu, Y. The elemental composition of atmospheric particles at Beijing during Asian dust events in Spring 2004. *Aerosol Air Qual Res* 2010, 10, 67–75.

Zhang, X. Y.; Gong, S. L.; Zhao, T. L.; Arimoto, R.; Wang, Y. Q.; Zhou, Z. J. Sources of Asian dust and role of climate change versus desertification in Asian dust emission. *Geophys Res Lett* 2003, 30, 2272.

CHAPTER 3

CHLORINATED COMPOUNDS IN NATURAL AND BIOTECHNOLOGICAL PROCESSES: MERITS, RISKS, AND USES

SÁNDOR T. FORCZEK[1], JOSEF HOLÍK[1], LUDĚK REDERER[2], and MARTIN FERENČÍK[3]

[1]Isotope Laboratory, Institute of Experimental Botany, Academy of Sciences of the Czech Republic, Vídeňská 1083, Prague CZ-14220, Czech Republic

[2]Povodí Labe, State Enterprise, Víta Nejedlého 951, Hradec Králové CZ-50003, Czech Republic

[3]Institute of Environmental and Chemical Engineering, Faculty of Chemical Technology, University of Pardubice, Studentská 573, Pardubice CZ-53210, Czech Republic

CONTENTS

3.1 INTRODUCTION

Chlorine, similarly to other elements, undergoes a complex biogeochemical cycle that includes the formation, conversion, and degradation of different inorganic and organic forms of chlorine. Chlorinated compounds participate in natural processes, biological, and chemical processes forming volatile organochlorine compounds; human activities also change their presence in the environment. Chlorinated pollutants are in the center of interest due to their reactive nature, causing degradation of ozone in the atmosphere, and due to health concerns, as some chlorinated compounds are highly toxic. Some volatile chlorinated hydrocarbons (VCHs) are both reactive and toxic, such as chloroform. Moreover, chloroform has natural and anthropogenic sources and can be formed in abiotic and biotic processes. In this chapter, we discuss the role of VCHs in the natural environment and their anthropogenic impact. Furthermore, statutorily determinations of adsorbable organic halogens (AOX) and chloroform will be evaluated in connection with a sample study in a clean area Hamry, Czech Republic. The catchment has low human activity and the collected water in the Hamry water reservoir is used as a source of drinking water. The area contains many bogs and forests and a high content of dissolved organic carbon has therefore been determined in it. The high concentration of AOX and chloroform in the water source area is caused by the natural biological activity of the soil. The compounds cannot be found in the reservoir, as the AOX are diluted, while chloroform is evaporated during on their course to the reservoir.

3.2 CHLORINATED COMPOUNDS IN NATURAL PROCESSES

Chlorine is present in every part of the geosphere, including atmosphere, hydrosphere, lithosphere, and biosphere as well. Natural processes including several abiotic and biogenic ways yield chlorinated compounds. Organochlorines (Cl_{org}), in which chlorine is covalently bonded to an organic compound, participate in the complex biogeochemical cycle of chlorine involving the formation, conversion, and degradation of different inorganic and organic forms of chlorine. Natural sources, such as wild fires, volcanoes, and other geothermal processes account for a wide range of volatile organochlorines (VOCl) and inorganic chlorine (Lobert et al., 1999). Chlorine is present in volcanic gases not only as hydrogen chloride gas, but also as organic forms of chlorine, such as chloromethane, chloroform, tetrachloroethylene, carbon tetrachloride, and several chlorofluorocarbons (CFCs) such as Freons, which are chemicals formerly thought only to result from the human activities (Gribble, 1999).

It has been suggested that chlorine contributes to the decay of soil organic matter (SOM) leading to the formation of large molecules of chlorohumus (Asplund, 1995; Matucha et al., 2010). Degradation of chlorohumus leads to smaller intermediates, such as chlorinated acetic acids, anisol-, orcinol- and hydroquinone-based substances, which may be taken up by plants or fungi. In addition to insoluble compounds or smaller molecules present in soil and dissolved in pore water, VOCl are also present in soil. Some of the VOCls can also be found in hundreds of years old ice in Antarctica (Saito et al., 2006), demonstrating that volatile VOCls have been emitted long before anthropogenic activity altered the atmosphere of the Earth (Lovelock et al., 1973).

Production of Cl_{org} is widespread in every domain of living organisms, including *Archaea*, *Bacteria*, and *Eukarya* as well. Examples of production and emission are found in but not restricted to animals, almost all higher and lower plants, fungi, algae, and microorganisms (Yokouchi et al., 2002; Gribble, 2003). Many of the biogenic Cl_{org} are volatile and so can be emitted to the atmosphere, where relevant atmospheric reactions take place.

3.3   ATMOSPHERIC CYCLE OF CHLORINATED COMPOUNDS

Several reactions take place in the atmosphere between inorganic chloride and chlorinated compounds. Most of them are notoriously well known, as chlorinated compounds have long been considered as of anthropogenic origin only and studied in connection with ozone depletion (Gay et al., 1976; Tuazon et al., 1988; Nelson et al., 1990; McCulloch and Midgley, 1996). Chlorine and other halogens enter the atmosphere in the form of sea salt spray as water particles are lofted into the atmosphere by the motion of ocean waves. These particles can undergo reactions with trace atmospheric gases and internal mixing with anthropogenic pollutants which are deposited on the surface of the particles. Several studies have shown that NaCl particles in the atmosphere are depleted in chloride and have attributed this to reactions with inorganic acids. These reactions, which are unique for aerosolized particles, were previously overlooked and are important in atmospheric chemistry models (Laskin et al., 2012). The reactions in atmospheric particles containing sea salt and organic acids liberate HCl (g) and promote the formation of organic salts in the particle phase. In the atmospheric environment, the released HCl (g) may result in consecutive acidification of coexisting neighboring particles, and trigger additional acid–base reactions, especially with alkaline components of mineral dust (Sullivan et al., 2007). The uptake of chlorine species onto dust particles modifies the chlorine chemistry budget in the marine boundary layer, thus altering the overall HCl cycle and the cycles of important species involved in these heterogeneous reactions (such as SO_x, NO_y, ClO_x, and O_3). These processes will therefore affect the chemical budgets of ClO_x, NO_x, and SO_x species in the marine boundary layer (Sullivan et al., 2007; Faxon and Allen, 2013). The chemically reactive gases ClO_x, NO_x, and SO_x can form organically bound chlorine in the atmosphere (Gay et al., 1976; Tuazon et al., 1988; Itoh et al., 1994; Sidebottom and Franklin, 1996).

Precipitation washes out chloride and water-soluble chlorinated compounds from the atmosphere. Chloride therefore does not stay in the atmosphere, but it is transferred by wet and dry deposition. The halide concentrations of real seawater are: Cl^- 19000 mg/L, Br^- 65 mg/L, I^-, IO_3^{3-} 0.060 mg/L, while in wet deposition, chloride concentration can be still significant for terrestrial ecosystems; under continental climate it can reach values around 1 mg/L. In the form of precipitation, Cl^- ends up on the soil surface or on the vegetation, wherefrom it is taken up by plants or washed down into the topsoil by tree throughfall and stem-flow. Chloride in aerosols is directly captured by vegetation, ranging from several kg/ha/year in continental areas (e.g., 250 mg/m^2/year in Central Europe) up to tens of kg/ha/year in coastal areas (e.g., 4000 mg/m^2/year in Western France) depending on geographical situation and weather conditions (Delalieux et al., 2006). Estimates of the yearly production of atmospheric marine salt range from 10^9 to 10^{10} metric tons, and a realistic value is most probably somewhere in between (Blanchard, 1985; Erickson and Duce, 1988). Estimates of relative contributions of dry and wet removal processes to total sea-salt removal vary largely, from an amount of 67% salt brought down by rain to 33% dry deposition (Blanchard, 1985; Erickson and Duce, 1988).

3.4   FORMATION OF ORGANOCHLORINE COMPOUNDS

Chloride which enters the terrestrial ecosystems becomes dissolved in precipitation, running, or soil water, where it can react with soil organic substances, leading to the formation of halogenated compounds (Johansson et al., 2000; Godwin et al., 2003; Matucha et al., 2010).

Formation of halogenated organic compounds can happen in soil in an abiotic way (Keppler et al., 2000) or through the action of exoenzymes produced by microorganisms and plants. Myneni (2002) reported that stable, chlorinated hydrocarbons with aliphatic and aromatic groups are formed rapidly at the expense of inorganic Cl^- during weathering and humification of plant material. The magnitude of VOCl production in soil is difficult to estimate, and the achieved emission rates have so far not been used to estimate a global emission rate, but the results show that the terrestrial ecosystem has an enormous potential to release CH_3Cl, CH_3Br, and CH_3I into the atmosphere through an abiotic process.

In addition to the abiotic formation, the emission of VOCl by living organisms was also confirmed. The emissions of chloroform, carbon tetrachloride, or chloromethane in addition to numerous brominated and iodinated organic compounds from some marine and terrestrial sources, such as marine macroalgae, coastal salt marshes, forest vegetation, and soil, are well documented (Laturnus et al., 1996; Laturnus et al., 1998; Giese et al., 1999; Rhew et al., 2000; Hoekstra et al., 2001; Svensson et al., 2007; Laturnus et al., 2010). The pathways of formation of some chlorinated compounds have been elucidated while others are still to be revealed. Formation of haloacetic acids and chloroform is closely interconnected in the soil, and both are intensively studied. Both compounds are found to be produced during degradation of resorcinolic structures, which are common structural elements of humic material. In the course of their formation, the aromatic rings are chlorinated and ring cleavage occurs, followed by haloform reaction with aliphatic chains and final cleavage into trichloroacetic acid (TCA) and chloroform. The former is in prevalence in acidic soils while the latter dominates in basic soils with a pH over eight (Hoekstra et al., 1999a; Hoekstra et al., 1999b). Haloacetic acids are readily taken up by vegetation (Forczek et al., 2004), or can be formed from volatile halogenated hydrocarbons inside the plants due to detoxification processes (Schröder et al., 2003; Weissflog et al., 2007). Active movement and metabolism of halogen containing xenobiotics have been intensively studied, and the role of glutathione in plants was elucidated (Schröder and Wolf, 1996; Schröder et al., 2007). Plants possess an elaborate enzyme-based detoxification system for organic xenobiotics and agrochemicals, comprising a metabolic cascade proceeding in three phases. In phase I, the xenobiotics are activated by P450 monooxygenases, peroxidases, or similar enzymes that catalyze oxidation, reduction, or hydrolysis of the foreign compound. Detoxification in the strict sense of the word proceeds in phase II, where the xenobiotic is conjugated to biomolecules, that is, sugars via glycosyltransferases or to the tripeptide glutathione via glutathione S-transferases. Glutathione conjugates are less toxic and lipophilic than the parent xenobiotics, but they might still carry some unwanted properties rendering them problematic for the cell. It is therefore generally accepted that xenobiotic conjugates are sequestered from the cytosol in higher plants in phase III by the activity of tonoplast MRP transporters, which belong to the wide family of ABC-transporters (Schröder et al., 2007).

VOCl emission by plants is theoretically possible: (1) during the detoxification of halogenated compounds; (2) due to the emission of already taken up VOCl from the environment, or; (3) due to *de novo* synthesis. To answer the first possibility, chloroacetic acids were studied as model compounds in plants by radiotracer methods (Blanchard, 1954; Forczek et al., 2004). No degradation products were found in most cases, which points to total mineralization by dechlorination and CO_2 emission, whereas in some cases trichloromethyl compounds were found (Mayer, 1957). The main degradation product during thermal decomposition of TCA is chloroform, which was however never found in plants (Forczek et al., 2004). In the second place, the re-emission of previously absorbed VOCl has also been studied, and it is a completely reversible process in soils (Chen and Dural, 2002). Plants are also capable of absorbing VOCl, and based on this capability, some lichens are used as bio-indicators of air pollution (Conti and Cecchetti, 2001). Re-emission is therefore also possible in plants. In the third place, the VOCl emission by plants was determined by field and laboratory experiments (Yokouchi et al., 2000; Forczek et al., 2015). The emitted VOCl are proved to be the products of *de novo* synthesis of plant enzymes.

The enzymes responsible for the formation of VOCl in plants can be heme peroxidases (EC 1.11.1.X) such as chloroperoxidase (EC 1.11.1.10), or non-heme peroxidases such as vanadate-dependent peroxidases (EC 1.11.1.7), perhydrolases, and $FADH_2$-dependent halogenases (Walter and Ballschmiter, 1992; Asplund et al., 1993; Wever and Hemrika, 2001; van Pée, 2003); most of them ubiquitously present also in soil by microorganisms and fungi (e.g., *Caldariomyces* sp., *Pseudomonas* sp., *Streptomyces* sp.).

The halogenating enzymes containing the heme group or vanadium in their molecule require hydrogen peroxide for their halogenating activity. Perhydrolases and $FADH_2$-dependent halogenases do not contain either a heme group or any metal ions but the former also require hydrogen peroxide for their halogenating activity. The reaction mechanisms of halogenating enzymes vary. Perhydrolases act in the presence of H_2O_2. Perhydrolysis of the acyl-enzyme intermediate results in the formation of peracids, which can oxidize halide ions to hypohalous acids which then act as the halogenating agent. On the other hand, heme-type haloperoxidases and vanadate-dependent peroxidases produce free hypohalous acid as the halogenating agent also in the presence of H_2O_2 and halide ions. Thus, heme-and vanadium-containing haloperoxidases and also perhydrolases are halogenating enzymes without any substrate specificity and regioselectivity. The halogenation reactions in some biosynthetic pathways like 7-chlorotetracyclin biosynthesis are catalyzed by enzymes with high substrate specificity and regioselectivity (van Pée, 2003). $FADH_2$-dependent halogenase enzymes (EC 1.14.14.7) are therefore responsible for specific halometabolite synthesis, whereas haloperoxidases and perhydrolases are probably involved in defense mechanisms in which unspecific halogenation occurs.

3.5 VOLATILE CHLORINATED COMPOUNDS IN THE SOIL

Sodium chloride, the most common compound of chlorine, primarily comes to terrestrial ecosystems by atmospheric transport from the oceans and secondarily by weathering of bedrocks, which is negligible (Winterton, 2000; Öberg, 2003). Some anthropogenic sources/activities such as irrigation, road salting, or coal combustion locally increase the concentration of salt in soil, and can cause immense damage. In terms of environmental impact, chloride was considered a chemically inert substance and its role in the forest ecosystem has been considered negligible. After degradation of plant residues, forest soil comprises mainly lignin, humic and fulvic acids, and a variety of other substances resulting from their further degradation, altogether called humus. Halogenation in soil by chlorine radicals can produce chlorinated SOM (chlorohumus) containing a wide range of chlorinated compounds such as large chlorinated humic and fulvic acids, and a number of low-molecular chlorinated compounds. The formation of chlorinated aromatic compounds derived from lignin degradation is also anticipated. These substances are present in the runoff from forest ecosystems. Low-molecular chlorinated compounds, such as chloroacetic acids and chloroform are relatively well detectable. The formation of chloroacetic acids has been demonstrated in forest soil by investigations during long-term model experiments with the help of chlorine 36 (^{36}Cl) and carbon 14 (^{14}C) isotopes (Matucha et al., 2007a, 2007b) and formation of VCHs was proved (Rohlenová et al., 2009).

The chlorination achieved through abiotic formation and by living organisms, described in previous sections, produces a wide spectrum of volatile and non-volatile halogenated compounds from inorganic halides. The above-mentioned chlorinating enzymes are present in the forest ecosystem not only as a product of plants, but also due to the action of fungi and microorganisms, and are located either inside the cells of living organisms, or as extracellular enzymes (exoenzymes) emitted into the soil. Chlorination occurs fairly fast, chlorine is bound to organic molecules and the biogeo-chemical cycle of chlorine continues in soil (Winterton, 2000; Öberg, 2003). Through volatilization and mineralization, chlorine gets back into the atmosphere. Most of the chlorine in the atmosphere in an organic form is chloromethane, chloroform, and carbon tetrachloride (Butler et al., 1999; Gribble, 1999; Montzka et al., 1999).

Formation of chloromethane by fungal methylation of chloride was reported by Harper (1985), and reaction of chloride with pectin (Hamilton et al., 2003) leads to the same VOCls as those that cause damage to the ozone layer. Myneni (2002) and Leri et al. (2006) showed also chlorination of plant litter material.

Vinyl chloride found in the atmosphere also has natural sources, contrary to previous considerations that dedicated all of it to anthropogenic sources only. Soil air and ambient air from a rural area in Northern Germany were investigated for volatile chlorinated halocarbons, and the concentrations of vinyl chloride in the soil air were significantly enhanced as compared to ambient air, indicating a natural formation of this compound in the soil. A series of laboratory experiments using different soils and model compounds clearly proved that vinyl chloride can be produced during soil processes (Keppler et al., 2000; Keppler et al., 2002). The authors propose that this highly reactive compound can be formed during the oxidative degradation of organic matter in soil, for example in a reaction between humic substances, chloride ions, and an oxidant (ferric ions or hydroxyl radicals). The redox-sensitive aromatic compounds in soil such as catechols and o-quinones can be degraded to CO_2, accompanied by the release of vinyl chloride and other volatile chlorinated compounds (Keppler et al., 2000).

3.6   FORMATION OF CHLORINATED COMPOUNDS BY CYANOBACTERIA

Cyanobacteria, very simple prokaryotic oxygenic phototroph microorganisms found in nearly every conceivable habitat on earth, also produce chlorinated compounds. Some of these are simple compounds well known in plant physiology, such as halogenated hydrocarbons (Gribble, 2003), whereas others are specific secondary metabolites such as chlorinated cyanotoxins. The blue-green algae *Nostoc* and *Oscillatoria* are two taxa rich in both types of organohalogen compounds. For example, blue-green algae produce aeruginosins, found predominantly in the genera *Microcystis* and *Planktothrix*. These toxins can hold chlorine in their molecule in more positions (Fig. 3.1)

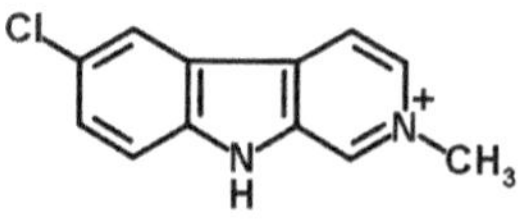

Nostocarboline

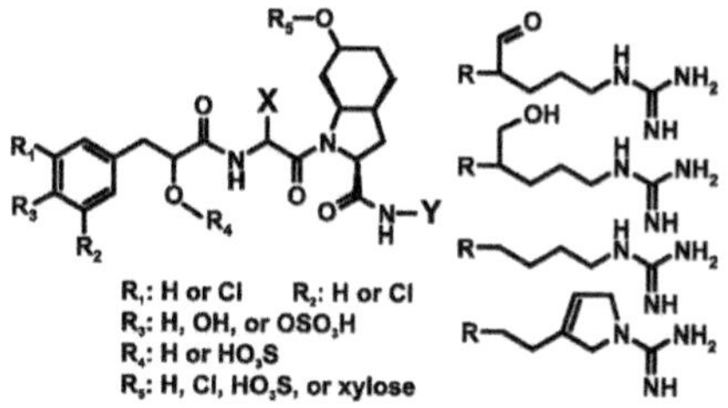

R₁: H or Cl   R₂: H or Cl
R₃: H, OH, or OSO₃H
R₄: H or HO₃S
R₅: H, Cl, HO₃S, or xylose

Aeruginosins, where X denotes different amino acids that can occupy the second position: Leu, Ile, Phe, Tyr, and homotyrosine. Y denotes the C-terminal variable moieties as shown to the right (from top): argininal (linear tautomer), argininol, agmatine, and 1-amino-2-(N-amidino-Δ3-pyrrolinyl)-ethyl (Ishida et al., 2009).

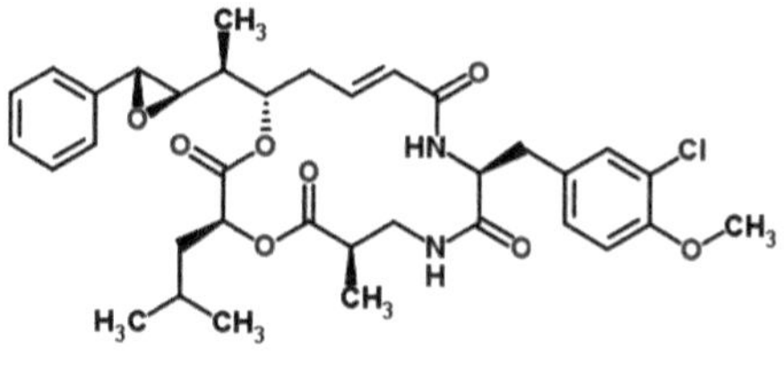

Cryptophycin A

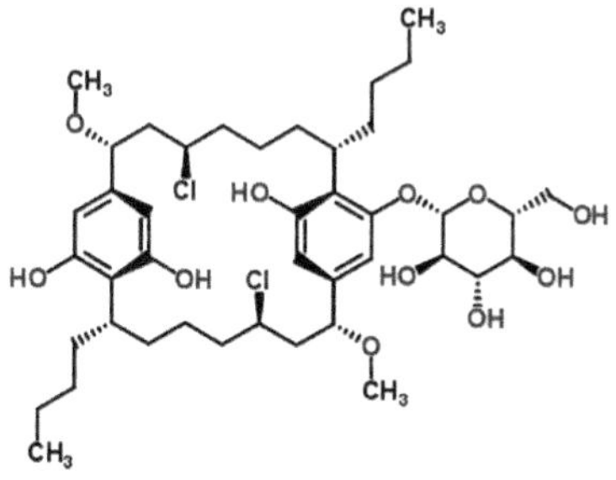

Nostocyclophane B

FIGURE 3.1 Chlorinated natural compounds isolated from blue-green algae.

(Ishida et al., 2009). The potent anticancer drug candidate cryptophycin A (Fig. 3.1) was also isolated from cultures of a blue-green alga (*Nostoc* sp.), and the structurally novel nostocyclophane (Fig. 3.1) is produced by *Nostoc linckia* (Gribble, 2003). These special secondary metabolites are encoded in specific gene clusters, aimed by the algae to produce highly specific compounds. As these genes were identified, it will be possible to search for and identify with the aid of genome-mining approaches organisms with similar genes, which can encode promising new natural products.

Although some cyanobacteria actually produce halogenated compounds, their role in the global halocarbon production is not well established. Studies on distributions and sea-to-air fluxes of VOCls were conducted only very recently (Roy et al., 2011) and only poor correlations were found between concentrations of halogenated compounds with various marker pigments. The results support the existence of multiple sources and sinks of halogenated compounds, which might obscure the relationship between halocarbons and phytoplankton composition. In another study, however, He et al. (2013a, 2013b) found a positive correlation between chlorophyll-a, salinity, and nutrients on the distributions of gaseous halocarbons in the coastal shelf of the Yellow Sea and the East China Sea. Elevated levels of chloroform ($CHCl_3$), trichloroethylene (C_2HCl_3), tetrachloroethylene (C_2Cl_4), chlorodibromomethane ($CHBr_2Cl$), and other simple halogenated hydrocarbons were found in sea water, whereas a decreasing trend was found with the distance from the coast, with low values found in the open sea, bringing evidence that phytoplankton is related with halocarbon emission. These results indicate that the coastal shelf contributes significantly to the global oceanic emissions of gaseous halocarbons, and especially phytoplankton is responsible for its formation, but they do not state explicitly the role of cyanobacteria.

The formation of chlorinated compounds in blue-green algae is known to have allelopathic effects to affect other organisms living in the same aquatic environment. Active compounds are synthesized to fight the harmful cyanobacterial water bloom. Some chemicals are derived from natural compounds, such as those found and identified from decomposition of straw, which show toxicity toward cyanobacteria. Several chemicals were derived and are tested from anthraquinone (found in lignin) and stilbenes, which are produced by plants in response to stress. Further compounds found to have cyanocide properties are the alkaloids of carbolines such as nostocarboline (Fig. 3.1), isolated from the cyanobacterium *Nostoc* (Jancula et al., 2011). Nostocarboline, which is a novel neurochemical to fight Alzheimer's disease, also contains chlorine in the molecule (Becher et al., 2005). Many of the chemicals used as natural or synthetic herbicides are chlorine-containing compounds. Their metabolites and degradation products may differ in many aspects from the original compounds. In addition to changes of physico-chemical properties, their subsequent biological and toxicological properties are modified. The change of solubility and hydrophobicity defines new hazards present in the novel compounds.

3.7 VOLATILE CHLORINATED COMPOUNDS OF ANTHROPOGENIC ORIGIN

In addition to naturally produced organohalogens, human activity also releases halogenated compounds to the environment, which then pose risk on human health. Halogenated organic compounds constitute the largest fraction of priority pollutants as designated by the U.S. Environmental Protection Agency (U.S. EPA), with over 50 % of the compounds containing chlorine or another halogen. Quite often, the mentioned pollutants are volatile, and that is why they reach the natural environment.

As a simple organochlorine, chloroform can serve for example to estimate human activities. Anthropogenic sources of chloroform and other VOCls are industrial production processes, chlorination of drinking, swimming pool, and cooling water, pulp and paper bleaching (de Fouw, 1994), and waste incineration (Jay and Stieglitz, 1995). Other sources of chloroform include hazardous waste sites and sanitary landfills. Chloroform is one of the most strictly controlled compounds in drinking water among other trihalogenated methanes (THM) that were found to be carcinogenic in animal tests. U.S. EPA has classified chloroform as a group B2, probable human carcinogen, and the U.S. Food and Drug Administration has banned its use in consumer products in 1976. Chloroform in the drinking water is therefore regulated. To protect drinking water from pathogens, water suppliers often add a disinfectant, such as chlorine, to raw water. However, disinfectants can react with naturally occurring materials present in the water to form halogenated byproducts, which may pose health risks. Hazardous disinfection byproducts include THM and haloacetic acids formed during chlorination of drinking water, wastewater, and swimming pools, where dissolved organic materials react with chlorine, chloramine, and bromine.

Nowadays chloroform is primarily used as a solvent in pharmaceutical industry and as an important raw material for producing dyes, pesticides, and hydrochlorofluorocarbons (HCFCs). The U.S. EPA has posed a control on the maximum contaminant level of chloroform in drinking water to 70 μgL^{-1}, whereas in European countries the guideline for maximum acceptable concentration level is 200 µg L^{-1}. Several European countries set a lower level for example, in the Czech Republic the maximum acceptable concentration level of chloroform is 30 µgL^{-1} or 50 µgL^{-1} of THM.

Several methods are used to remove anthropogenic pollution from natural environment *in situ*. They are applied according to the nature of the pollutant, such as air stripping, carbon adsorption, soil venting or other aeration technologies, and biological remediation. Biotechnological methods based on biological systems, living organisms, and their derivatives, though still in the research stage can help to clean environmental pollutions by chlorinated products in natural ecosystems. In bioremediation, contamination or hazardous substances are turned into less toxic or non-toxic form by removal or breakdown of these substances. Some examples of bioremediation-related technologies are phytoremediation, bioventing, bioleaching, landfarming, bioreactor, composting, bioaugmentation, rhizofiltration, and biostimulation. Biodegradation of contaminants can be achieved by dechlorinating microorganisms having enzymes which remove halogens by hydrolytic nucleophilic substitution (Thompson et al., 2005; Huang et al., 2014). Some enzymes work in both directions, not only dehalogenating, but also achieving halogenation whereas other enzymes degrade different parts of the organic unit. Still other enzymes conjugate the compounds to xenobiotic substrates. Cyanobacteria and other microorganisms utilized in biotechnological applications therefore can produce VOCl, which can have adverse effects on health and on the environment (Huang et al., 2014).

During bioremediation, chloroform can also be emitted as a degradation product, for example in the course of bioaugmentation of carbon tetrachloride remediation (Wang et al., 2002). Bioremediation of chloroform-contaminated sites is considered a favorable alternative to physical and chemical approaches such as air stripping or sorption onto activated carbon that transfer contaminants from one medium to another rather than causing contaminant destruction (Cappelletti et al., 2012).

3.8 CHLOROFORM

Although chloroform alone may play a minor role in the global chlorine cycle, it may be worth considering more seriously when taken together with other naturally produced chlorocarbons. To highlight chloroform among the above-mentioned facts on VOCl, we can summarize that it is formed in many different ways naturally and through anthropogenic action. The major anthropogenic sources of chloroform are fossil-fuel combustion, waste incineration, and industrial releases primarily due to pulp and paper manufacturing and water treatment (Laturnus et al., 2002). Chloroform can be transformed from other VCHs produced by human activity. Other uses of chloroform, such as industrial solvent, surgical anesthetic or in cough syrups, and toothpastes are banned or replaced by different solvents.

Chloroform is one of the most complex compounds in terms of its sources. According to the hypothesis of Hoekstra et al. (1999), chloroform cannot be produced by decarboxylation of TCA in soil (Hoekstra et al., 1999a; Hoekstra et al., 1999b), while Schöler et al. (2003) claim that moderately elevated temperature (50–70 °C, 30 min) already leads to thermal decomposition of TCA to chloroform. In general, elevated temperature is not expected in soil, whereas it can happen in plant leaves, due to strong irradiance from full sunlight (Dickey et al., 2005). Plant temperatures generally follow the diurnal pattern of air temperatures, but they may be higher (sometimes 10 °C or more); this supports Schöler's hypothesis (Schöler et al., 2003) on partial thermal degradation of TCA. Chloroform and methyl-chloroform are released from biomass burning (Rudolph et al., 2000), rice fields (Khalil et al., 1998), termite mounds (Khalil et al., 1990), fungi (Hoekstra et al., 1998a, 1998b), and spruce forests (Haselmann et al., 2000a; Haselmann et al., 2000b). Chloroform represents the most abundant halocarbon in the atmosphere (Harper, 2000; Cappelletti et al., 2012) but the anthropogenic flux of chloroform into the environment is too much low to account for observed background concentrations. Therefore, it is clear that natural sources must make major contributions.

Natural abiotic ways of chloroform formation include, but are not limited to, emission of chloroform from volcanic sites, wild fires, and geothermal processes and abiotic formation in the soil. Biogenic formation of chloroform is also fairly diverse. The oceans have been found to be a major source for naturally produced volatile halogenated compounds, including chloroform. Marine organisms, mainly phytoplankton and macroalgae have been found to produce chloroform. Budget calculations indirectly indicate that the terrestrial ecosystem may also be an important source of volatile halocarbons, thereby contributing to the global halogen cycle. Though terrestrial natural sources of chloroform are still poorly investigated, some sources have already been identified. Khalil et al. (1998) found a positive flux of chloroform liberated into the atmosphere during the investigation of rice fields. Dimmer et al. (2001) discovered a flux of chloroform from Irish peatland ecosystems in globally significant amounts. Chloroform in soil can be formed from TCA, especially in alkaline media (Haiber et al., 1996; Schöler et al., 2003), through intracellular microbial activity and via the activity of exoenzymes in the soil. Metabolism of TCA in plants can also produce chloroform, while in soil, chloroform is produced by fungi as the result of the chlorination of natural organic matter by naturally produced hypochlorous acid (Hoekstra et al., 1998a, 1998b). Concentration patterns of TCA and chloroform in leaves suggest that TCA might be partly decarboxylated to the intermediate trichloromethanide which forms chloroform, but this was never confirmed in experiments (Schöler et al., 2003; Forczek et al., 2004). Due to its low tendency to sorb to soil organic carbon, chloroform has a high mobility in aquifers (Cappelletti et al., 2012).

3.9   ADSORBABLE ORGANIC HALOGENS

Until the 1980s, halogenated compounds present in nature were considered exclusively as industrial products. To detect these possibly harmful compounds in natural environment, the AOX method was introduced, originally developed to quantify anthropogenic pollution of surface waters such as various herbicides (e.g., 2,4-dichlorophenoxyacetic acid, atrazine, and TCA), pesticides (DDT, lindane, etc.), chlorinated waste products (emitted e.g., during cellulose bleaching, or in PCB manufacturing) or chlorinated solvents (perchloroethylene, chloroform). The method to determine adsorbable water-soluble organohalogen compounds in water on charcoal was proposed by Kühn in 1976 (Müller, 2003) and has been included in 1985 into the German DIN-Norm 38409 as a sum parameter. During AOX (where X stands for chlorine, bromine and iodine) determination, halogenated compounds are adsorbed on activated carbon in a slightly acidic aqueous solution, and combusted thereafter along with the carbon, and then the halogen content is determined by microcoulometric titration. AOX therefore are determined as a group parameter, without further characterization of the compounds.

Investigations conducted since then have clearly shown that plants growing in marine, limnic, and terrestrial environments are a natural source of organohalogens, which are also measured as AOX (Müller, 2003). During the development of monitoring of environmental quality and pollution studies of anthropogenic activities, high levels of AOX have been revealed in Scandinavia and elsewhere in the world in pristine natural areas (Silk et al., 1997). In these vast forest areas, no herbicides or pesticides were ever applied or no industrial enterprises were present whereas in surface waters tens to hundreds of µg/L AOX were determined.

In natural terrestrial ecosystems, the most common halogen is chlorine and halogenated substances are therefore also containing mostly chlorine.

Further research proved that chlorinated substances of natural origin are arising from chlorination of SOM. It has been shown that the chlorination is carried out in the forest soil and mainly catalyzed by microbial exoenzymes, specifically by chloroperoxidases (Asplund et al., 1993). Studies of Cl_{org} in forest ecosystems and watercourses were conducted primarily by a group from Linköping University (Öberg, 2002; Johansson et al., 2003; Öberg, 2003; Öberg et al., 2005; Öberg and Sandén, 2005; Bastviken et al., 2007). The resulting Cl_{org} have not been characterized in detail. The results indicated only the presence of chlorinated humic and fulvic acids, and additional chlorinated substances of low molecular weight (LMW), possibly chloroacetic acids, chloroform, chlorophenol, chloroanisole, and other chlorinated aromatic compounds, some of which could be toxic even at low concentrations.

The Cl_{org} formed in the forest soil and determined as AOX usually fall between 200 and 1000 mg Cl/kg soil and its amount is proportional to the total carbon content of the soil and often exceeds its free inorganic chloride (Cl^-) content. The formation of various types of chlorinated compounds in forest soils leads to more easily degradable and better water-soluble soil components. The dissolved Cl_{org} in effluent waters from the forest ecosystem can end up in reservoirs of drinking water, where these substances then contribute to increased concentration of AOX. According to the AOX indicator, the water quality is then deteriorated. It is important to understand the biogeochemical cycle of chlorine and to separate natural processes from anthropogenic influences in forest ecosystems to identify the threats stemming from human activities. Water catchments with high concentrations of total organic carbon (TOC) and Cl_{org} in the water are common worldwide; AOX levels can pose health risk, which raises concerns on the consumers. One such water source can be found in the catchment area of the Hamry water reservoir (Czech Republic).

3.10   SAMPLE STUDY

The Povodí Labe s.p. (Elbe River Basin, Czech State Enterprise) has extensively monitored the level of AOX in the catchment area of the Hamry water supply reservoir between 2006 and 2008 (Ferenčík et al., 2008). Although the reservoir is situated in a lightly populated and anthropogenically non-loaded area, elevated concentrations of AOX have occasionally been found in the surface waters. High concentration of TOC has also been found here, because humic substances are leached from the forest and peatland areas in the catchment. No xenobiotics were found during the detailed analyses of water samples, where 80 LMW anthropogenic organohalogen compounds (especially herbicides and pesticides) were tested individually. Haloacetic acids were not found in the water either. The absence of natural haloacetic acids in the analyzed samples was not expected due to their rapid microbial degradation in soil and soil water (Matucha et al., 2003). The chloroacetic acids are supposed to be degraded very rapidly, without the formation of VOCls, especially without chloroform and chlorinated methanes (Matucha et al., 2003, 2007a, 2007b). The findings gathered during the 2009–2010 monitoring indicate that the occasionally elevated AOX values, which reach or exceed the specified national limit for sources of drinking water, were not caused by human activity in the area but by natural soil processes. Similar findings were reported in Sweden (Öberg, 2002; Öberg, 2003; Johansson et al., 2003; Öberg and Sandén, 2005) showing that AOX measures both natural and anthropogenic Cl_{org}, whereas in some cases AOX determination does not include important compounds such as chloroform (which is evaporated during the preparation of soil samples in the course of drying). This conclusion was also confirmed in directed experiments performed by our team. Therefore, the question arose whether substances of natural origin can pollute surface waters like in this case, transporting compounds to the Hamry reservoir from forest and peatlands in the catchment by streams.

The measured concentrations of AOX crossed the hygienic threshold for sources of drinking water (30 µg/L) which, according to the present legislation in effect, indicated deterioration of drinking water. In this case, water cannot be used as raw material for the preparation of drinking water, or special precautions must be taken to remove the pollutants. This procedure can be a financial burden and sometimes technically impossible or unavailable. Recent publications in the literature are addressing the fate of chlorine in a forest ecosystem, but monitoring studies including organochlorine are still unique. The study also confirms the conclusions of Muller that AOX in the original form cannot be simply used for indication of anthropogenic pollution by halogenated compounds (Müller, 2003). This is especially the case when bogs are present in the watershed, where halogenated SOM, chlorinated humic acids, and other chlorinated compounds are being formed (Holík et al., 2010).

The source area of one of the tributaries to Krejcarský stream, bearing large fields of peatlands with low drainage, contained higher concentration of humic substances and many metabolites than running waters outside the area. Humic substances leached from forest and peatland areas of the Hamry water reservoir are responsible for higher TOC and AOX content of the surface waters. The found values of AOX and TOC in the sample study showed a correlation between the concentrations of AOX and humic substances. These findings are in agreement with the natural origin of the determined AOX. This confirms the original assumption that Cl_{org} are produced naturally in the forest ecosystem during biotic and abiotic chlorination of complex organic matter present in the soil, such as fulvic and humic substances, lignin, and a wide variety of their chlorinated degradation products.

Although concentration of AOX in several cases exceeds the specified national limit for drinking water in the Hamry catchment area, the water in the reservoir itself, which is used to prepare drinking water, never reaches the AOX hygienic limit.

During the monitoring, the study focused on the explanation of the presence and the origin of AOX in the catchment area of the Hamry water reservoir, which is an important source of drinking water in the area (Pardubice region, Czech Republic) supplying approximately 20 000 people with water. Within the framework of cooperation in the years 2010–2011, the Institute of Experimental Botany, Academy of Sciences of the Czech Republic; the Pardubický kraj (Pardubice region, Czech Republic); the Povodí Labe, státní podnik (Elbe River Basin, Czech State Enterprise); the Vodárenská společnost Chrudim, a.s. (Waterworks Co. Chrudim plc.) and the Vodovody a kanalizace Chrudim, a.s. (Water Utilities Co. Chrudim plc.) participated in studying the Hamry water catchment. The study revealed the primary importance of natural processes in the formation of AOX, and assessed the potential health risks to humans using water reservoirs such as Hamry as sources of drinking water, which was found to be negligible. At the same time, the study can endorse the most appropriate water treatment technologies for preparation of drinking water. In this case, the removal of humic substances and/or TOC would be sufficient. Regarding the presence of naturally occurring chloroform, which was determined before the water treatment as a model compound for the occurrence of volatile chlorinated compounds in raw water, it can be treated as anthropogenic chloroform, also normally occurring during water chlorination.

To further characterize dissolved macromolecules, we separated a LMW fraction of 0–500 daltons (Da) by ultrafiltration. This way, large molecules are separated from smaller molecules which can usually pose a health risk. Some of the low molecular compounds such as chloroacetic acids or chloroform are rapidly decomposed in water samples or evaporated during sample preparation. Some of these low molecular compounds in the LMW fractions such as chloroform, chloroacetic acids, chlorophenol, chloroanisole, chlorinated dicarboxylic acid, and so forth can pose a health risk in increased concentrations. For this reason, they are regulated in the drinking water by national laws in many countries. The AOX and TOC concentration were determined in the original water samples and in LMW fractions. In order to analyze and characterize compounds in the LMW fractions, we used a pre-concentration method using a DEAE-cellulose packed column. The use of DEAE-cellulose for the concentration of humic substances has been described previously (Hiraide et al., 1994; Ibrahim et al., 2008). Fractions with high absorbance (indicating high concentration of humic acids) were selected for further concentration, acidified with hydrochloric acid to pH 1, NaCl was added and then extracted with methyl t-butyl ether. The ether extracts were concentrated by rotary evaporator, derivatized with diazomethane and analyzed by GC-MS, as recommended for chlorinated phenols.

According to our expectations, UV–VIS spectrum analysis of ultra-filtrated fractions had a much lower overall extinction than the original water samples. Moreover, the samples showed a distinct plateau in the region of 260–280 nm, which can correspond to the presence of organic molecules with aromatic ring or heterocycle in their molecules. It can be assumed that these substances might arise from the degradation of humic substances and lignin degradation. From these results, we can conclude that abiotic and biotic chlorination enables high-molecular substances in the forest ecosystem to facilitate their degradation, as their fractions are more water-soluble.

Hydrolysis of large humic substances during their degradation, along with chlorination, and chlorination of SOM can produce chlorinated organic substances some of which are volatile. The chlorinated volatile substances may partially increase the measured AOX level of water, but their presence in the samples is not controlled and can change to various extents during sample transport and preparation. Hence, volatile compounds have to be determined separately from non-volatile compounds. During our monitoring campaigns, we determined the levels of chloroform in water samples in addition to AOX determinations. The results indicate that chloroform production is largely due to biotic chlorination of microorganisms in the soil and is dependent on the season and temperature. The level of chloroform in surface waters is also dependent on the weather as chloroform level shows a close correlation with water temperature.

AOX and TOC analysis of original water samples and ultrafiltrated fractions with molecular size under 500 Da showed that the AOX to TOC ratio in the LMW fractions is higher than in the original samples. This means that LMW fraction is either more halogenated or faster degraded than large molecules. By definition, compounds quantified by the AOX method are dissolved in water and not well soluble in apolar solvents. Therefore, their lipophilicity is low, meaning that they have a low potential to bioaccumulation or biomagnification.

Analyses of organochlorine compounds by gas chromatography-mass spectrometry (GC-MS) in preconcentrated fractions of ultrafiltrates showed that the achieved concentration factor does not yet allow detailed analysis of the present compounds. The determinations led us to suspect that chlorophenols and chloroanisoles might be present, but only in concentrations which are at the limit of detection of the method used. We are not yet able to distinguish whether the traces of chloroanisole come from studied samples, or if they are produced by methylation of chlorophenol, which theoretically can happen by natural processes. In model experiments using radioactive chlorine Cl-36, where forest soil was incubated with radiolabelled chloride, traces of radioactive 4-chlorophenol and radioactive chloroacetic acid were present in the water extracts (Matucha et al., 2007a, 2007b; Rohlenová et al., 2009).

3.11 CONCLUSION

Our study was based on a monitoring campaign which was carried out in 2006–2008 by the Povodí Labe s.p., in the catchment of the Hamry water reservoir. During the first monitoring campaign, elevated levels of AOX were found in the catchments area, which only occasionally exceeded the specified national limit (30 µg/L) for sources of drinking water. This limit has not been exceeded in the water reservoir itself. The continued monitoring in 2009–2010 was aimed to elucidate the AOX levels of water samples and find out whether it is caused by anthropogenic contaminants. Our study confirmed the deduction set out in the conclusions of monitoring conducted by Povodí Labe s.p., that is that substances collectively described by the parameter AOX are in this case of natural origin. This was also found in a study of the presence and origin of chlorinated organic substances in surface waters in southeastern Sweden, which was originally also designed to detect anthropogenic substances. From the research conducted in the Hamry water reservoir it is evident that the observed AOX are of natural origin and that the chlorinated compounds arise mainly during microbial chlorination of SOM. SOM is part of humus formed from the biodegradation of dead plant material.

The AOX concentration limit for sources of drinking water (i.e., 30 µg/L) was exceeded in some cases and at some locations during the second monitoring campaign in 2009–2010, for example in stagnant peat waters and their draining stream waters. These waters were carrying compounds from peat-land humic substances and many chlorinated metabolites of natural origin. The chlorinated metabolites are measured by the AOX method, but are not necessarily toxic. Running waters outside the area diluted these compounds, and volatile compounds are evaporated in their course to the water reservoir. The found AOX values were proportional to TOC values as well as to low-molecular substances, indicating that LMW fraction is either more halogenated or faster degraded than larger molecules. Experiments on microbial chlorination of SOM using radiotracer techniques confirmed the formation of LMW substances, which have been identified as dichloro- and TCAs, chloroform, and other chlorinated substances. These latter substances, suspected to be present in trace amounts and are presumed to be chlorophenols and chloroanisoles, are still waiting to be identified. Unidentified chlorophenol compounds are of special interest because in low concentrations they may have an impact on human health. The attempts to identify and determine the content of LMW substances in surface waters is still in the research stage.

The concentrations of chloroform in natural waters were much lower (0.1–0.8 µg/L) than the maximum permissible value (30 µg/L) for drinking water since its volatility further reduces its concentration in raw water. Some naturally produced organohalogens (such as chloroform) cannot be distinguished from anthropogenic counterparts, and have to be treated in the same way, whereas chlorinated humic substances probably do not represent harm to human health and can be easily removed from raw water.

ACKNOWLEDGMENTS

The results of the 2010–2011 monitoring were obtained within the framework of cooperation of the Academy of Sciences of the Czech Republic with the Pardubický kraj (Pardubice region, Czech Republic), the Povodí Labe, státní podnik (Elbe River Basin, Czech State Enterprise), the Vodárenská společnost Chrudim, a.s. (Waterworks Co. Chrudim plc.), and the Vodovody a kanalizace Chrudim, a.s. (Water Utilities Co. Chrudim plc.). The authors also gratefully acknowledge financial support provided by the Grant Agency of the Czech Republic (Grant Number 13-11101S).

KEYWORDS

- Adsorbable organic halogens
- Biogeochemical cycle of chlorine
- Bioremediation
- Chlorinated organic compounds
- Chlorinated plant products
- Chloroacetic acids
- Chloroform
- Chlorohumus
- Chloroperoxidases
- Chlorophenols
- Cyanobacteria
- Hygienic threshold levels
- Natural bioactive compounds
- Natural chlorination
- Radiotracer studies
- Sodium chloride
- Soil organic matter
- Trihalomethanes
- Volatile organochlorines
- Water disinfection byproducts

REFERENCES

Asplund, G. Origin and occurrence of halogenated organic matter in soil. In *Naturally-produced organohalogens*. Grimvall, A., de Leer, E. W. B., Eds.; Kluwer Academic Publishers: Dordrecht, 1995, pp 35–48.

Asplund, G.; Christiansen, J. V.; Grimvall, A. B. A chloroperoxidase-like catalyst in soil detection and characterization of some properties. *Soil Biol Biochem* 1993, 25, 41–46.

Bastviken, D.; Thomsen, F.; Svensson, T.; Karlsson, S.; Sandén, P.; Shaw, G.; Matucha, M.; Öberg, G. Chloride retention in forest soil by microbial uptake and by natural chlorination of organic matter. *Geochim Cosmochim Ac* 2007, 71, 3182–3192.

Becher, P. G.; Beuchat, J.; Gademann, K.; Jüttner, F. Nostocarboline: Isolation and synthesis of a new cholinesterase inhibitor from Nostoc 78-12A. *J Nat Prod* 2005, 68, 1793–1795.

Blanchard, D. C. The oceanic production of atmospheric sea salt. *J Geophys Res* 1985, 90, 961–963.

Blanchard, F. A. Uptake, distribution and metabolism of carbon-14 labelled trichloroacetate in corn and pea plants. *Weeds* 1954, 3, 274–278.

Butler, J. H.; Battle, M.; Bender, M. L.; Montzka, S. A.; Clarke, A. D.; Saltzman, E. S.; Sucher, C. M.; Severinghaus, J. P.; Elkins, J. W. A record of atmospheric halocarbons during the twentieth century from polar firn air. *Nature* 1999, 399, 749–755.

Cappelletti, M.; Frascari, D.; Zannoni, D.; Fedi, S. Microbial degradation of chloroform. *Appl Microbiol Biotechnol* 2012, 96, 1395–1409.

Chen, C. H.; Dural, N. H. Chloroform adsorption on soils. *J Chem Eng Data* 2002, 47, 1110–1115.

Conti, M. E.; Cecchetti, G. Biological monitoring: Lichens as bioindicators of air pollution assessment – a review. *Environ Pollut* 2001, 114, 471–492.

De Fouw, J. *Chloroform, environmental health criteria series, Vol. 163*. WHO: Geneva, 1994, pp 31–34.

Delalieux, F.; van Grieken, R.; Potgieter, J. H. Distribution of atmospheric marine salt depositions over Continental Western Europe. *Mar Pollut Bull* 2006, 52, 606–611.

Dickey, C. A.; Heal, K. V.; Cape, J. N.; Stidson, R. T.; Reeves, N. M.; Heal, M. R. Addressing analytical uncertainties in the determination of trichloroacetic acid in soil. *J Environ Monit* 2005, 7, 137–144.

Dimmer, C. H.; Simmonds, P. G.; Nickless, G.; Bassford, M. R. Biogenic fluxes of halomethanes from Irish peatland ecosystems. *Atmos Environ* 2001, 35, 321–330.

Erickson, D. J.; Duce, R. A. On the global flux of atmospheric sea salt. *J Geophys Res* 1988, 93, 14079–14088.

Faxon, C. B.; Allen, D. T. Chlorine chemistry in urban atmospheres: A review. *Environ Chem* 2013, 10, 221–233.

Ferenčík, M.; Rederer, L.; Pešava, J. *Adsorbovatelné organické halogeny (AOX) a organické látky (CHSK$_{Mn}$) v povodí vodárenské nádrže Hamry*. Povodí Labe: Hradec Králové, 2008, pp 1–8.

Forczek, S. T.; Laturnus, F.; Doležalová, J.; Holík, J.; Wimmer, Z. Emission of climate relevant volatile organochlorines by plants occurring in temperate forests. *Plant Soil Environ* 2015, 61, 103–108.

Forczek, S. T.; Uhlířová, H.; Gryndler, M.; Albrechtová, J.; Fuksová, K.; Vágner, M.; Schröder, P. Trichloroacetic acid in Norway spruce/soil-system. II. Distribution and degradation in the plant. *Chemosphere* 2004, 56, 327–333.

Gay, B. W. J.; Hanst, P. L.; Bufalini, J. J.; Noonan, R. C. Atmospheric oxidation of chlorinated ethylenes. *Environ Sci Technol* 1976, 10, 58–67.

Giese, B.; Laturnus, F.; Adams, F. C.; Wiencke, C. Release of volatile iodinated C(1)–C(4) hydrocarbons by marine macroalgae from various climate zones. *Environ Sci Technol* 1999, 33, 2432–2439.

Godwin, K. S.; Hafner, S. D.; Buff, M. F. Long-term trends in sodium and chloride in the Mohawk River, New York: The effect of fifty years of road-salt application. *Environ Pollut* 2003, 124, 273–281.

Gribble, G. W. The diversity of naturally produced organohalogens. *Chemosphere* 2003, 52, 289–297.

Gribble, G. W. Organochlorines - natural and anthropogenic. In *Environmental health*, Mooney, L., Bate, R., Eds.; Butterworth Heinemann: Oxford, 1999, pp 161–176.

Haiber, G.; Jacob, G.; Niedan, V.; Nkusi, G.; Schöler, H. F. The occurrence of trichloroacetic acid (TCAA) - Indications of a natural production? *Chemosphere* 1996, 33, 839–849.

Hamilton, J. T. G.; McRoberts, W. C.; Keppler, F.; Kalin, R. M.; Harper, D. B. Chloride methylation by plant pectin: An efficient environmentally significant process. *Science* 2003, 301, 206–209.

Harper, D. B. Halomethane from halide ion - a highly efficient fungal conversion of environmental significance. *Nature* 1985, 315, 55–57.

Harper, D. B. The global chloromethane cycle: Biosynthesis, biodegradation and metabolic role. *Nat Prod Rep* 2000, 17, 337–348.

Haselmann, K. F.; Ketola, R. A.; Laturnus, F.; Lauritsen, F. R.; Gron, C. Occurrence and formation of chloroform at Danish forest sites. *Atmos Environ* 2000a, 34, 187–193.

Haselmann, K. F.; Laturnus, F.; Svensmark, B.; Gron, C. Formation of chloroform in spruce forest soil – results from laboratory incubation studies. *Chemosphere* 2000b, 41, 1769–1774.

He, Z.; Yang, G. P.; Lu, X. L. Distributions and sea-to-air fluxes of volatile halocarbons in the East China Sea in early winter. *Chemosphere* 2013a, 90, 747–757.

He, Z.; Yang, G. P.; Lu, X. L.; Zhang, H. H. Distributions and sea-to-air fluxes of chloroform, trichloroethylene, tetrachloroethylene, chlorodibromomethane and bromoform in the Yellow Sea and the East China Sea during spring. *Environ Pollut* 2013b, 177, 28–37.

Hiraide, M.; Shima, T.; Kawaguchi, H. Separation and determination of dissolved and particulate humic substances in river water. *Microchim Acta* 1994, 113, 269–276.

Hoekstra, E. J.; de Leer, E. W. B.; Brinkman, U. A. T. Findings supporting the natural formation of trichloroacetic acid in soil. *Chemosphere* 1999a, 38, 2875–2883.

Hoekstra, E. J.; de Leer, E. W. B.; Brinkman, U. A. T. Mass balance of trichloroacetic acid in the soil top layer. *Chemosphere* 1999b, 38, 551–563.

Hoekstra, E. J.; de Leer, E. W. B.; Brinkman, U. A. T. Natural formation of chloroform and brominated trihalomethanes in soil. *Environ Sci Technol* 1998a, 32, 3724–3729.

Hoekstra, E. J.; Duyzer, J. H.; de Leer, E. W. B.; Brinkman, U. A. T. Chloroform – concentration gradients in soil air and atmospheric air, and emission fluxes from soil. *Atmos Environ* 2001b, 35, 61–70.

Hoekstra, E. J.; Verhagen, F. J. M.; Field, J. A.; de Leer, E. W. B.; Brinkman, U. A. T. Natural production of chloroform by fungi. *Phytochemistry* 1998, 49, 91–97.

Holík, J.; Forczek, S. T.; Matucha, M.; Rohlenová, J. Studium adsorbovatelných organicky vázaných halogenů (AOX) neantropogenního původu v povodí vodárenské nádrže (Study of the adsorbable organically bound halogens (AOX) from non-anthropogenic origin in the catchment of water reservoir Hamry), Conference Orlicko - Kladsko 2010, Jablonné nad Orlicí, Czech Republic, 2010, pp 100–105.

Huang, B.; Lei, C.; Wei, C.; Zeng, G. Chlorinated volatile organic compounds (Cl-VOCs) in environment - sources, potential human health impacts, and current remediation technologies. *Environ Int* 2014, 71, 118–138.

Ibrahim, M. B. M.; Moursy, A. S.; Bedair, A. H.; Radwan, E. K. Comparison of XAD-8 and DEAE for isolation of humic substances from surface water. *J Environ Sci Technol* 2008, 1, 90–96.

Ishida, K.; Welker, M.; Christiansen, G.; Cadel-Six, S.; Bouchier, C.; Dittmann, E.; Hertweck, C.; Tandeau de Marsac, N. Plasticity and evolution of aeruginosin biosynthesis in cyanobacteria. *Appl Environ Microb* 2009, 75, 2017–2026.

Itoh, N.; Kutsuna, S.; Ibusuki, T. A product study of the OH radical initiated oxidation of perchloroethylene and trichloroethylene. *Chemosphere* 1994, 28, 2029–2040.

Jancula, D.; Marsalek, B. Critical review of actually available chemical compounds for prevention and management of cyanobacterial blooms. *Chemosphere* 2011, 85, 1415–1422.

Jay, K.; Stieglitz, L. Identification and quantification of volatile organic components in emissions of waste incineration plants. *Chemosphere* 1995, 30, 1249–1260.

Johansson, E.; Krantz-Rülcker, C.; Zhang, B. X.; Öberg, G. Chlorination and biodegradation of lignin. *Soil Biol Biochem* 2000, 32, 1029–1032.

Johansson, E.; Sanden, P.; Oberg, G. Organic chlorine in deciduous and coniferous forest soils in southern Sweden. *Soil Sci* 2003, 168, 347–355.

Johansson, E.; Sandén, P.; Öberg, G. Spatial patterns of organic chlorine and chloride in Swedish forest soil. *Chemosphere* 2003, 52, 391–397.

Keppler, F.; Borchers, R.; Pracht, J.; Rheinberger, S.; Schöler, H. F. Natural formation of vinyl chloride in the terrestrial environment. *Environ Sci Technol* 2002, 36, 2479–2483.

Keppler, F.; Eiden, R.; Niedan, V.; Pracht, J.; Schöler, H. F. Halocarbons produced by natural oxidation processes during degradation of organic matter. *Nature* 2000, 403, 298–301.

Khalil, M. A. K.; Rasmussen, R. A.; French, J. R.; Holt, J. A. The influence of termites in atmospheric trace gases: CH_4, CO_2, $CHCl_3$, N_2O, CO, H_2, and light hydrocarbons. *J Geophys Res* 1990, 95, 3619–3634.

Khalil, M. A. K.; Rasmussen, R. A.; Shearer, M. J.; Chen, Z.; Yao, H.; Yang, J. Emissions of methane, nitrous oxide, and other trace gases from rice fields in China. *J Geophys Res* 1998, 103, 25241–25250.

Laskin, A.; Moffet, R. C.; Gilles, M. K.; Fast, J. D.; Zaveri, R. A.; Wang, B.; Nigge, P.; Shutthanandan, J. Tropospheric chemistry of internally mixed sea salt and organic particles: Surprising reactivity of NaCl with weak organic acids. *J Geophys Res* 2012, 117, D15302.

Laturnus, F.; Adams, F. C.; Wiencke, C. Methyl halides from Antarctic macroalgae. *Geophys Res Lett* 1998, 25, 773–776.

Laturnus, F.; Haselmann, K. F.; Borch, T.; Gron, C. Terrestrial natural sources of trichloromethane (chloroform, $CHCl_3$) - an overview. *Biogeochemistry* 2002, 60, 121–139.

Laturnus, F.; Svensson, T.; Wiencke, C. Release of reactive organic halogens by the brown macroalga *Saccharina latissima* after exposure to ultraviolet radiation. *Polar Res* 2010, 29, 379–384.

Laturnus, F.; Wiencke, C.; Klöser, H. Antarctic macroalgae - sources of volatile halogenated organic compounds. *Mar Environ Res* 1996, 41, 169–181.

Leri, A. C.; Hay, M. B.; Lanzirotti, A.; Rao, W.; Myneni, S. C. B. Quantitative determination of absolute organohalogen concentrations in environmental samples by X-ray absorption spectroscopy. *Anal Chem* 2006, 78, 5711–5718.

Lobert, J. M.; Keene, W. C.; Logan, J. A.; Yevich, R. Global chlorine emissions from biomass burning: Reactive chlorine emissions inventory. *J Geophys Res* 1999, 104, 8373–8389.

Lovelock, J. E.; Maggs, R. J.; Wade, R. J. Halogenated hydrocarbons in and over the Atlantic. *Nature* 1973, 241, 194–196.

Matucha, M.; Clarke, N.; Lachmanová, Z.; Forczek, S. T.; Fuksová, K.; Gryndler, M. Biogeochemical cycles of chlorine in the coniferous forest ecosystem: Practical implications. *Plant Soil Environ* 2010, 56, 357–367.

Matucha, M.; Forczek, S. T.; Gryndler, M.; Uhlířová, H.; Fuksová, K.; Schröder, P. Trichloroacetic acid in Norway spruce/soil-system. I. Biodegradation in soil. *Chemosphere* 2003, 50, 303–309.

Matucha, M.; Gryndler, M.; Forczek, S. T.; Schröder, P.; Bastviken, D.; Rohlenová, J.; Uhlířová, H.; Fuksová, K. A chlorine-36 and carbon-14 study of the role of chlorine in the forest ecosystem. *J Label Compd Radiopharm* 2007a, 50, 437–439.

Matucha, M.; Gryndler, M.; Schröder, P.; Forczek, S. T.; Uhlířová, H.; Fuksová, K.; Rohlenová, J. Chloroacetic acids - degradation intermediates of organic matter in forest soil. *Soil Biol Biochem* 2007b, 39, 382–385.

Mayer, F. Zur Wirkungsweise von Trichloroacetat auf die höhere Pflanze. Z. *Naturforschg* 1957, 12B, 336–346.

McCulloch, A.; Midgley, P. M. The production and global distribution of emissions of trichloroethene, tetrachloroethene and dichloromethane over the period 1988–1992. *Atmos Environ* 1996, 30, 601–608.

Montzka, S. A.; Butler, J. H.; Elkins, J. W.; Thompson, T. M.; Clarke, A. D.; Lock, L. T. Present and future trends in the atmospheric burden of ozone-depleting halogens. *Nature* 1999, 398, 690–694.

Müller, G. Sense or no-sense of the sum parameter for water soluble "adsorbable organic halogens" (AOX) and "absorbed organic halogens" (AOX-S18) for the assessment of organohalogens in sludges and sediments. *Chemosphere* 2003, 52, 371–379.

Myneni, S. C. B. Formation of stable chlorinated hydrocarbons in weathering plant material. *Science* 2002, 295, 1039–1041.

Nelson, L.; Shanahan, I.; Sidebottom, H. W.; Treacy, J.; Nielsen, O. J. Kinetics and mechanism for the oxidation of 1,1,1-trichloroethane. *Int J Chem Kinet* 1990, 22, 577–590.

Öberg, G. The biogeochemistry of chlorine in soil. In *The Handbook of Environmental Chemistry*, *Vol. 3*, Gribble, G. W., Ed.; Springer-Verlag: Berlin, Heidelberg, 2003, pp 43–62.

Öberg, G. The natural chlorine cycle - fitting the scattered pieces. *Appl Microbiol Biotechnol* 2002, 58, 565–581.

Öberg, G.; Holm, M.; Sanden, P.; Svensson, T.; Parikka, M. The role of organic-matter-bound chlorine in the chlorine cycle: A case study of the Stubbetorp catchment, Sweden. *Biogeochemistry* 2005, 75, 241–269.

Öberg, G.; Sandén, P. Retention of chloride in soil and cycling of organic matter-bound chlorine. *Hydrol Process* 2005, 19, 2123–2136.

Rhew, R. C.; Miller, B. R.; Weiss, R. F. Natural methyl bromide and methyl chloride emissions from coastal salt marshes. *Nature* 2000, 403, 292–295.

Rohlenová, J.; Gryndler, M.; Forczek, S. T.; Fuksová, K.; Handová, V.; Matucha, M. Microbial chlorination of organic matter in forest soil: Using [36]Cl-chloride and its methodology. *Environ Sci Technol* 2009, 43, 3652–3655.

Roy, R.; Pratihary, A.; Narvenkar, G.; Mochemadkar, S.; Gauns, M.; Naqvi, S. W. A. The relationship between volatile halocarbons and phytoplankton pigments during a *Trichodesmium* bloom in the coastal eastern Arabian Sea. *Estuar Coast Shelf Sci* 2011, 95, 110–118.

Rudolph, J.; von Czapiewski, K.; Koppmann, R. Emissions of methyl chloroform (CH_3CCl_3) from biomass burning and the tropospheric methyl chloroform budget. *Geophys Res Lett* 2000, 27, 1887–1890.

Saito, T.; Yokouchi, Y.; Aoki, S.; Nakazawa, T.; Fujii, Y.; Watanabe, O. A method for determination of methyl chloride concentration in air trapped in ice cores. *Chemosphere* 2006, 63, 1209–1213.

Schöler, H. F.; Keppler, F.; Fahimi, I. J.; Niedan, V. W. Fluxes of trichloroacetic acid between atmosphere, biota, soil and groundwater. *Chemosphere* 2003, 52, 339–354.

Schröder, P.; Matucha, M.; Forczek, S. T.; Uhlířová, H.; Fuksová, K.; Albrechtová, J. Uptake, translocation and fate of trichloroacetic acid in Norway spruce/soil system. *Chemosphere* 2003, 52, 437–442.

Schröder, P.; Scheer, C. E.; Diekmann, F.; Stampfl, A. How plants cope with foreign compounds - translocation of xenobiotic glutathione conjugates in roots of barley (*Hordeum vulgare*). *Environ Sci Pollut Res* 2007, 14, 114–122.

Schröder, P.; Wolf, A. E. Characterization of glutathione S-transferases in needles of Norway spruce trees from a forest decline stand. *Tree Physiol* 1996, 16, 503–508.

Sidebottom, H.; Franklin, J. The atmospheric fate and impact of hydrochlorofluorocarbons and chlorinated solvents. *Pure Appl Chem* 1996, 68, 1757–1769.

Silk, P. J.; Lonergan, G. C.; Arsenault, T. L.; Boyle, C. D. Evidence of natural organochlorine formation in peat bogs. *Chemosphere* 1997, 35, 2865–2880.

Sullivan, R. C.; Guazzotti, S. A.; Sodeman, D. A.; Tang, Y.; Carmichael, G. R.; Prather, K. A. Mineral dust is a sink for chlorine in the marine boundary layer. *Atmos Environ* 2007, 41, 7166–7179.

Svensson, T.; Laturnus, F.; Sanden, P.; Oberg, G. Chloroform in runoff water - a two-year study in a small catchment in Southeast Sweden. *Biogeochemistry* 2007, 82, 139–151.

Thompson, I. P.; van der Gast, C. J.; Ciric, L.; Singer, A. C. Bioaugmentation for bioremediation: The challenge of strain selection. *Environ Microbiol* 2005, 7, 909–915.

Tuazon, E. C.; Atkinson, R.; Aschmann, S. M.; Goodman, M. A.; Winer, A. M. Atmospheric reactions of chloroethenes with the OH radical. *Int J Chem Kinet* 1988, 20, 241–265.

van Pée, K. -H.; Unversucht, S. Biological dehalogenation and halogenation reactions. *Chemosphere* 2003, 52, 299–312.

Walter, B.; Ballschmiter, K. Formation of C_1/C_2-bromo-/chloro-hydrocarbons by haloperoxidase reactions. *Fresen J Anal Chem* 1992, 342, 827–833.

Wang, X. P.; Gordon, M. P.; Strand, S. E. Mechanism of aerobic transformation of carbon tetrachloride by poplar cells. *Biodegradation* 2002, 13, 297–305.

Weissflog, L.; Krüger, G. H. J.; Forczek, S. T.; Lange, C. A.; Kotte, K.; Pfennigsdorff, A.; Rohlenová, J.; Fuksová, K.; Uhlířová, H.; Matucha, M.; Schröder, P. Oxidative biotransformation of tetrachloroethene in the needles of Norway spruce (*Picea abies* L.). *S Afr J Bot* 2007, 73, 89–96.

Wever, R.; Hemrika, W. Vanadium Haloperoxidases. In *Handbook of Metalloproteins*. Messerschmidt, A., Huber, R., Poulos, T., Wieghardt, K., Eds.; John Wiley and Sons: Chichester, 2001, pp 1417–1428.

Winterton, N. Chlorine: The only green element - towards a wider acceptance of its role in natural cycles. *Green Chem* 2000, 2, 173–225.

Yokouchi, Y.; Ikeda, M.; Inuzuka, Y.; Yukawa, T. Strong emission of methyl chloride from tropical plants. *Nature* 2002, 416, 163–165.

Yokouchi, Y.; Noijiri, Y.; Barrie, L. A.; Toom-Sauntry, D.; Machida, T.; Inuzuka, Y.; Akimoto, H.; Li, H. J.; Fujinuma, Y.; Aoki, S. A strong source of methyl chloride to the atmosphere from tropical coastal land. *Nature* 2000, 403, 295–298.

CHAPTER 4

ENVIRONMENTAL IMPACT OF PESTICIDE USE ON MICROBIAL COMMUNITIES AND SOIL BIOPROCESSES: A PHYSIOLOGICAL, BIOCHEMICAL, AND MOLECULAR PERSPECTIVE

JEYABALAN SANGEETHA[1], MUNISWAMY DAVID[2], DEVARAJAN THANGADURAI[3], ETIGEMANE RAMAPPA HARISH[2], JADHAV SHRINIVAS[2], PRATHIMA PURUSHOTHAM[3], and KARTHEEK RAJENDRA MALOWADE[2]

[1]Department of Environmental Science, Central University of Kerala, Kasaragod, Kerala 671316, India

[2]Department of Zoology, Karnatak University, Dharwad 580003, Karnataka, India

[3]Department of Botany, Karnatak University, Dharwad 580003, Karnataka, India

CONTENTS

4.1   INTRODUCTION

Pesticides are agrochemicals which are in use to prevent pests infesting crops. Usage of pesticides largely increased from the past few decades which even started from pre-sowing stage. In agricultural fields, soil is the most important component as it is the site for biological interactions. Repeated application of pesticides ultimately contaminates the soil via spray drift during foliage treatment, ground water through seepage, and surface water through runoff. Ultimately, pesticide contamination adversely affects the soil ecosystem function by affecting microflora, macroflora, fauna, and physicochemical properties of the soil. Finally, it leads to the degradation of soil fertility. Nutrient cycle, roots, plants, and soil biological activities normally found in the top 20–30 cm (8–12 in), known as the rhizosphere. The most important microflora of the rhizosphere soil are nitrogen fixing bacteria, phosphate solubilizing bacteria, and mineral solubilizers. In addition, many types of bacteria produce plant-growth promoting substances. These microorganisms are responsible for the overall fertility of the soil (Asadu et al., 2015). The excess and frequent application of pesticides may adversely affect these kinds of agriculturally beneficial microorganisms and ultimately spoil fertility of the soil.

The production and utilization of agro-based chemicals has been widely undertaken by low to average income countries. Plants, soil matrix, and soil organisms are having interrelationship as triad in the rhizosphere region, where one compound in triad affected by external sources will have the impact on other two compounds (Coleman et al., 2004). In general, broad spectrum of pesticides acts against pests, insects, or weeds and kill them but also affects the non-target beneficial organisms including microbes. Thus, use of pesticides shows severe threat to microbial diversity of the agricultural soil. Mode of action of the pesticides differs based on the active chemical compounds. However, during their application majority of the pesticides are affecting non-target organisms in addition to the pests. Hence, the repeated application of the pesticides ultimately leads to loss of biodiversity.

In particular, assessment of side effects of chemical or biological compounds is a complex and problematic in environmental systems (Rebecchi et al., 2000). Bioavailability of pesticides to soil organisms, including rhizosphere microorganisms is of paramount importance to the expected effect (Gevao et al., 2000). Bioavailability of the pesticide residues basically depends on the percentage of pesticide used, soil type, leachate, and degradation of the compound. A wide variety of pesticides are known to persist in soil environment without degradation and are further found to pollute water table as well as soil environment. Consequently, it influences micro- and macroorganisms through bioaccumulation in food chain (Macdonald et al., 2000; Gill and Garg, 2014). In general, continuous application of pesticides leads to several negative impacts on the environment which cannot be ignored. Directly or indirectly, pesticides may affect the fundamental biochemical reaction of soil ecosystem such as organic matter mineralization, nitrogen fixation, nitrification, denitrification, ammonification, sulfur and phosphorous solubilization (Kinney et al., 2004; Menon et al., 2005; Hussain et al., 2009). Many of the agriculturally important bacterial species are more sensitive to pesticides. Filimon et al. (2015) proved the inhibition of nitrification due to the sulfonylurea herbicides. Fungicides particularly, chlorothalonil and dinitrophenyl have shown adverse effects on the rhizosphere microorganisms which are response for the nitrification and denitrification processes (Niewiadomska and Klama, 2005; Lang and Cai, 2009). It has been reported by earlier researchers that residual concentration of pesticides like pentachlorophenol, DDT, and methyl parathion from soil are known to affect signaling in leguminous plants such as alfalfa, peas, and soybeans with symbiotic soil bacteria.

This phenomenon is comparable to endocrine disruption of pesticides in human and animals, thereby significantly disrupting N_2 fixation (Fox et al., 2007; Mnif et al., 2011). Organochlorine pesticides are more persistent and its metabolites are more toxic than other types of pesticides, hence it has been banned to use in agriculture in many of the developed countries (Ravikumar et al., 2013).

Pesticide residues are found in many agricultural products due to continuous use of pesticides in agricultural fields. On the basis of global concern and widespread criticism, certain pesticides like DDT, dieldrin, endosulfan, and lindane have been banned citing the potential threat on ecosystem. These pesticides are having long half life, for example, half life of DDT in soil is from 22 to 33 years, toxaphene up to 14 years, mirex about 12 years, dieldrin about 7 years, and chlordecone up to 30 years; hence, it posses high persistence in the environment. Additionally, many pesticides are potential to accumulate in the fatty tissues of living organisms due to its water phobic or low water solubility nature (Cone, 2005). Hence, pesticides are great threat to the biodiversity including macro- and microorganisms.

4.2   PESTICIDES AND ITS APPLICATIONS

The term pesticide is used to define certain group of chemicals that are synthetically produced yet, are of biological origin, and are used to counter pests of plants and animals that could harm the productivity, processing, and transport of food and agricultural products (Arthur et al., 2000; Mushobozi and Santacoloma, 2010; Saini et al., 2014). The term synthetic pesticide comprises the wide range of compounds which include organochlorine, organophosphate, pyrethroids, and carbamates having a property to act as an insecticide, fungicide, herbicide, rodenticide, molluscicide, nematicide, and plant growth regulators (Aktar et al., 2009). Pesticides are important because of their tremendous benefits to the civilized people, through integral part of the process by reducing the losses of yield from the weeds, disease and insect pests that can markedly reduce risk in forestry, public health and the domestic sphere and, of course, in agriculture (Vinita and Veena, 2015).

The present scenario of agriculture is dependent on the wide range of pesticides which include insecticides, fungicides, and herbicides (Lopez et al., 2002). An ideal pesticide should be toxic to only target species and nontoxic to other organisms including humans. Additionally, higher crop yield by the sage of pesticides should bring additional revenue to the people that lead to wealth of the country (Aktar et al., 2009).

About 38% of global land are terrestrial biomes which have been utilized for agricultural practices such as the cultivation of food, pasture, feed, and fodder crops (Tilman et al., 2001; Foley, 2011). During the past six decades, expansion and intensification of agricultural practices led to the rapid increase in pesticide production (>75%) with the worldwide market of about US $50 billion (Tilman et al., 2001; Stehle and Schulz, 2015). Traditional agriculture has witnessed a considerable crop and economic losses in several parts of third world countries, wherein the industrial production of pesticides and their steady use as part of green revolution in the past have shown potential increase in crop productivity and plant protection (Warren, 1998; Webster et al., 1999).

Since pesticides play a significant role in an agricultural production and as such farmers invest significant sums in management of pests. Therefore, it is imperative that the quality of pesticides is assured for which the system of collection of samples requires improvement, infrastructure for analysis of pesticides samples particularly in the synthetic pesticides and plants growth regulators. List of common pesticides used for different crops, and its target pests are given in Table 4.1.

TABLE 4.1 List of Common Pesticides Used in Different Crops and Its Target Organisms.

Pesticide	Target and Chemical Class	Crops Used	References
Phorate	Insect, Organo-phosphate (OP)	Bajra, barley, maize, paddy, sorghum, wheat, black gram, green gram, pigeon pea	Rajendran (2003); Indiradevi (2010); Bhushan et al. (2013)
Mancozeb	Fungicide, Dithiocarbamate	Potato, tomato, wheat, maize, paddy, jowar, chilies, onions, tapioca, ginger, sugarbeet, cauliflower, groundnut, grapes, guava, banana, apple, cumin, tobacco, mustard, black pepper, pearlmillet, cucumber	Bhushan et al. (2013)
Methyl Parathion	Insect, Acaricide (OP)	Paddy, cotton, black gram, green gram, soybean, mustard, groundnut	Rajendran (2003); Bhushan et al. (2013)
Cypermethrin	Insect, Pyrethroid	Brinjal, cotton, cabbage okra, sugarcane, wheat, sunflower, rice	Rajendran (2003); Bhushan et al. (2013)
Carbendazim	Fungicide, Carbamates	Paddy, wheat, barley, tapioca, cotton, jute, groundnut, sugarbeet, beans, cucurbits, brinjal, apples, grapes, walnut, rose, ber, mango	Bhushan et al. (2013)
Monocroto-phos	Insect, OP	Paddy, maize, bengal gram, green-gram, pea, red gram, sugarcane, cotton, castor, mustard, citrus fruits, mango, coffee, cardamom	Bhushan et al. (2013)
Malathion	Insect, OP	Paddy, sorghum, soybean, cotton, castor, groundnut, mustard, sunflower, okra, cauliflower, radish, turnip, tomato, apple, grape, mango	Rajendran (2003); Indiradevi (2010); Bhushan et al. (2013)
Quinalphos	Insect, Organo-thiophosphate	Chilies, paddy, sugarcane, sorghum, okra, cotton, brinjal, tomato, tea, tur, groundnut, wheat, bengal gram, blackgram, red gram, french bean, soybean, jute, mustard, sesame, cabbage, cauliflower, onion, apple, banana, citrus fruits, mango, pomegranate, cardamom, coffee, gram, safflower	Bhushan et al. (2013)
Carbofuran	Insect, Carbomates	Barley, bajra, sorghum, jute, groundnut, french bean, potato, tomato, apple, citrus fruits, maize, paddy, mustard, soybean, sugarcane, bhindi, chilies, cabbage, wheat, brinjal, banana, peach, mandarins, cotton, pea, tea, sweet pepper	Bhushan et al. (2013)

TABLE 4.1   *(Continued)*

Pesticide	Target and Chemical Class	Crops Used	References
Phosalone	Insect, OP	Barley, paddy, sorghum, cotton, jute, groundnut, bhindi, brinjal, cabbage, chilies, tomato, tea, mustard	Bhushan et al. (2013)
Chlorpyrifos	Insect, OP	Rice, beans, gram, sugarcane, cotton, groundnut, mustard, brinjal, cabbage, onion, apple, ber, citrus fruits	Rajendran (2003); Bhushan et al. (2013)
Dimethoate	Insect, OP	Bajra, maize, sorghum, red gram, cotton, castor, groundnut, mustard, safflower, bhindi, brinjal, cabbage, cauliflower, chilies, onion, potato, tomato, apple, apricot, banana, citrus fruits, fig, mango, rose	Bhushan et al. (2013)
Dichlorvos	Insect, OP	Paddy, wheat, soybean, sugarcane, castor, groundnut, mustard, sunflower, cucurbits, cashew	Bhushan et al. (2013)
Paraquat Dichloride	Weed, Herbicides	Tea, cotton, potato, rubber, rice, wheat, maize, grapes, apple, aquatic weeds	Rajendran (2003); Bhushan et al. (2013)
Zineb	Fungicide, Organosulfur compound	Jowar, paddy, wheat, ragi, tobacco, onion, potato, tomato, chilies, brinjal, cucurbits, cauliflower, cumin, apple, citrus fruits, cherries, grapes, guava	Bhushan et al. (2013)
Captan	Fungicide, Organochlorides (OC)	Chilies, potato, apple, cherry, grapes, cabbage, cauliflower, brinjal, beans, tomato, citrus fruits, rose, paddy, tobacco	Rajendran (2003); Bhushan et al. (2013)
2,4–D	Weed, Chloro-phenoxy acid	Paddy, maize, wheat, sorghum, potato, sugarcane, citrus fruits, grapes	Rajendran (2003); Bhushan et al. (2013)
Fenvalerate	Insecticide, Organochlorides	Cotton, cauliflower, brinjal, okra	Bhushan et al. (2013)
Triazophos	Sucking and chewing insects, OP	Cotton, rice, soybean, brinjal	Bhushan et al. (2013)
Acephate	Insect, OP	Cotton, safflower, rice	Bhushan et al. (2013)

4.3   ENVIRONMENTAL IMPACT OF PESTICIDES

Throughout the world, farmers are following a number of pest controlling measures by using different chemical pesticides to control pest infestation and minimize crop yield losses. Pesticides are highly biologically active substances that can threaten the ecological integrity of aquatic and terrestrial ecosystem; hence, there is a need to incorporate good agricultural practices (De Leo and Levin, 1997). In most parts of old world, pesticides were simply applied without proper understanding on ecology of pests, cost of pesticides, cost-benefit analysis, mode of application, type of pesticide to be used, crop specificity and different developmental stages of the crop plants, which are all potential attributes for heavy crop as well as economic losses (Ombe, 2014). The level of pesticide consumption is comparatively less (Indiradevi, 2010) in several developing and poorly developed countries like Sri Lanka, Lebanon, India, China, Bangladesh, Philippines, Mali, Ecuador, Zimbabwe, and Vietnam than that of many other developed and industrialized nations (Rola and Pingali, 1993; Van Der Hoek et al., 1998; Dung and Dung, 1999; Wilson, 2000; Huang et al., 2001; Ajayi, 2002; Maumbe and Swinton, 2003; Rahman, 2003; Yanggen et al., 2003; Gupta, 2004; Salameh et al., 2004).

Recently, Stehle and Scuze (2015) for the first time reported at the global scale that, more than 50% of detected insecticide concentrations exceed regulatory threshold levels, thereby posing threat to aquatic biodiversity because of surface water pollution due to modern farming practices. Angelini et al. (2013) and Anzuay et al. (2015) indicates that the pesticide application caused a decrease on the number of cultivated nitrogen fixing population of peanut soils of Cordoba as well on the nitrogen fixing ability of these soil. According to World Health Organization (WHO), one million cases of pesticide poisoning occur every year and consequently there are 20000 deaths globally. The most damaging ecological disturbance of injudicious use of pesticides is the existence of high concentration of pesticide residues in food chain, including cereals, pulses, vegetables, fruits, milk and milk products (including mother's milk), fishes, poultry, meat products, and water (Jeyanthi and Kombairaju, 2005).

The use of pesticides and production of synthetic pesticides doubles in future due to increased cash-crop and plantation-style farming in the modern agricultural activity. Also, the rate of individual risk cases of intentional and unintentional acute poisoning may increase over the next decade despite a decrease in the proportion of the overall population directly involved in agricultural production (Aktar et al., 2009). Hence, revisions regarding the present regulatory methods and process of pesticide usage are necessary to minimize the damage to environment due to exploitation of synthetic chemicals for agricultural purpose. To prevent any harmful effects on the environment, the level of pesticide residues in air, soil, and water should be monitored regularly. The monitoring of pesticide residues in food is one of the most important approaches in minimizing the potential hazard to human health. When unacceptable levels of pesticides are found, appropriate steps should be taken to identify the cause and to prevent recurrence.

4.4   AGRICULTURALLY IMPORTANT MICROORGANISMS

The era of green revolution in the 20th century was the result of intense agricultural activity tagged with modern scientific technologies. This was accompanied with surprising cost to the ecological system which further led to environmental unsustainability and extensive threat to biodiversity (Vance, 1998). As a result, immense global response was received toward the deteriorating environment by suggesting certain measures which could support the development of sustainable agriculture and increased productivity without compromising the environmental health (Vance, 2001; Noble and Ruaysoongnern, 2010).

One of the feasible technologies to achieve this condition was to avoid the application of hazardous agrochemicals which included synthetic fertilizers and toxic pesticides, instead to proceed with eco-friendly methodologies like employing symbiotic microorganisms which could help in attaining good growth of crops and better health of livestock in addition to their protection from pests and additionally impart resistance to environmental stresses (Higa and Parr, 1994; Yang et al., 2009). The interaction of microbes with plants is a well-established process due to the habitat availability shown by plants to different microbes. The wide range of habitats for microbial association include, phyllosphere which refers to aerial plant part, endosphere referring to internal transport system and rhizosphere; zone of influence of root system (Lynch, 1990; Lindow et al., 2002). Certain groups of free-living bacteria that are present in soil are known to promote the growth of plant by colonizing the plant roots, and are hence known as plant growth promoting rhizobacteria (PGPR) (Kloepper et al., 1989; Cleyet-Marcel et al., 2001).

The plant microbe interaction is of primary importance in order to carry out transformation and solubilization of nutrients from soil which is in limited fashion under the absence of microbes. This in turn is in crucial proposition for plants to realize their genetic ability to utilize nutrients and grow. Considering biological aid for better production of crops is becoming more acceptable by wide range of farmers around the globe for its ability to induce higher growth rate and in addition avoid the hazardous conditions which would arise upon making use of synthetic chemical fertilizers. With this scenario, PGPR has attained a pivotal role for its potential to increase crop yield along with sustainable environment (Sturz et al., 2000; Shoebitz et al., 2009). The occurrence of agriculturally important microbes is available under a cluster of symbiotic (*Rhizobium* sp.) as well as non-symbiotic (*Azotobacter*, *Azospirullum*, *Bacillus* and *Klebsiella* sp.) bacterial species which are being used to ensure maximum productivity of agricultural crops (Burd et al., 2000; Cocking, 2003).

Environmental factors are known to play a crucial role in determining the composition of bacterial communities that are known to exist with constant interaction with plant species. One among these factors is identified to be the nature of soil which can determine the microbial existence in association with the plant roots (Lundberg et al., 2012; Bulgarelli et al., 2013). At the same time, reports have suggested that the host plant is known to play a crucial role in determining the microbiota of its root habitat (Marschner et al., 2005; Doornbos et al., 2011) especially endophytic bacterial communities (Haichar et al., 2008). The enormous contribution could be seen by the important microbes that are known to establish their association with plants and are often known to play a critical role in determining environmental stability. Analysis of these microbes under molecular levels have revealed that more than 4000 species are known to be present in one gram of soil. However, many of these microorganisms have known to be non-cultivable which include a plant symbiotic vesicular-arbuscular mycorrhiza (VAM) endomycorrhizae which are often seen in angiosperms and gymnosperms or are viable yet in non-cultivable conditions (Colwell and Grimes, 2000).

The accessibility of nutrient pool in an ecosystem is highly dependent on ability of microbes existing in that very environment and their capacity to uptake the nutrients from soil. Retention of nitrogen by microbes has been identified as a temporary supply of nitrogen in few terrestrial ecosystem (Zogg et al., 2000; Bardgett et al., 2003) thereby potentially limiting the export of nitrogen to adjacent ecosystems and groundwater (Brooks et al., 1998).

Understanding the localization of nitrogen due to seasonal variations is important for acquiring nitrogen by plants that are more particular to ecosystems with poor nitrogen availability. This is where microbial species tend to localize maximum amounts of nitrogen thereby helping the plants during autumn and preserve it throughout winter until the arrival of spring during which it could be released for utilization of plants (Zogg et al., 2000; Bardgett et al., 2005).

The genus *Frankia* constitutes an important group of microorganisms that have immense agricultural importance. They are filamentous, sporulating, and Gram-positive bacteria that are capable of fixing atmospheric nitrogen through establishment of symbiotic nodules in different dicotyledonous plants (Susamma et al., 2002). The presence of *nif* genes in *Frankia* makes it possible to carry out nitrogen fixation (Lois et al., 1999; Joel et al., 2002). One of the important and well-known relationships between legume-*Rhizobium* is seen between few plants and microbes. The rapid growing nature of *Rhizobium* makes it easy to be isolated it in pure culture. *Frankia* on other end is also a strong symbiotic organism that is known for its alliance with plants like *Alnus* and *Casurina*. As compared with the other species of symbiotic groups, *Frankia* constitute highly active strains (Ganesh et al., 1994). It is known to give rise to root nodules in which nitrogen is actively converted to ammonia. They are highly diverse and they differ in morphology when compared to actinomycete genera. They are known to form vegetative hyphae, sporangia, and vesicles (Tjepkema et al., 1980). Vesicles are characterized to protect nitrogenase, as it is highly reactive to O_2. The establishment of *Frankia* in to plants is known to be gained by root hair infection. After gaining the entry through root hair infection, *Frankia* is known to form nodules on lateral roots with cortical cylinder of vascular tissue (Ganesh et al., 1994). Vegetative mycelia are the active state of *Frankia* that is seen in root nodules. Thus, most of the nitrogen requirement that a plant is in need of is supplied by *Frankia* species. The process of fixation of nitrogen begs the need of large amount of energy and this is in turn provided by the host plants in the form of organic carbon. The primary significant role is played by the soil condition which is more or less a deciding factor for establishment of *Frankia* with the plants in a symbiotic relationship (Reddel et al., 1986; Smolander et al., 1988).

The identification of *Rhizobium* as source of fixing nitrogen in root nodules is considered as not less than a milestone in the field of agricultural microbiology (Fisher and Newton, 2004). The process of nitrogen fixation is carried out in the root nodules wherein the free nitrogen reacts with H molecules resulting in the formation of NH_3 (Zahran, 1999). This process is carried out by *Rhizobium* symbiotically during which it is gaining the energy from host plant (Kondorosi et al., 2013). *Rhizobium* is identified as a Gram-negative organism which is known to live freely in soil (Trinick, 1973). The uniqueness of *Rhizobium* is due to its potential to fix atmospheric nitrogen thereby helping the host plant for maintenance of its normal physiology (Lancelle and Torrey, 1984). However, certain strains of rhizobia-legume relations are constrained and not similar as above. Few species are known to exist as an endosymbiotic microorganism which is known to gain entry in to root of legumes through root hair thereby forming the root nodule. One of the prominent natures of rhizobia is that it has been found to be associated with development of shoot and root growth in rice plants (Yanni et al., 1997; Yanni and Abd-El-Fattah, 1999).

Azotobacter is a genus which is known to exist in symbiotic association with plants under aerobic soil conditions. It is a Gram-negative organism and is known for fixing the nitrogen (Lakshminarayana, 1993). They are known to be either free living or in association with plants (Gandora et al., 1998; Martyniuk and Martyniuk, 2003). The distribution of *Azotobacter* sp. is complex and is mainly dependent on critical environmental factors in-cluding soil condition and its constituents. The species of *Azotobacter* has been used as a crucial biofertilizer in association with cereals and rice crop by employing seedling dip method and seed dipping technique (Singh et al., 1999; Rüttimann et al., 2003). Species of *Azotobacter* are known to be highly specialized toward the nitrogen fixation process. Since the nitrogen fixation is known to be highly sensitive to O_2, the specialty of *Azotobacter* to counter the O_2 inside cells perhaps makes it feasible to fix the nitrogen com-fortably (Yu et al., 2005). O_2 is known to have antagonistic approach toward nitrogenase enzyme, however, the high rate of respiration in *Azotobacter* is known for utilizing the free O_2 thereby protecting the nitrogenase enzyme as well. One of the crucial factors involved in nitrogen fixation process is the participation of homocitrate ions and C source is the chief requirement to supply energy to microbes (Kanungo et al., 1997). *A. vinelandis* is known for its high nitrogenase activity under *in vitro* conditions (Schubert et al., 1976; Smith et al., 1976). Few reports indicating its involvement in growth of wheat crop are also evident (Kader et al., 2002).

Azolla is a fern that is known to be free-floating. It is capable of fixing atmospheric nitrogen under the collaborative approach with nitrogen fixing cyanobacterium *Anabaena azollae*. *Azolla* encompasses seven species and is a critical symbiotic complex in the aquatic ecosystem (Mian, 2002). This organism is known to reside in dorsal lobes of *Azolla* leaves endophytically and known for supplying nitrogen to rice crop. The lone plant cyanobacterial complexity exists between *Azolla-Anabaena* that is used as a biofertilizer in agricultural scenario. The remarkable ability of Azolla to retrieve soils there-by dejecting harmful invasions by weed species is known to be well evident in rice crops. In addition, its use in waste-water treatment and degradation of certain heavy metals has also been well documented. Besides these, its use in the form of aquaculture feed has also been practiced and the terminology "Azobiofer" has been used to describe this (Hove and Lejeune, 1996).

Cyanobacteria, commonly known as blue green algae are an important class of bacteria that are known to derive energy through the process of photosynthesis (Stewart, 1980). The bacteria are known to be widely distributed with its characteristic presence seen as phototrophic biofilms that are seen in fresh water and marine waters as well. The *Cyanobacteria* are found to be highly resistant to climatic conditions and under unfavorable environmental conditions are known to pose special structures called as heterocysts. The heterocyst is known to contain nitrogenase which has a pivotal role to play in the process of nitrogen fixation. The heterocyst forming *cyanobacteria* are given special importance for their ability to fix the atmospheric nitrogen and efficiently converting it to ammonia, nitrites, and nitrate which the plants can readily absorb for fulfilling their energy requirement. Ample studies indicating the use of *Cyanobacteria* have been presented under rice field conditions and is known to reduce the usage of chemical urea (Uma and Kannaiyan, 1999). Their contribution to the field of soil chemistry is also immense, as they are known to prevent the process of soil erosion. *Cyanobacteria* is found to highly benefit the rice plants by the production of certain growth promoters, that is followed by elevating the phosphorus availability by the process of excreting organic acids which is further speculated to contribute in preventing soil erosion ultimately.

Rhizosphere microorganisms are playing a vital role to supply nutrients to the plants, providing growth promoting substances to enhance the plant growth, protecting plants from pathogens and also to maintain the soil fertility (Ahmad et al., 2008; Ahemad and Khan, 2011). The plant-microbe interactions are important for the solubilization, mobilization, and transformation of minerals to the plants (Shoebitz et al., 2009; Hayat et al., 2010). Many researchers have studied the effect of pesticides on root colonizing microorganisms in various crop fields (Anderson et al., 2004; Fox et al., 2007; Datta et al., 2011).

4.5 PESTICIDE IMPACT ON SOIL MICROBIAL DIVERSITY

Pesticides greatly influence soil bioprocesses carried out by microorganisms. These also affects the bioavailability of organic compounds and even affect the process by which microbes convey organic compounds in soil to their mineralized forms and thus biotransformation (Demanou et al., 2004; Kinney et al., 2005). However, recent advances has lead to the development of molecular tools and techniques which can be used in understanding the impact of pesticides on microbial community structures and functions

(Widenfalk et al., 2008). Various environmental factors and properties of soil play a major role in determining the effect of pesticide on soil microflora (Ecobichon, 1991). Due to indiscriminate use of pesticides, a large deviation in standard quality of organic matter in soil affecting diverse soil microbial community in it. Since microbial biomass one or the other way is linked with various nutrient recycling and biotransformation, any alteration in their population could seriously alter soil fertility. Many pesticides are intended to ward off certain insect pests. However, more than 90% of foliar application reach soil and thereby affect non-target microorganisms altering their vital biochemical processes, thus cell division and molecular composition (De Lorenzo et al., 2001). Initial pesticide exposure decreases the microbial diversity, but prolonged persistency leads to resistance or tolerance. Ryan (1999) reported that routine agricultural practices involving chemical pesticide may alter some groups of soil organisms, but the overall integrity of community structure would remain constant. However on the contrary, Kalia and Gosal (2011) studied overall soil microbial community structure from rice and wheat cultivating lands under pesticide stress and concluded that average population in all soil microbes considered for study decreased drastically with pesticide application.

Soil properties and chemical nature of a pesticide decides the fate of pesticide in soil; as transformations like degradation, transit, and adsorption/desorption. According to Singh and Walker (2006), soil microorganism and their metabolites often react with pesticides and thus alter the physiological and biochemical properties of soil microbes. However, available microbial activity and biomass in a given soil determines soil health and fertility, as the major nutrient transformations are linked with microorganisms in soil. Many recent studies shows nugatory effect of some pesticides on soil microbial population thus soil fertility (Pampulha and Oliveira, 2006; Zhou et al., 2006). More precisely, decrease in soil fertility implies reduction in microbial activity and increase implies induction in soil microbial activity.

On the contrary, some consortia of microorganisms are capable of utilizing administered pesticides for their energy requirements to grow, develop, and multiply. There are reports stating administered pesticides reduced the microbial diversity but helped to increase the functional diversity of microbial community (Pampulha and Oliveira, 2006; Wang et al., 2006). However, pesticides may outnumber some group of microorganisms by removing and outnumbering them from competition. For instance, fungicide application as reported by Chen et al. (2001) removed the activity of certain fungi leading to rapid elevation in bacterial population. Similar activity was reported by

López et al. (2006) with increase in heterotrophic mesophilic and psychrophilic aquatic bacteria when applied with herbicide simazine.

Despite several other parameters, numerous environmental factors determine the effect of pesticide on soil microorganisms. One of such major factor is the bioavailability of pesticide in soil environment. Several adsorption/desorption reactions modulate the concentration of pollutant in the vicinity of the soil and hence its bioavailability (Katagi, 2008). Further, Menon et al. (2004) have reported inhibitory activity of chlorpyrifos and quinalphos are differed with sandy loam, and loamy sand as the bioavailability differed with respect to different soil environments. Gundi et al. (2005) observed synergetic effects of three insecticides (monocrotophos, quinalphos, and cypermethrin) at lower level in black clay soil and harmful effect at higher level of concentration. On the contrary, Widenfalk et al. (2004) reported toxic effect of pesticides on fresh water sediment microbial community even under predicted environmentally safe concentrations (Table 4.2).

TABLE 4.2   Effect of Some Pesticides on Soil Microorganisms.

Pesticide	Target Microbe and Effects	References
Methamidophos	Soil bacteria (decreased biomass)	Wang et al. (2006)
Metalaxyl	Soil microflora (decreased biomass)	Sukul and Spiteller (2001)
Mefenoxam, Bensulfuron methyl	Inhibition in N-fixing bacteria	Monkiedje et al. (2002)
Isoproturon	Actinomycetes and fungi (suppress growth)	Nowak et al. (2004)
Captan	*Rhizobium ciceri* (decrease in viable count)	Kyei-Boahen et al. (2001)
Atrazine, Metribuzin	*Bradyrhizobium* sp. (adverse effect)	Khan et al. (2006)
Agroxone, Atranex, 2,4-D amine	*Rhizobium phaseoli*, *Azotobacter vinelandii* (most toxic)	Das and Mukherjee (1998)

Further, apart from bioavailability, certain other parameters like soil texture, organic matter, and available vegetation may also influences pesticide toxicity toward microbial biomass. The resistance to pesticides may get enhanced by the additional supplementation of carbon sources (glucose, acetate, amino acids) (Mishra and Pandey, 1989). This effect could be well documented in the soil which has been well tilled. Agronomically important microorganisms such as bacteria as well as arbuscular mycorrhiza (AM) form a symbiosis together. This interaction is often affected by pesticide application resulting in stress and deviation from normal conditions (Sainz et al., 2006). Fungi are the most affected at this function as they are prone to growth, development, nitrogen fixation, and other metabolic activities. However, further work is needed in understanding comparative sensitivity of fungi to many pesticides (Ma et al., 2004).

4.6 IMPACT OF PESTICIDES ON SOIL BIOPROCESSES

Biotransformation is one of the key aspects in microbial biochemical processes. They have the ability to transform various biomolecules like nitrogen (N), phosphorus (P), sulfur (S) and carbon (C). Biochemical reaction such as mineralization of organic matter, nitrogen fixation, and ammonification are usually affected by pesticides directly or indirectly either by acting upon microbes alone or by acting on enzymes from microbial metabolism (Kinney et al., 2005; Menon et al., 2005). Biological nitrogen fixation (BNF) is one of the major components of nitrogen fixation which is estimated about twice (about 175 million tons) the amount of nitrogen fixation from non-biological sources (Tate, 1995). However, pesticides are known to affect the root nodulation and BNF in legumes. Niewiadomska (2004) and Niewiadomska and Klama (2005) have studied and reported the adverse effects of carbendazime, thiram (fungicides), and imazetapir (herbicide) on nitrogenase activity of *Rhizobium leguminosarum*, *Sinorhizobium meliloti*, and *Bradyrhizobium* sp., and clover, lucerne, and serradella plants.

Biological activities are more pronounced in black soil than red soil. This is because black soil contains more organic matter than red soil (Shrinivas and David, 2015). Hence, organic matter is the most crucial part of the soil which determines fertility of soil. Quality of organic matter and dynamics is often controlled by biological activities of the soil and rate at which nutrient recycling takes place. Some researchers have thrown light on how pesticides deteriorates nutrient recycling and biological decomposition of organic matter under grassland area, forest ecosystem and desert place (Weary and Merriam, 1978; Perfect et al., 1981; Santos and Whitford, 1981). Whereas, others have documented beneficiary effect of pesticide on mineralization processes. Sukul (2006) reported significant decrease in C and N content of the soil with 30-day exposure to metalaxyl.

Pesticides when applied can interfere with bacterial communities affecting many biological processes such as nitrification, denitrification, and ammonification. Kinney et al. (2005) studied the effect of fungicides and herbicides (mancozeb, chlorothalonil, prosulfuron) on nitrifying bacteria and concluded toxic potentials of these pollutants on nitrification (N_2O) and denitrification (NO) which are supposed to be environmentally significant trace gases. Similar observation was made by Ogunseitan and Odeyemi (1985) in tropical soil with lindane, captan, and malathion on nitrification, phosphate solubilization and sulfur oxidation. They observed nitrification and phosphate solubilization were deprived in 30 days exposure period with all three pesticides. It was even reported that malathion increased sulfur oxidation while lindane and capta affected the reaction adversely.

Soil also contains certain enzymes in free form, immobilized and within microbial cells and represent normal/abnormal health of soil. Certain foliar applied pesticides however reach soil system thereby interacting with soil microbes and may affect enzymatic activities (Shrinivas and David, 2015). Monkiedje and Spiteller (2002) have reported negative impact of pesticides on hydrolases, oxidoreductases, and dehydrogenase activities in the soil. Whereas, there are reports advocating the increased enzyme activities and ATP contents in the soil due to certain pesticide applications (Shukla, 1997). Malkomes (1997) attributed such differences to the dual behavior of pesticides (both harmful and beneficial for soil enzymes), diversity and various stages of the processes taking place in soil that are frequently overlapped. Enzyme activity in soils reflects not only enzymes in soil solution and living tissue, but also enzymes bound to soil colloids and humic substances (Nannipieri et al., 1990).

However, on a lighter note there is no comprehensive information to understand the role of pesticides on soil microbes as well bioprocesses despite of extensive researches conducted all over. This is because some pesticides act as energy sources by providing carbon and nitrogen to microbes and are easily get degraded in return. Whereas, some act as recalcitrant and adversely affect soil microflora and bioprocesses. Soil enzymatic activity acts as a 'biological index' to measure the soil fertility and biological process in soil and is also reduced by the pesticides (Monkiedje et al., 2002; Antonious, 2003). The mineralization of organic compounds and biotransformation of nutrients by soil microorganisms are also adversely affected by the pesticides (Demanou et al., 2004; Niewiadomska, 2004; Kinney et al., 2005; Mahiá et al., 2008). Therefore, it is naive to give a definite conclusion on pesticide effect on soil micropopulation. The same is true in case of soil biochemical processes and soil enzymatic activities.

4.7 MOLECULAR ASSESSMENT OF PESTICIDE IMPACT ON MICROBIAL DIVERSITY

In this scientific era, molecular techniques are most widely used to study the effect of agrochemicals on the function and structure of microbial communities, as this technique is culture independent. However, the impact of agrochemicals on the genetic structure and degradation ability of soil microbial communities have been less investigated (Hussain et al., 2009). Reports by Schloss and Handelsman (2004) have indicated the availability of not less than 50 bacterial phyla of which 50% of bacterial communities have been represented on the basis of molecular sequencing. Culturable microorganisms constitute <1% of all microbial species (Hugenholtz, 2002). For example, from soil samples most of the microbial species belong to one of four phyla (Proteobacteria, Firmicutes, Bacteroidetes, and Actinobacteria) due to their ease of cultivation in laboratory conditions. Conversely, Acidobacteria comprise approximately 20% of soil bacterial communities which are represented by few genera; such bacterial species are difficult to cultivate in laboratory conditions (Schloss and Handelsman, 2004). These findings suggest that in depth characterization of environmental microbial population needs molecular techniques for isolation, cultivation, and characterization.

The introduction of culture independent methods such as 16S rRNA and DNA based methods to assess the total microbial diversity has found the solution for the limitations of culture dependent methods. Assessment of microbial diversity in environmental samples is possible through molecular tools (Sharma et al., 2014). Since last two decades, amplification and sequencing of 16S rRNA has been in use for the assessment and identity of abundance and taxonomic variation of microbial species in the environment (Pace, 2009). Gradually, 16S rRNA observed shows variation among the strains of same species (Acinas et al., 2004). Thus, molecular tools have been developed as a new approach for revealing the microbial diversity in different environments.

There are apparent improvements in the microbial diversity study through several scientific and technological developments in nucleic acid sequencing methods (Rastogi and Sani, 2011). The advent and effect of sequencing techniques has changed the assessment of microbial diversity and its functional pattern in an ecosystem (Chistoserdova, 2010; Manickam et al., 2010; Morales and Holben, 2011). For the assessment of environmental risk, environmental impact, and public health, culture independent methods would be used rather than culture dependent methods, so as to infer the significance of a specific taxonomic group in a community (Vaz-Moreira et al.,

2011). There are two types of molecular tools that are available to assess the microbial diversity such as whole community analysis and partial community analysis (Fig. 4.1).

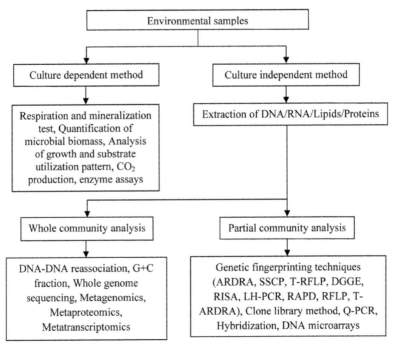

FIGURE 4.1 Molecular methods to characterize the diversity of microorganisms from pesticide contaminated soil samples (Rastogi and Sani, 2011).

The strategy of partial community analysis is based on polymerase chain reaction (PCR) methods, where the characterization of the microbial species can be achieved by the total RNA/DNA extracted from the environmental samples. The PCR product might have the mixture of genetic sequences of all microorganisms including viable but non-culturable (VBNC) microorganisms present in the sample (Rastogi and Sani, 2011). PCR amplification of 16S rRNA from an environmental sample has been used for all types of sample due to its ubiquitous nature, structurally and functionally conserved, having variable, and highly conserved regions (Hugenholtz, 2002). In microbial diversity studies, 16S rRNA technology is a prime choice through its suitable gene size and increasing number of 16S rRNA sequences available in database for comparison. Using 16S rRNA sequencing methods the phylogenetic relatedness for the known microorganism can be estimated, in which the closest affiliation of a new isolate is assigned. According to Ghebremedhin et al. (2008), RNA polymerase beta subunit (*rpoB*), gyrase beta subunit (*gyrB*), recombinase A (*recA*) and heat shock protein (*hsp60*) have also been used in microbial ecology to differentiate the microbial species. These amplified PCR products further analyzed either by (i) genetic finger printing such as Denaturing- or Temperature Gradient Gel Electrophoresis (DGGE/TTGE), Single-Strand Conformation Polymorphism (SSCP), Random Amplified Polymorphic DNA (RAPD), DNA Amplification Fingerprinting (DAF), Amplified Ribosomal DNA Restriction Analysis (ARDRA), Terminal Restriction Fragment Length Polymorphism (T-RFLP), Length Heterogeneity PCR (LH-PCR), and Ribosomal Intergenic Spacer Analysis (RISA), (ii) clone library method, or (iii) DNA microarray or using a combination of these three techniques (Fromin et al., 2002; Haack et al., 2004; de Figueiredo et al., 2007; Rastogi and Sani, 2011).

Sequence analysis of 16S rRNA is generally used by researchers in microbial ecology using conserved gene sequence regions. Though, this technique does not provide adequate resolution at species and strain level (Konstantinidis et al., 2006). In this view, whole community analysis provides an inclusive outlook of genetic variability and diversity of microbial population compared to partial community analysis. In principle, complete genetic information present in the total DNA extracted from the environmental sample is thoroughly analyzed by whole community analysis and GC content (Rastogi and Sani, 2011). DNA-DNA hybridization (DDH) offers promising, accurate, efficient, and reliable results for the comparison among the whole genome of different organisms. Moreover, Guanine-Cytosine (G+C) content of DNA is varying within different prokaryotes at 3–5%, which can also be used to compare the genomic composition of phylogenetically related bacterial community present in samples from extreme environments (Nüsslein and Tiedje, 1999).

Shotgun cloning method is also used to sequence the whole microbial genome. Assessing the microbial systems through whole-genome analysis is a wide-ranging and incorporated approach to understand the structural and functional pattern of microbial ecology (Huson et al., 2007). Through the genome sequencing technique, massive amount of information gathered and deposited in the searchable databases that could be incorporated with other bioinformatics tools which available in the Integrated Microbial Genomes (IMG) web server (Markowitz et al., 2010). In addition, metagenomics is commonly used to analyze the collective microbial genomes from an environmental samples (Riesenfeld et al., 2004). Metagenomics is also functioning based on the principle of sequencing whole genome to sequencing and analyzing the entire genetic composition of environmental microbial communities (Rastogi and Sani, 2011; Imfeld and Vuilleumier, 2012).

4.8   CONCLUSION AND FUTURE PERSPECTIVES

The unscientific and uncontrolled excessive and indiscriminate use of pesticides in the agricultural sector results in many human health problems and environmental pollution. Hazards arising during the application of pesticides are mainly due to lack of information, knowledge, awareness, poor supervision during spraying, absence of proper legislation or of enforcement of legislation, and sale on the open market of highly toxic pesticides. Hence, it is necessary to monitor the occupational health of workers in the agricultural sector, with appropriate surveillance and record-keeping (Roy, 2015). Food and Agricultural Organization (FAO) introduce the word for safe use of pesticide in the environment and it define it as 'Good Agricultural Practice' (GAP) in the use of pesticide as "the officially recommended or authorized usage of a pesticide under practical conditions at any stage of production, storage, transport, distribution and processing of food and other agricultural commodities, bearing in mind the variations in requirements within and between regions, and which takes into account the minimum quantities necessary to achieve adequate control, the pesticides being applied in a manner so as to leave a residue which is the smallest amount practicable and which is toxicologically acceptable" (FAO, 1977). In developing countries, it is necessary to train the general public regarding health problems in the agricultural sector with respect to quantities of pesticides to be used for the respected crops, scientific way of pesticide application, the current state of health of workers in relation to pesticide exposure, coordination between hospital physicians, occupational health specialists, the number, sex, and age of the workers exposed, and so forth. Comprehensive occupational health histories should be obtained from all workers adversely affected by pesticide exposure (Roy, 2015).

The use of pesticide is beneficial for plant growth by protecting it from the damage by the pests and insects, but decreasing the overall soil fertility and contaminating the environment. Soil microorganisms are playing a crucial role in biogeochemical cycle, which maintains soil fertility for plant growth and productivity. According to many research findings, continuous application of pesticides poses adverse effect on soil especially rhizosphere microorganisms (Sethi and Gupta, 2013). The effects of pesticides on microbial community and soil bioprocesses are based on the nature of the chemical compound in pesticide, its metabolites, and quantity of the pesticide application. Elucidating the risk of pesticides on microbial diversity and soil biological processes can be achieved only through the molecular assessment of whole microbial community in pesticide contaminated-environment (Hussain et al., 2009). The field of microbial ecology is enduring the exceptional desirable changes along with the development and application of molecular genomic tools.

Our knowledge and understanding on microbial composition and nutrient dynamics in agricultural soil ecosystem is still insufficient. In future, several issues related to the impact of pesticides on microbial diversity is need to be addressed and widely acceptable risk assessment frame work in agro-ecosystems with microbial indicators should be prepared. Currently, research focus is shifting toward the exploration of the soil microbial responses to pesticide contamination. A complete analysis is necessary for the global patterns of microbial species diversity which can be obtained from sequencing methods and its alteration by the pesticide chemical exposure.

KEYWORDS

- **Agrochemicals**
- **Azobiofer**
- **Biological index**
- **Biologically active substances**
- **Biotransformation**
- **Endosphere**
- **Environmental impact**
- ***Frankia***
- **Microbial indicators**
- **Nitrogen fixation**
- **Pesticides**
- **Plant growth regulators**
- ***Rhizobium***
- **Rhizosphere**
- **Soil microbial diversity**
- **Synthetic pesticide**

REFERENCES

Acinas, S. G.; Marcelino, L. A.; Klepac-Ceraj, V.; Polz, M. F. Divergence and redundancy of 16S rRNA sequences in genomes with multiple *rrn* operons. *J Bacteriol* 2004, 186, 2629–2635.

Ahemad, M.; Khan, M. S. Functional aspects of plant growth promoting rhizobacteria: recent advancements. *Insight Microbiol* 2011, 1, 39–54.

Ahmad, F.; Ahmad, I.; Khan, M. S. Screening of free-living rhizospheric bacteria for their multiple plant growth promoting activities. *Microbial Res* 2008, 163(S2), 173–181.

Ajayi, O. *Socio-economic assessment of pesticide use in Mali.* University of Hanover: Germany, 2002.

Aktar, W.; Sengupta, D.; Chowdhury, A. Impact of pesticides use in agriculture: their benefits and hazards. *Interdisc Toxicol* 2009, 2(1), 1–12.

Anderson, L. M.; Stockwell, V. O.; Loper, J. E. An extracellular protease of *Pseudomonas fluorescens* inactivates antibiotics of *Pantoea agglomerans*. *Phytopathology* 2004, 94, 1228–1234.

Angelini, J.; Ghio, S.; Taurian, T.; Ibanez, F.; Tonelli, M. I.; Valetti, I.; Anzuay, M. S.; Luduena, L.; Munoz, V.; Fabra, A. The effects of pesticides on bacterial nitrogen fixers in peanut-growing area. *Arch Microbiol* 2013, 195, 683–892.

Antonious, G. F. Impact of soil management and two botanical insecticides on urease and invertase activity. *J Environ Sci* 2003, 38, 479–488.

Anzuay, M. S.; Frola, O.; Angelini, J. G.; Luduena, L. M.; Ibanez, F.; Fabra, A.; Taurian, T. Effect of pesticides application on peanut (*Arachis hypogae* L.) associated phosphate solubilizing soil bacteria. *Appl Soil Ecol* 2015, 95, 31–37.

Arthur, J. R.; Lavilla-Pitogo, C. R.; Subasinghe, R. P. *Use of chemicals in aquaculture in Asia.* Southeast Asian Fisheries Development Center: Tigbauan: Philippines, 2000, pp 1–235.

Asadu, C. L. A.; Nwafor, I. A.; Chibuike, G. U. Contributions of microorganisms to soil fertility in adjacent forest, fallow and cultivated land use types in Nsukka, Nigeria. *Int J Agric For* 2015, 5(3), 199–204.

Bardgett, R. D.; Bowman, W. D.; Kaufmann, R.; Schmidt, S. K. Linking aboveground and belowground communities: a temporal approach. *Trends Ecol Evol* 2005, 20, 634–641.

Bardgett, R. D.; Streeter, T.; Bol, R. Soil microbes compete effectively with plants for organic nitrogen inputs to temperate grasslands. *Ecology* 2003, 84, 1277–1287.

Bhushan, C.; Bhardwaj, A.; Misra, S. S. *State of pesticide regulations in India.* Centre for Science and Environment: New Delhi, 2013.

Brooks, P. D.; Williams, M. W.; Schmidt, S. K. Inorganic nitrogen and microbial biomass dynamics before and during spring snowmelt. *Biogeochemistry* 1998, 43, 1–5.

Bulgarelli, D.; Schlaeppi, K.; Spaepen, S.; Ver Loren van Themaat, E.; Schulze-Lefert, P. Structure and functions of the bacterial microbiota of plants. *Annu Rev Plant Biol* 2013, 64, 807–838.

Burd, G.; Dixon, D. G.; Glick, B. R. Plant growth promoting bacteria that decrease heavy metal toxicity in plants. *Can J Microbiol* 2000, 46, 237–245.

Chen, S. K.; Edwards, C. A.; Subler, S. Effect of fungicides benomyl, captan and chlorothalonil on soil microbial activity and nitrogen dynamics in laboratory incubations. *Soil Biol Biochem* 2001, 33, 1971–1980.

Chistoserdova, L. Recent progress and new challenges in metagenomics for biotechnology. *Biotechnol Lett* 2010, 1351–1359.

Cleyet-Marcel, J. C.; Larcher, M.; Bertrand, H.; Rapior, S.; Pinochet, X. Plant growth enhancement by rhizobacteria. In *Nitrogen assimilation by plants: Physiological, biochemical and molecular aspects*, Morot Gaudry, J. F., Eds.; Science Publishers: Plymouth, 2001, pp 185–197.

Cocking, E. C. Endophytic colonization of plant roots by nitrogen-fixing bacteria. *Plant Soil* 2003, 252(1), 169–175.

Coleman, D. C.; Crossley Jr, D. A.; Hendrix, P. F. *Fundamentals of soil ecology*. Elsevier Academic Press: California, 2004.

Colwell, R. R.; Grimes, D. J. *Nonculturable microorganisms in the environment*. American Society for Microbiology: Washington, 2000.

Cone, M. *Silent snow: The slow poisoning of the arctic*. Grove Press: New York, 2005.

Das, A. C.; Mukherjee, D. Insecticidal effects on soil microorganisms and their biochemical processes related to soil fertility. *World J Microbiol Biotechnol* 1998, 14, 903–909.

Datta, M.; Palit, R.; Sengupta, C.; Pandit, M. K.; Banerjee, S. Plant growth promoting rhizobacteria enhance growth and yield of chilli (*Capsicum annuum* L.) under field conditions. *Aust J Crop Sci* 2011, 5(5), 531–536.

de Figueiredo, D. R.; Pereira, M. J.; Moura, A.; Silva, L.; Barrios, S.; Fonseca, F.; Henriques, I.; Correia, A. Bacterial community composition over a dry winter in meso- and eutrophic Portuguese water bodies. *FEMS Microb Ecol* 2007, 59, 638–650.

De Leo, G. A.; Levin, S. The multifaceted aspects of ecosystem integrity. *Conserv Ecol* 1997, 1(1), 3.

De Lorenzo, M. E.; Scott, G. I.; Ross, P. E. Toxicity of pesticides to aquatic microorganisms: a review. *Environ Toxicol Chem* 2001, 20, 84–98.

Demanou, J.; Monkiedje, A.; Njine, T.; Foto, S. M.; Nola, M.; Serges, H.; Togouet, Z.; Kemka, N. Changes in soil chemical properties and microbial activities in response to the fungicide Ridomil gold plus copper. *Int J Environ Res Public Health* 2004, 1, 26–34.

Doornbos, R. F.; van Loon, L. C.; Bakker, P. A. H. M. Impact of root exudates and plant defense signalling on bacterial communities in the rhizosphere: a review. *Agron Sustain Dev* 2011, 32, 227–243.

Dung, N. H.; Dung, T. T. T. Economic and health consequences of pesticide use in paddy production in the Mekong Delta, Vietnam. EEPSEA Research Report Series, IDRC, Singapore, 1999.

Ecobichon, D. J. Toxic effects of pesticides. In *Casarett and Doull's Toxicology*, Amdur, M. O., Donl, J., Klassen, C. D., Eds.; Pergamon Press: New York, 1991, pp 2–18.

FAO. Ad-hoc government consultation on international standardization of pesticide registration requirements. Food and Agricultural Organization, Rome, Italy, 1977, pp 12–33.

Filimon, M. N.; Vlad, D. C.; Verdes, D.; Dumitrascu, V.; Popescu, R. Enzymatic and biological assessment of sulfonylurea herbicide impact on soil bacterial communities. *Afr J Agric Res* 2015, 10(14), 1702–1708.

Fisher, K.; Newton, W. E. Nitrogen fixation: an historical perspective. In *Catalysts of nitrogen fixation: Nitrogenases, relevant chemical models and commercial processes*, Smith, B. E., Richards, R. L., Newton, W.E., Eds.; Kluwer: Netherlands, 2004, pp 1–31.

Foley, J. A. Solutions for a cultivated planet. *Nature* 2011, 478(7369), 337–342.

Fox, J. E.; Gulledge, J.; Engelhaupt, E.; Burow, M. E.; McLachlan, J. A. Pesticides reduce symbiotic efficiency of nitrogen-fixing rhizobia and host plants. *Proc Natl Acad Sci USA* 2007, 104(24), 10282–10287.

Fromin, N.; Hamelin, J.; Tarnawski, S.; Roesti, D.; Jourdain-Miserez, K.; Forestier, N.; Teyssier-Cuvelle, S.; Gillet, F.; Aragno, M.; Rossi, P. Statistical analysis of denaturing gel electrophoresis (DGE) fingerprinting patterns. *Environ Microbiol* 2002, 4(11), 634–643.

Gandora, V.; Gupta, R. D.; Bhardwaj, K. K. R. Abundance of azotobacter in great soil groups of north-west Himalayas. *J Ind Soc Soil Sci* 1998, 46(3), 379–383.

Ganesh, G.; Misra, A. K.; Chapelon, C.; Normand, P. Morphological and molecular characterization of *Frankia* sp. isolates from nodules of *Alnus nepalensis* Don. *Arch Microbiol* 1994, 161, 152–155.

Gevao, B.; Mordaunt, C.; Semple, K. T.; Piearce, T. G.; Jones, K. C. Bioavailability of non-extractable (bound) pesticide residues to earthworms. *Environ Sci Technol* 2000, 35(3), 501–507.

Ghebremedhin, B.; Layer, F.; König, W.; König, B. Genetic classification and distinguishing of *Staphylococcus* species based on different partial gap, 16S rRNA, *hsp60*, *rpoB*, *sodA*, and *tuf* gene sequences. *J Clin Microbiol* 2008, 46, 1019–1025.

Gill, H. K.; Garg, H. Pesticides: environmental impacts and management strategies. In *Pesticides - toxic aspects*, Soloneski, S., Ed.; InTech: Croatia, 2014, pp 187–230.

Gundi, V. A. K. B.; Narasimha, G.; Reddy, B. R. Interaction effects of insecticides on microbial populations and dehydrogenase activity in a black clay soil. *J Environ Sci Health B* 2005, 40, 269–283.

Gupta, P. Pesticide exposure - Indian scene. *Toxicology* 2004, 198(1–3), 83.

Haack, S. K.; Fogarty, L. R.; West, T. G.; Alm, E. W.; McGuire, J. T.; Long, D. T.; Hyndman, D. W.; Forney, L. J. Spatial and temporal changes in microbial community structure associated with recharge-influenced chemical gradients in a contaminated aquifer. *Environ Microbiol* 2004, 6(5), 438–448.

Haichar, F. E. Z.; Marol, C.; Berge, O.; Rangel-Castro, J. I.; Prosser, J. I.; Balesdent, J.; Heulin, T.; Achouak, W. Plant host habitat and root exudates shape soil bacterial community structure. *ISME J* 2008, 2, 1221–1230.

Hayat, R.; Ali, S.; Amara, U.; Khalid, R.; Ahmed, I. Soil beneficial bacteria and their role in plant growth promotion: a review. *Ann Microbiol* 2010, 60, 579–598.

Higa, T.; Parr, J. F. *Beneficial and effective microorganisms for a sustainable agriculture and environment*. International Nature Farming Research Center: Atami, 1994, pp 1–16.

Hove, C. V.; Lejeune, A. *Biological nitrogen fixation associated with rice production*. Kluwer: Dordrecht, 1996, pp 83–94.

Huang, J.; Qiao, F.; Zhang, L.; Rozelle, S. *Farm pesticides, rice production, and human health in China*. Economy and Environment Program for Southeast Asia: Tanglin, Singapore, 2001, pp 1–58.

Hugenholtz, P. Exploring prokaryotic diversity in the genomic era. *Genome Biol* 2002, 3(2), reviews0003.1–0003.8.

Huson, D. H.; Auch, A. F.; Qi, J.; Schuster, S. C. MEGAN analysis of metagenomic data. *Genome Res* 2007, 17, 77–386.

Hussain, S.; Siddique, T.; Saleem, M.; Arshad, M.; Khalid, A. Impact of pesticides on soil microbial diversity, enzymes, and biochemical reactions. In *Advances in Agronomy*, Sparks, D. L., Ed.; Elsevier: New York, 2009, pp 159–200.

Imfeld, G.; Vuilleumier, S. Measuring the effects of pesticides on bacterial communities in soil: a critical review. *Eur J Soil Biol* 2012, 49, 22–30.

Indiradevi, P. Pesticides in agriculture - a boon or a curse? A case study of Kerala. *Econ Polit Wkly* 2010, 26(27), 199–207.

Jeyanthi, H.; Kombairaju, S. Pesticide use in vegetable crops: frequency, intensity and determinant factors. *Agric Econ Res Rev* 2005, 18, 209–221.

Joel, W. R.; Glenn, D. K.; Matthew, S. C.; Louis, S. T. Heavy metal resistance patterns of *Frankia* strains. *Appl Environ Microbiol* 2002, 68, 923–927.

Kader, M. A.; Miar, M. H.; Hoque, M. S. Effects of Azotobacter inoculants on the yield and nitrogen uptake by wheat. *J Biol Sci* 2002, 2(4), 259–261.

Kalia, A.; Gosal, S. K. Effect of pesticide application on soil microorganisms. *Arch Agron Soil Sci* 2011, 57(6), 569–596.

Kanungo, P. K.; Ramakrishnan, B.; Rao, V. R. Placement effect of organic sources on nitrogenases activity and nitrogen-fixing bacteria in flooded rice soils. *Biol Fertl Soils* 1997, 25, 103–108.

Katagi, T. Surfactant effects on environmental behaviour of pesticides. *Rev Environ Contam Toxicol* 2008, 194, 1–177.

Khan, M. S.; Zaidi, A.; Rizvi, P. Q. Biotoxic effects of herbicides on growth, nodulation, nitrogenase activity, and seed production in chickpeas. *Commun Soil Sci Plant Anal* 2006, 37, 1783–1793.

Kinney, C. A.; Mandernack, K. W.; Mosier, A. R. Laboratory investigations into the effects of the pesticides mancozeb, chlorothalonil and prosulfuron on nitrous oxide and nitric oxide production in fertilized soil. *Soil Biol Biochem* 2005, 37, 837–850.

Kinney, C. A.; Mosier, A. R.; Ferrer, I.; Furlong, E. T.; Mandernack, K. W. Effects of the herbicides prosulfuron and metolachlor on fluxes of CO_2, N_2O and CH_4 in a fertilized Colorado grassland soil. *J Geophys Res* 2004, 109, D05304.

Kloepper, J. W.; Lifshitz, R.; Zablotowicz, R. M. Free-living bacterial inocula for enhancing crop productivity. *Trends Biotechnol* 1989, 7, 39–43.

Kondorosi, E.; Mergaert, P.; Kereszt, A. A paradigm for endosymbiotic life: cell differentiation of *Rhizobium* bacteria provoked by host plant factors. *Annu Rev Microbiol* 2013, 67, 611–628.

Konstantinidis, K. T.; Ramette, A.; Tiedje, J. M. The bacterial species definition in the genomic era. *Philos Trans R Soc B* 2006, 361, 929–1940.

Kyei-Boahen, S.; Slinkard, A. E.; Walley, F. L. Rhizobial survival and nodulation of chickpea as influenced by fungicide seed treatment. *Can J Microbiol* 2001, 47, 585–589.

Lakshminarayana, K. Influence of Azotobacter on nutrition of plant and crop productivity. *Proc Ind Natl Sci Acad* 1993, 59, 303–308.

Lancelle, S. A.; Torrey, J. G. Early development of Rhizobium-induced root nodules of *Parasponia rigida*. I. infection and early nodule initiation. *Protoplasma* 1984, 123, 26–37.

Lang, M.; Cai, Z. Effects of chlorothalonil and carbendazim on nitrification and denitrification in soils. *J Environ Sci* 2009, 21(4), 458–467.

Lindow, S. E.; Hecht-Poinar, E. I.; Elliot, V. J. *Phyllosphere microbiology*. American Phytopathological Society Press: Minnesota, 2002.

Lois, S. T.; Matthew, S. C.; Glenn, D. K.; Joel, R. Antibiotic resistance patterns of *Frankia* strains. *Can J Bot* 1999, 77(9), 1257–1260.

Lopez, L.; Pozo, C.; Gomez, M. A.; Calvo, C.; Gonzalez Lopez, J. Studies on the effects of the insecticide aldrin on aquatic microbial populations. *Biodeterior Biodegrad* 2002, 50, 83–87.

López, L.; Pozo, C.; Rodelas, B.; Calvo, C.; González-López, J. Influence of pesticides and herbicides presence on phosphatase activity and selected bacterial microbiota of a natural lake system. *Ecotoxicology* 2006, 15, 487–493.

Lundberg, D. S.; Lebeis, S. L.; Paredes, S. H.; Yourstone, S.; Gehring, J.; Malfatti, S.; Tremblay, J.; Engelbrektson, A.; Kunin, V.; del Rio, T. G.; Edgar, R. C.; Eickhorst, T.; Ley, R. E.; Hugenholtz, P.; Tringe, S. G.; Dangl, J. L. Defining the core *Arabidopsis thaliana* root microbiome. *Nature* 2012, 488, 86–90.

Lynch, J. M. Introduction: consequences of microbial rhizosphere competence for plant and soil. In *The Rhizosphere*, Lynch, M., Ed.; Wiley: New York, 1990, pp 1–10.

Ma, J.; Lin, F.; Wang, S.; Xu, L. Differential sensitivity of green algae to herbicides: acute toxicity of 20 herbicides to *Scenedesmus quadricauda*. *Bull Environ Contam Toxicol* 2004, 72, 1164–1171.

Macdonald, R. W. L. A.; Barrie, T. F.; Bidleman, M. L.; Diamond, D. J.; Gregor, R. G.; Semkin, W. M. J.; Strachan, Y. F.; Li, F.; Wania, M.; Alaee, S.; Backus, M.; Bewers, C.; Gobeil, C.; Halsall, J.; Hoff, L.; Lockhart, D.; Mackay, M. D.; Pudykiewicz, J.; Reimer, K.; Smith, J.; Stern, G.; Schroeder, W.; Wagemann, R.; Yunker, M. Contaminants in the Canadian Arctic: five years of progress in understanding sources, occurrence and pathways. *Sci Total Environ* 2000, 254, 93–234.

Mahiá, J.; Cabaneiro, A.; Carballas, T.; Díaz-Raviña, M. Microbial biomass and C mineralization in agricultural soils as affected by atrazine addition. *Biol Fertil Soils* 2008, 45, 99–105.

Malkomes, H. P. Applications of ecotoxicity tests to assess side effects of pesticides in soils. In *Soil ecotoxicology*, Tarradellas, J., Bitton, G., Rossel, D., Eds.; Lewis Publishers: Boca Raton, 1997, pp 319–343.

Manickam, N.; Pathak, A.; Saini, H. S.; Mayilraj, S.; Shanker, R. Metabolic profiles and phylogenetic diversity of microbial communities from chlorinated pesticides contaminated sites of different geographical habitats of India. *J Appl Microbiol* 2010, 109, 1458–1468.

Markowitz, V. M.; Chen, I. M.; Palaniappan, K.; Chu, K.; Szeto, E.; Grechkin, Y.; Ratner, A.; Anderson, I.; Lykidis, A.; Mavromatis, K.; Ivanova, N. N.; Kyrpides, N. C. The integrated microbial genomes system: an expanding comparative analysis resource. *Nucl Acids Res* 2010, 38, 382–390.

Marschner, P.; Grierson, P. F.; Rengel, Z. Microbial community composition and functioning in the rhizosphere of three *Banksia* species in native woodland in Western Australia. *Appl Soil Ecol* 2005, 28, 191–201.

Martyniuk, S.; Martyniuk, M. Occurrence of *Azotobacter* spp. in some Polish soils. *J Environ Stud* 2003, 12(3), 371–374.

Maumbe, B. M.; Swinton, M. S. Hidden health costs of pesticide use in Zimbabwe's small holder cotton growers. *Soc Sci Med* 2003, 57(15), 59–71.

Menon, P.; Gopal, M.; Parsad, R. Effects of chlorpyrifos and quinalphos on dehydrogenase activities and reduction of Fe^{3+} in the soils of two semi-arid fields of tropical India. *Agric Ecosyst Environ* 2005, 108, 73–83.

Menon, P.; Gopal, M.; Parsad, R. Influence of two insecticides, chlorpyrifos and quinalphos, on arginine ammonification and mineralizable nitrogen in two tropical soil types. *J Agric Food Chem* 2004, 52, 7370–7376.

Mian, M. H. Azobiofer: a technology of production and use of *Azolla* as biofertiliser for irrigated rice and fish cultivation. In *Biofertilisers in action*, Kennedy, I. R., Choudhury, A. T. M. A., Eds.; Rural Industries Research and Development Corporation: Canberra, 2002, pp 45–54.

Mishra, A. K.; Pandey, A. B. Toxicity of three pesticides to some nitrogen fixing cyanobacteria. *Ecotoxicol Environ Saf* 1989, 17, 236–246.

Mnif, W.; Hassine, A. I. H.; Bouaziz, A.; Bartegi, A.; Thomas, O.; Roig, B. Effect of endocrine disruptor pesticides: a review. *Int J Environ Res Public Health* 2011, 8(6), 2265–2303.

Monkiedje, A.; Ilori, M. O.; Spiteller, M. Soil quality changes resulting from the application of the fungicides mefenoxam and metalaxyl to a sandy loam soil. *Soil Biol Biochem* 2002, 34, 1939–1948.

Monkiedje, A.; Spiteller, M. Effects of the phenylamide fungicides, mefenoxam and metalaxyl, on the biological properties of sandy loam and sandy clay soils. *Biol Fertil Soils* 2002, 35, 393–398.

Morales, S. E.; Holben, W. E. Linking bacterial identities and ecosystem processes: can 'omic' analyses be more than the sum of their parts? *FEMS Microb Ecol* 2011, 75, 2–16.

Mushobozi, W. L.; Santacoloma, P. *Good Agricultural Practices (GAP) on horticultural production for extension staff in Tanzania.* Food and Agriculture Organization of the United Nations: Rome, 2010, pp 1–186.

Nannipieri, P.; Ceccanti, B.; Gregos, S. Ecological significance of the biological activity in soil. In *Soil biochemistry*, Bollag, J. M., Stotzky, G., Eds.; Marcel Dekker: New York, 1990, pp 293–355.

Niewiadomska, A. Effect of carbendazim, imazetapir and thiram on nitrogenase activity, the number of microorganisms in soil and yield of red clover (*Trifolium pratense* L.). *Pol J Environ Stud* 2004, 13, 403–410.

Niewiadomska, A.; Klama, J. Pesticide side effect on the symbiotic efficiency and nitrogenase activity of Rhizobiaceae bacteria family. *Pol J Microbiol* 2005, 54, 43–48.

Noble, A. D.; Ruaysoongnern, S. The nature of sustainable agriculture. In *Soil microbiology and sustainable crop production*, Dixon, R., Tilston, E., Eds.; Springer Science and Business Media: Berlin, 2010, pp 1–25.

Nowak, A.; Nowak, J.; Klodka, D.; Pryzbulewska, K.; Telesinski, A.; Szopa, E. Changes in the microflora and biological activity of the soil during the degradation of isoproturon. *J Plant Dis Protec* 2004, 19, 1003–1016.

Nüsslein, K.; Tiedje, J. M. Soil bacterial community shift correlated with change from forest to pasture vegetation in a tropical soil. *Appl Environ Microbiol* 1999, 65, 3622–3626.

Ogunseitan, O. A.; Odeyemi, O. Effects of lindane, captan and malathion on nitrification, sulphur oxidation, phosphate solubilisation and respiration in a tropical soil. *Environ Pollut Ecol Biol* 1985, 37(4), 343–354.

Ombe, G. M. W. *Risk of agrochemicals on the environment and human health. In Mukaro location, Nyeri County, Kenya.* MSc Thesis, Kenyatta University, Nairobi, Kenya, 2014.

Pace, N. R. Mapping the tree of life: progress and prospects. *Microbiol Mol Biol Rev* 2009, 73, 565–576.

Pampulha, M. E.; Oliveira, A. Impact of an herbicide combination of bromoxynil and prosulfuron on soil microorganisms. *Curr Microbiol* 2006, 53, 238–243.

Perfect, T. J.; Cook, A. G.; Critchley, B. R.; Smith, A. R. The effect of crop protection with DDT on the microarthropod population of a cultivated forest soil in the sub-humid tropics. *Pedobiologia* 1981, 21, 7–18.

Rahman, S. Women's employment in Bangladesh agriculture: composition, determinants and scope. *Journal of Rural Studies* 2000, 16, 497–507.

Rajendran, S. Environment and health aspects of pesticides use in Indian agriculture. In *Proceedings of the third International conference on environment and health*, Bunch, M. J., Suresh, V. M., Kumaran, T. V., Eds.; University of Madras: Chennai, 2003, pp 353–373.

Rastogi, G.; Sani, R. K. Molecular techniques to assess microbial community structure, function, and dynamics in the environment. In *Microbes and microbial technology: Agricultural and environmental applications*, Ahmad, I., Ahmad, F., Pichtel, J., Eds.; Springer: New York, 2011, pp 29–57.

Ravikumar, C. H.; Srinivas, P.; Seshaiah, K. Determination of organochlorine pesticide residues in rice by gas chromatography tandem mass spectrometry. *J Chem Pharm Res* 2013, 5(1), 361–366.

Rebecchi, L.; Sabatini, M. A.; Cappi, C.; Grazioso, P.; Vicari, A.; Dinelli, G.; Bertolani, R. Effects of a sulfonylurea herbicide on soil microarthropods. *Biol Fertil Soils* 2000, 30, 312–317.

Reddell, P.; Bowen, G. C.; Robson, A. D. Nodulation of Casuarinaceae in relation to host species and soil properties. *Aust J Bot* 1986, 34, 435–444.

Riesenfeld, C. S.; Schloss, P. D.; Handelsman, J. Metagenomics: genomic analysis of microbial communities. *Annu Rev Genet* 2004, 38, 525–552.

Rola, A. C.; Pingali, P. L. *Pesticides, rice productivity, and farmers' health: an economic assessment*. International Rice Research Institute and World Resources Institute: Philippines, 1993.

Roy, S. Pesticide, health and economics. *Int J Res Sci Technol*. 2015, 5(2), 85–101.

Rüttimann, J. C.; Rubio, L. M.; Dean, D. R.; Ludden, P. W. VnfY is required for full activity of the vanadium-containing dinitrogenase in *Azotobacter vinelandii*. *J Bacteriol* 2003, 185, 2383–2386.

Ryan, M. Is an enhanced soil biological community relative to conventional neighbours a consistent feature of alternative (organic and biodynamic) agricultural systems? *Biol Agr Hortic* 1999, 17(2), 131–144.

Saini, R. K.; Yadav, G. S.; Kumari, B. *Novel approaches in pest and pesticide management in agro-ecosystem*. CCS Haryana Agricultural University: Hisar, 2014, pp 1–330.

Sainz, M. J.; González-Penalta, B.; Vilariňo, A. Effects of hexachlorocyclohexane on rhizosphere fungal propagules and root colonization by arbuscular mycorrhizal fungi in *Plantago lanceolata*. *Eur J Soil Sci* 2006, 57, 83–90.

Salameh, P. R.; Baldi, I.; Brochar, P. Pesticides in Lebanon: a knowledge, attitude and practice study. *Environ Res* 2004, 94, 1–6.

Santos, P. F.; Whitford, W. G. The effects of microarthropods on litter decomposition in a Chihuahuan desert ecosystem. *Ecology* 1981, 62, 654–663.

Schloss, P. D.; Handelsman, J. Status of the microbial census. *Microbiol Mol Biol Rev* 2004, 68, 686–691.

Schubert, K. R.; Evans, H. J. Hydrogen evolution: a major factor affecting the efficiency of nodulated symbionts. *Proc Natl Acad Sci USA* 1976, 73, 1207–1212.

Sethi, S.; Gupta, S. Impact of pesticides and biopesticides on soil microbial biomass carbon. *Univers J Environ Res Technol* 2013, 3(2), 326–330.

Sharma, B.; Narzary, D.; Jha, D. K. Culture independent diversity analysis of soil microbial community and their significance. In *Bacterial diversity in sustainable agriculture, sustainable development and biodiversity*, Maheshwari, D. K., Ed.; Springer International Publishing: Switzerland, 2014, pp 305–340.

Shoebitz, M.; Ribaudo, C. M.; Pardo, M. A.; Cantore, M. L.; Ciampi. L.; Curá, J. A. Plant growth promoting properties of a strain of *Enterobacter ludwigii* isolated from *Lolium perenne* rhizosphere. *Soil Biol Biochem* 2009, 41, 1768–1774.

Shrinivas, S. J.; David, M. Modulatory impact of flubendiamide on enzyme activities in tropical black and red agriculture soils of Dharwad (North Karnataka), India. *Int J Agri Food Sci* 2015, 5(2), 43–49.

Shukla, A. K. Effect of herbicides butachlor, fluchloralin, 2,4-D and oxyfluorfen on microbial population and enzyme activities of rice field soil. *Ind J Ecol* 1997, 24, 189–192.

Singh, B. K.; Walker, A. Microbial degradation of organophosphorus compounds. *FEMS Microbiol Rev* 2006, 30, 428–471.

Singh, M. S.; Devi, R. K. T.; Singh, N. I. Evaluation of methods for *Azotobacter* application on the yield of rice. *Ind J Hill Farm* 1999, 12, 22–24.

Smith, L. A.; Hill, S.; Yates, M. G. Inhibition by acetylene of conventional hydrogenase in nitrogen-fixing bacteria. *Nature* 1976, 262, 209–210.

Smolander, A.; Van Dijk, C.; Sundman, V. Survival of *Frankia* strains introduced into soil. *Plant and Soil* 1988, 106, 65–72.

Stehle, S.; Schulz, R. Agricultural insecticides threaten surface waters at the global scale. *Proc Natl Acad Sci USA*, 2015, 112(18), 5750–5755.

Stehle, S.; Schulz, R. Pesticide authorization in the EU – environment unprotected. *Environ Sci Pollut Res* 2015, 22, 19632–19647.

Stewart, W. D. P. Some aspects of structure and function in N fixing cyanobacteria. *J Bacteriol* 1980, 137, 321.

Sturz, A. V.; Christie, B. R.; Novak, J. Bacterial endophytes: potential role in developing sustainable system of crop production. *Crit Rev Plant Sci* 2000, 19, 1–30.

Sukul, P. Enzyme activities and microbial biomass in soil as influenced by metalaxyl residues. *Soil Biol Biochem* 2006, 38, 320–326.

Sukul, P.; Spiteller, M. Persistence, fate and metabolism of ^{14}C-metalaxyl in typical Indian soils. *J Agric Food Chem* 2001, 49, 2352–2358.

Susamma, V.; Misra, A. K. Frankia-actinorhizal symbiosis with special reference to host-microsymbiont relationship. *Curr Sci* 2002, 83(4), 404–408.

Tate, R. L. *Soil microbiology (symbiotic nitrogen fixation)*. Wiley: New York, 1995, pp 307–333.

Tilman, D.; Fargione, J.; Wolff, B.; D'Antonio, C.; Dobson, A.; Howarth, R.; Schindler, D.; Schlesinger, W. H.; Simberloff, D.; Swackhamer, D. Forecasting agriculturally driven global environmental change. *Science* 2001, 292(5515), 281–284.

Tjepkema, J. D.; Ormerod, W.; Torrey, J. G. Vesicle formation and acetylene reduction activity in *Frankia* sp. CP11 cultured in defined nutrient media. *Nature* 1980, 287, 633–635.

Trinick, M. J. Symbiosis between *Rhizobium* and the nonlegume *Trema aspera*. *Nature* 1973, 244, 459–460.

Uma, D.; Kannaiyan, S. Studies on salt stress on growth, ammonia excretion and nitrogen fixation by the cyanobacterial mutant *Anabaena variabilis*. *J Microb World* 1999, 1, 9–18.

Van Der Hoek, K.; Konradsen, F.; Athukoral, K.; Wanigadewa, T. Pesticide poisoning: a major health problem in Sri Lanka. *Soc Sci Med* 1998, 46, 495–504.

Vance, C. P. Legume symbiotic nitrogen fixation: agronomic aspects. In *The Rhizobiaceae: Molecular biology of model plant-associated bacteria,* Spaink, H. P., Kondorosi, A., Hooykaas, P. J. J., Eds.; Kluwer: Dordrecht, 1998, pp 509–530.

Vance, C.P. Symbiotic nitrogen fixation and phosphorus acquisition: plant nutrition in the world of declining renewable resources. *Plant Physiol* 2001, 127, 390–397.

Vaz-Moreira, I.; Egas, C.; Nunes, O. C.; Manaia, C. M. Culture-dependent and culture-independent diversity surveys target different bacteria: a case study in a freshwater sample. *Antonie Van Leeuwenhoek* 2011, 100(2), 245–57.

Vinita, B.; Veena, S. Impact of pesticides use in agriculture. *Int J Develop Res Engin* 2015, 2(2), 1–21.

96

Wang, M. C.; Gong, M.; Zang, H. B.; Hua, X. M.; Yao, J.; Pang, Y. J.; Yang, Y. H. Effect of methamidophos and urea application on microbial communities in soils as determined by microbial biomass and community level physiological profiles. *J Environ Sci Health B* 2006, 41, 399–413.

Warren, G. F. Spectacular increases in crop yields in the United States in the Twentieth century. *Weed Tech* 1998, 12, 752–760.

Weary, G. C.; Merriam, H. G. Litter decomposition in a red maple woodlot under natural conditions and under insecticide treatment. *Ecology* 1978, 59, 180–184.

Webster, J. P. G.; Bowles, R. G.; Williams, N. T. Estimating the economic benefits of alternative pesticide usage scenarios: wheat production in the United Kingdom. *J Crop Prod* 1999, 18, 1–83.

Widenfalk, A.; Bertilsson, S.; Sundh, I.; Goedkoop, W. Effects of pesticides on community composition and activity of sediment microbes - responses at various levels of microbial community organization. *Environ Pollut* 2008, 152, 576–584.

Widenfalk, A.; Svensson, J. M.; Goedkoop, W. Effects of the pesticides captan, deltamethrin, isoproturon and pirimicarb on the microbial community of a freshwater sediment. *Environ Toxicol Chem* 2004, 23, 1920–1927.

Wilson, C. Environmental and human costs of commercial agricultural production in South Asia. *Int J Soc Econ* 2000, 27, 816–884.

Yang, J.; Kloepper, J. W.; Ryu, C. M. Rhizosphere bacteria help plants tolerate abiotic stress. *Trends Plant Sci* 2009, 14, 1–4.

Yanggen, D.; Cole, D.; Crissman, C.; Sherwood, S. *Human health, environmental and economic effects of pesticide use in potato production in Ecudor.* International Potato Center: Lima, Peru, 2003.

Yanni, Y. G.; Abd-El-Fattah, F. K. Towards integrated biofertilization management with free living and associative dinitrogen fixers for enhancing rice performance in the Nile delta. *Symbiosis* 1999, 27, 319–331.

Yanni, Y. G.; Rizk, R. Y.; Corich, V.; Squartini, A.; Ninke, K.; Philip-Hollingsworth, S.; Orgambide, G.; de Bruijn, F.; Stoltzfus, J.; Buckley, D.; Schmidt, T. M.; Mateos, P. F.; Ladha, J. K.; Dazzo, F. B. Natural endophytic association between *Rhizobium leguminosarum* bv. *trifolii* and rice roots and assessment of its potential to promote rice growth. *Plant Soil* 1997, 194, 99–114.

Yu, S.; Demin, O.; Bogachev, A. V. Respiratory protection nitrogenase complex in *Azotobacter vinelandii. Succ Biol Chem* 2005, 45, 205–234.

Zahran, H. H. Rhizobium-legume symbiosis and nitrogen fixation under severe conditions and in an arid climate. *Microbiol Mol Biol Rev* 1999, 63, 968–989.

Zhou, Y.; Liu, W.; Ye, H. Effects of pesticides metolachlor and S-metolachlor on soil microorganisms in aquisols. II. soil respiration. *Ying Yong Sheng Tai Xue Bao* 2006, 17, 1305–1309.

Zogg, G. P.; Zak, D. R.; Pregitzer, K. S.; Burton, A. J. Microbial immobilization and the retention of anthropogenic nitrate in a northern hardwood forest. *Ecology* 2000, 81, 1858–1866.

CHAPTER 5

RECENT ADVANCES IN APPLICATIONS OF NANOMATERIALS FOR WATER REMEDIATION

KALIYAPERUMAL RANI[1] and BARINDRA SANA[2]

[1]*Division of Bioengineering, School of Chemical and Biomedical Engineering, Nanyang Technological University, 62 Nanyang Drive 637459, Singapore*

[2]*p53 Laboratory, Agency for Science, Technology and Research (A*STAR), 1 Fusionopolis Way, #20-10 Connexis North Tower 138632, Singapore*

CONTENTS

5.1 INTRODUCTION

According to National Nanotechnology Initiative, nanotechnology (NT) refers to engineering functional systems at the nanoscale (about 1–100 nm size) where unique phenomena facilitate innovative applications. This emerging transdisciplinary field opens novel perspectives in all sectors including medicine, biology, cosmetics, optical science, semi conductor, memory and storage technologies and electronic and automotive industry. Nanotechnology plays an inevitable role in improving the environmental sustainability by means of detecting, preventing, and removing pollutants and also by developing environmental friendly products (Figs. 5.1 and 5.2).

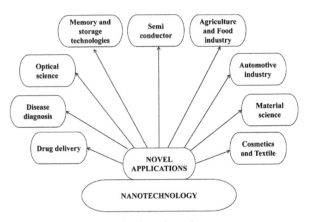

FIGURE 5.1   Applications of nanotechnology in various sectors.

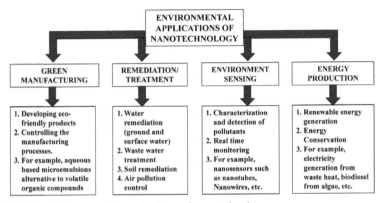

FIGURE 5.2   Environmental applications of nanotechnology.

Nanoremediation refers to the application of nanomaterials for remediation. These nanomaterials are highly reactive and effectively transform and extenuate the pollutants through oxidation or reduction. The standout features of nanomaterials such as large surface area, well-defined structure, and easy dispersability make them superior to conventional remediation technologies. They could be readily tailored according to the pollutants of concern. Further, nanoremediation technologies eliminate the necessities of ground water being drawn out above ground level and soil being shipped to different sites for treatments (Fig. 5.3) (Otto et al., 2008).

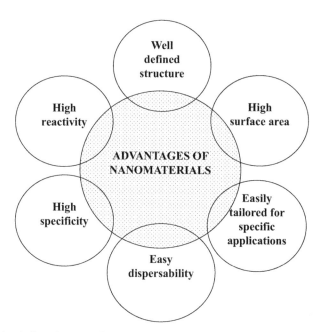

FIGURE 5.3   Salient features of nanomaterials.

Various nanotechnological platforms have been developed that could offer cost-effective solutions to numerous environmental concerns. For instance, nanoparticles containing zero-valent iron hold great promise for remediation of contaminated groundwater. Nanotitanium dioxide potentially degrades organic pollutants, which are deleterious to the environment. Carbon nanotube membranes are of great use in the process of desalination. Nanofilters are employed to clean up ground or surface water contaminated with chemicals and hazardous substances. Nanosensors are availed to detect waterborne contaminants. Nanosilver is applied to disinfect water owing to its antimicrobial potency. Overall, there is a multitude of promising environmental applications for nanotechnology; this chapter highlights some of the fascinating developments in the applications of following nanomaterials such as nanoscale zero-valent iron (nZVI), titanium dioxide (TiO_2), silver nanoparticles (AgNPs), nanomembranes, nanotubes, nanoclays, and nanocomposites.

5.2   NANOMATERIALS FOR WATER REMEDIATION

5.2.1   NANOSCALE ZERO-VALENT IRON (nZVI)

Numerous nanoparticles have been widely applied for remediation of aqueous contaminants; however, nanoscale iron (nZVI) gained a great deal of attention in the recent years and symbolizes a latest generation of environmental remediation technology. Nano zero-valent iron is a nanoscale material of 10 to 100 nm in diameter and is considered as a novel nanomaterial for soil and ground water remediation. They exhibit a typical core shell structure containing metallic iron core surrounded by a delicate amorphous oxide shell (Yan et al., 2010). In bimetallic nanoparticle, nZVI is often mixed with another metal such as palladium, silver, or copper, which acts as a catalyst. The inclusion of noble metal for instance palladium makes the remediation more effective as it catalyzes the dechlorination and hydrogenation processes more effectively (Dhakras, 2011).

nZVI has numerous advantages such as abundant surface area, which renders enhanced surface reactivity, exorbitant flexibility for *in situ* applications, efficiency to extenuate widespread environmental pollutants. In spite of these merits, nZVI bears certain site-specific requisites, which need to be fulfilled for efficient remediation. For instance, site characterization such as location, geologic conditions, concentration and types of contaminants, geochemical features, ground-water flow velocity, and depth are mandatory before nZVI is being applied. The most frequently used method of application of nZVI is injection of an nZVI suspension directly into an aquifer. Injection techniques include infiltration wells, sleeve pipe, push infiltration, or gravity infiltration. Another mode of application is by using a permeable reactive barrier, where the trench is loaded with nZVI particles (Mishra and Patel, 2009).

The nZVI appears to be very potent for degrading a wide a spectrum of organic pollutants including polychlorinated biphenyls (PCBs), chlorinated hydrocarbon (organochlorine) pesticides, and chlorinated organic solvents

(Zhang, 2003). The mechanisms through which chlorinated solvents such as trichloro ethane (TCE) are degraded by nZVI involve two pathways namely reductive dechlorination and β-elimination (Arnold and Roberts, 2000). The dechlorination reactions take place at the surface of nZVI and utilize the excess electrons formed from iron corrosion in water. This results in the production of nonchlorinated hydrocarbons such as ethane and ethene. However, the degradation of TCE by β-elimination pathway results in the formation of chloroacetylene, which is then converted to acetylene with the elimination of chlorine. Acetylene is further converted into ethane and ehthene (Arnold and Roberts, 2000). It has been demonstrated that nZVI is efficient in reducing inorganic anions such as nitrate, arsenate, arsenite, perchlorate, selenite, and chromate. When compared to granular iron, nZVI bears high catalytic rate and sorption potency. Further, nZVI is effective in eliminating heavy metals such as Pb and Ni present in aqueous solution by reducing them to zero-valent metals (USEPA, 2013).

nZVI particles have shown the potential to reduce perchlorate present in aqueous solution to chloride completely without forming any intermediates (Cao et al., 2005). It has been demonstrated that iron oxide nanoparticles are efficient in removing arsenic (As) through irreversible binding. Further, nZVI has been shown to catalyze the rapid transformation of As (V) to As (III) at neutral pH. However, for this reduction, nZVI was needed at larger quantity due to presence of phosphate, sulfate, and dissolved organic carbon (DOC) present in the solution (Kanel et al., 2006).

Chlorinated dense non-aqueous phase liquids (DNAPLs) remain as one the major constituents of groundwater contamination. nZVI acts as a remediation agent for DNAPL sources. As nZVI reduces the concentration of contaminants, it favors the formation of concentration gradient between DNAPLs and the aqueous phase, resulting in the translocation of DNAPLs to the dissolved aqueous phase. This in turn creates the necessity for DNAPLs to be treated again (Watlington, 2005). To overcome this issue, emulsified ZVI (eZVI) has been introduced. This contains iron particles in water surrounded by an oil-liquid membrane. The properties of eZVI's membrane (hydrophobicity) are similar to that of DNAPLs and hence are miscible with DNAPLs. Owing to the diffusion of halogenated hydrocarbons in DNAPL through the membrane into the interior aqueous phase where zero-valent iron presents, abiotic reductive dechlorination occurs. Moreover, the presence of hydrophobic membrane in eZVI protects the nZVI from other water constituents, thereby enhancing its availability for treating target contaminants (Hara et al., 2006).

5.2.2 TITANIUM DIOXIDE (TiO$_2$)

Nanocatalyst based photocatalysis is considered as a promising approach for treating contaminated water. The catalysts harness the UV radiation from sunlight and utilize the energy to degrade the pollutants. Photocatalysis involve in the degradation of wide spectrum of pollutants such as organic compounds and acids, inorganic compounds, pesticides, and dyes. There are several potential photocatalysts, among which titanium dioxide (TiO$_2$) is emerging as a promising option for water remediation owing to its semiconducting, energy converting, and gas sensing natures. They are very versatile and serve both as oxidative and reductive catalysts hence have been widely explored for oxidative transformation of organic and inorganic pollutants (Mayo et al., 2007). TiO$_2$ catalyzes the oxidation process through the formation of hydroxyl-free radicals and the reduction process through peroxide formation. The photocatalytic process catalyzed by TiO$_2$ is clearly demonstrated in Figure 5.4. In TiO$_2$ the electrons are confined to energy bands namely valence band and conduction band. When TiO$_2$ is illuminated by UV light, electrons are excited from the valence band to the conduction band. As a result, electron vacancies known as holes are produced in the valence band (Mills and Davies, 1993). As the holes are of positively charged, they form ·OH and H$^+$ ions upon binding with water molecules. Electrons reduce the dissolved oxygen to superoxide ions (O$_2$−). As these superoxide ions are highly reactive in nature, they bind with water molecules to produce peroxide radicals (·OOH) and hydroxide ions (OH−). Upon binding with H$^+$ ions, peroxide radicals produce ·OH and OH−. The hydroxide ions are oxidized by the holes to ·OH. Hence, ·OH ions formed during the photocatalysis directly attack the contaminants present in the aqueous mixture (Linsebigler et al., 1995) (Fig. 5.4).

TiO$_2$ nanoparticles are cost effective, photocatalytically active, chemically inert and are highly resistant to corrosion. The larger surface area of TiO$_2$ nanoparticles provides huge catalytic surface (Theron et al., 2008). These salient features make them to use as semiconductor photocatalyst in water purification systems. In the presence of UV light, TiO$_2$ nanoparticles are highly efficient in the removal of carbon from water contaminated with organic wastes (Chitose et al., 2003). However, they are unable to absorb visible light. Doping of transition metals and anionic nonmetals such as nitrogen, carbon, sulfur, or fluorine into TiO$_2$ enhances its optical absorbance to the visible region resulting in enhanced photocatalytic potency (Asahi et al., 2001; Noworyta and Augustynski, 2004; Livraghi et al., 2005; Ni et al., 2007). For instance, surface modification of TiO$_2$ nanoparticles with metals such as gold and platinum elevates the photocatalytic activity as the metals keep the electrons and holes from recombining. Further surface modification with chelating agents such as lauryl sulfate and arginine provides high surface area and thereby increases the active sites resulting in enhanced photocatalytic activity (Makarova et al., 2000). Similarly, nitrogen-doped TiO_2 nanoparticles were efficient in degrading azo dyes and Fe(III)-doped TiO_2 nanoparticles were efficient in the degradation of phenol (Liu et al., 2005; Nahar et al., 2006).

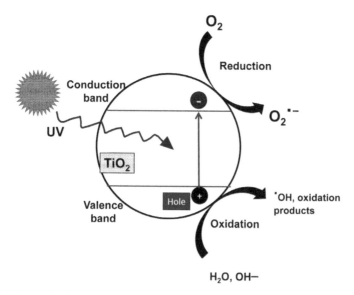

FIGURE 5.4 Mechanism of TiO_2 photocatalysis.

Microbial contaminants constitute a major part of water contamination. It has been demonstrated that nitrogen-doped TiO_2 nanoparticles are efficient in degrading microbial contaminants of water (Shalini et al., 2012). UV–visible light activated TiO_2 nanoparticles demonstrated enhanced bactericidal activity against *E. coli* (Ireland et al., 1993). Chemical oxygen demand (COD) (Kemnetz and Cody, 1996) test is often used to determine the concentration of organic compounds in water and hence acts as a useful measure of water quality. TiO_2 electrodes are of great use in detecting chemical oxygen demand of water, hence are employed as sensors for measuring the degree of water contamination (Kim et al., 2000).

Owing to the stable nature of TiO_2 in water, it could be immobilized onto thin films or membranes. The main goal of immobilizing TiO_2 is to eliminate the post separation predicaments related to the amorphous TiO_2. Apart from this, TiO_2 immobilization offers greater catalytic area, increased surface hydroxyl groups and enhanced adsorption efficiency (Zhu and Zou, 2009; Esparza et al., 2010; Jin and Dai, 2012). TiO_2 films are employed for filtration and bacterial inactivation purposes. TiO_2 films can be fabricated using different deposition techniques such as reactive sputtering, chemical vapor deposition and sol–gel process. The sol–gel method is the most popular approach because of its cost effectiveness and simplicity. The characteristics and the potential applications of TiO_2 films strongly rely on their synthesis procedures and relevant operational parameters (Sobczyk-Guzenda et al., 2013). It has been demonstrated that Ag/TiO_2 slurry solution and TiO_2 films fabricated by sol-gel method showed exceptional efficiency in the photocatalytic degradation of wastewater (Domínguez-Espíndola et al., 2013).

5.2.3 SILVER NANOPARTICLES

Silver nanoparticles (Ag NPs) possess unique optical, thermal and electrical properties and have been widely used as anti-bacterial agents. Ag NPs are mainly derived from silver nitrate and silver chloride. It has been reported that the antibacterial activities of Ag NPs mainly depend on the size and shape of the particles. Smaller Ag NPs exhibits enhanced bactericidal activity when compared to large size silver particles (Makhluf et al., 2005) and truncated triangular silver demonstrated exalted antibacterial potency than spherical and rod-shaped nanoparticles (Pal et al., 2007).

The exact mechanism through which Ag NP exert its antibacterial effect has not been clearly elucidated yet, however, various modes of actions have been proposed. It has been demonstrated that Ag NPs bind with the bacterial cell wall and cause structural changes in the membrane, which results in increased cell permeability and ultimately death (Sondi and Salopek-Sondi, 2004). It has also been suggested that silver ions released by the nanoparticles inactivate various enzymes of the bacterial cells by effectively binding with the SH functional groups of enzymes (Feng et al., 2000; Matsumura et al., 2003). Spectroscopy studies demonstrated that Ag NPs produce highly reactive free radicals, which in turn disrupt cell membrane resulting in cell death. These free radicals also attack the enzymes of the electron transport chain and inhibit electron transport system (Danilczuk et al., 2006; Kim et al., 2007). The potency of Ag NPs to bind with the sulfur and phosphorus components of DNA molecules provokes DNA damage and thus interferes with DNA replication (Hatchett and White, 1996). In gram-negative bacteria, Ag NPs cause dephosphorylation of tyrosine residues resulting in altered phosphotyrosine profile of bacterial peptides, which in turn interfere with signal transduction pathways and inhibit bacterial growth (Shrivastava et al., 2007) (Fig. 5.5).

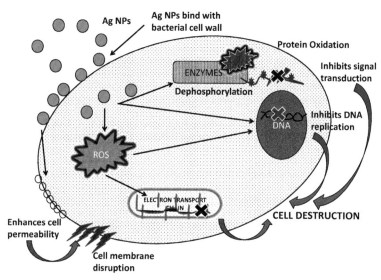

FIGURE 5.5   Antibacterial mechanisms of Ag NPs.

Immobilized Ag NPs showed greater antimicrobial activity, for instance, Ag NPs embedded with cellulose acetate fibers exhibited enhanced antimicrobial activity against both gram negative and gram-positive bacteria (Son et al., 2004). Further, Ag nanoparticles were incorporated into polysulfone membranes and used as nanofilters for water purification. The Ag NPs impregnated polysulfone membranes exhibited greater biofouling resistance (Zodrow et al., 2009). It has also been reported that these membranes possess excellent antimicrobial activities against *E. coli*, *Pseudomonas*, and so forth (Chou et al., 2005; Lee et al., 2007). Further, Ag NPs immobilized on amine-functionalized silica surface has been proven to possess enhanced bactericidal activity (Agnihotri et al., 2013).

5.2.4   NANOFILTERS, NANOMEMBRANES, AND NANOTUBES

Membrane technology plays an instrumental role in wastewater treatment and desalination of seawater. Membrane filtration (MF) involves the use of selective permeable membrane for the separation of dissolved solutes in aqueous mixture. One of the recent promising approaches in MF is the use of nanomembranes for filtration processes. NF membranes bear pore sizes between 0.2 and 4 nm. These pores are smaller than those of microfiltration and ultrafiltration membranes and are larger than those of membranes used in reverse osmosis process. Hence these membranes possess characteristics between those of ultra filtration (Makhluf et al., 2005) and reverse osmosis (RO) membranes. NF membranes are extensively employed in wastewater and potable water treatments. They are effective in removing turbidity, inorganic species such as Ca and Na and microbes. These membranes hold great promise in reducing water hardness and also in removing dissolved organic pollutants. NF-RO systems are employed in seawater desalination which primarily consisting of two steps where NF membranes are used for filtration prior to RO.

Recently carbon nanotube filters gained a great deal of attention in the field of water remediation. These filters posses numerous advantages such as hydrophobicity and high porosity. Despite the salient feature of CNT membranes, which could effectively filter water and retain Na^+/Cl^- ions, they are capable of desalinating seawater. The efficiency of CNT membrane in the removal of sodium chloride from seawater has been demonstrated recently by Srivastava et al. (2004). In addition, these membranes are of great use in removing bacteria such as *Escherichia coli* or nanometer sized poliovirus from water (Srivastava et al., 2004). In another approach, a novel carbonaceous nanofilter was synthesized using continuous spray pyrolysis process. For this, *n*-hexane was used as a carbon source and ferrocene served as a catalyst source. It has been demonstrated that at pressures of 8–11 bar, these filters efficiently remove viruses from contaminated water (Mostafavi et al., 2009). The added advantages of carbon nanotube filters are that they are easily cleaned up either by ultrasonication or autoclaving.

Carbon nanotubes are known to be effective against chemical contaminants including heavy metals such as Cr^{3+}, Zn^{2+} and Pb^{2+}, metalloids such as arsenic compounds, organics such as polycyclic aromatic organic compounds (PAH), and atrazine (Li et al., 2005; Peng et al., 2005; Di et al., 2006; Hedderman et al., 2006; Yang et al., 2006; Gotovac et al., 2007; Rao et al., 2007; Yan et al., 2008). Biological contaminants form a major portion of potable water contaminants and are mainly comprised of three categories namely microbes, biological toxins and natural organic matters (NOM). Carbon nanotubes possess exceptionally high bacterial adsorption capacities and serve as adsorbent media for eliminating biological contaminants. The standout features of carbon nanotubes such as high and selective adsorption of pathogens make them superior to conventional adsorbents (Upadhyayula et al., 2009).

The evolution of nanotechnology in the realm of ceramics has introduced innovative opportunities for the establishment of nanoceramic filters, which are of great use in removing viruses from blood stream. These filters possess exceptional chemical and thermal stability. Currently, nanoceramic membranes and filters are commercialized and used mainly for the removal of virus from blood. However, it only eliminates free viruses present in the blood stream (Zhao et al., 2008).

5.2.5 NANOCLAYS AND NANOCOMPOSITES

Nanoclays are one of the most affordable natural nanomaterials found in the clay fraction of soil. Owing to their outspread occurrence in nature, clays have long been availed as low cost flocculants and sorbents of environmental contaminants such as suspended particles, disease-bearing organisms, and other harmful pollutants (Churchman et al., 2006). The suitability of various nanoclays has been explored for water remediation over the past years. It has been reported that clays have a myriad of potential applications, which include the removal of oil and grease, heavy metals, chemicals and also reclaiming nitrogen from nitrogen-rich effluents (Kemnetz and Cody, 1996; Lo, 1996; Churchman, 2002a, 2002b; Yuan, 2004; Theng et al., 2006; Srinivasan, 2011).

As clays are layered minerals with space in between the layers, they readily adsorb water molecules and ions. They possess superior adsorption capacity, which make them attractive for applications in photocatalysis process (Chong et al., 2010). Clays also exchange adsorbed ions with their immediate environment. However, clays are catalytically inactive and lack the porosity. Hence, various researches have been carried out to manipulate them. The addition of materials such as metals, polymers with clays result in the formation of nanocomposites (clay composites). The salient features of montmorillonite clays such as excellent hydration and swelling capacity, durability, enlarged interlayer space, and immense reactivity make them excellent targets for manipulation. Modification of montmorillonite through pillaring with various polyoxy cations of Zr^{4+}, Si^{4+}, Al^{3+}, Fe^{3+}, Ti^{4+}, Ga^{3+} or Cr^{3+} and so forth resulting in enhanced metal adsorption capacity suggesting its potential in the elimination of heavy metals like Cd, Cu, Mn, Cr, Co, Ni, Fe, Pb, and Zn (Bhattacharyya and Gupta, 2008).

Recently, polymer-clay nanocomposites are emerging as promising options for remediation owing to their enhanced thermal, mechanical, and barrier performance features resulting from the large contact area between clay particles and polymer on a nanoscale. These nanocomposites have tremendous potential in removing organic pollutants, anionic pollutants, and herbicides especially atrazine, a well-known long-lived persistent herbicide (Breen, 1999; Churchman, 2002; Zadaka et al., 2009). For instance, chitosan-montmorillonite composites efficiently remove anionic pollutants (An et al., 2007; Li et al., 2009). It's efficiency in adsorbing selenium from water was exceptional when compared to commercially available adsorbents (Bleiman and Mishael, 2010). Chitosan-coated montmorillonite clay has been reported to be more efficacious compared to natural clay in the purge of tungsten (Gecol et al., 2006). In a batch test performed by Yuan et al. (2009), montmorillonite-supported magnetite nanoparticles were proven to be effective in removing hexavalent chromium. Ti-pillared montmorillonite has tremendous potential in the elimination of arsenite and arsenate from aqueous mixture (McNally et al., 2010).

Nitrates remain as the major pollutants of groundwater contamination in many parts. Hydrochloric acid activated montmorillonite exhibited greater nitrate adsorption competency. Recently, binary clay composites have been prepared and investigated for the removal of metals and other common water pollutants. For instance, the combination of carbon and sodium montmorillonite binary clay demonstrated increased sorption of lead compared to carbon alone (Ake et al., 2001).

Bentonite is well known for its competency to adsorb and remove toxins, impurities, heavy metals, and chemicals. Bentonite clay bears a strong negative charge while toxins bear positive charge; hence, the clay particles adsorb toxins easily resulting in the efficient removal of heavy metals such as lead and zinc from water (Mishra and Patel, 2009). Addition of iron oxide to bentonite clay has been shown to enhance competency of bentonite in adsorbing metal ions such as Ni^{2+}, Cu^{2+}, Cd^{2+}, and Zn^{2+} (Oliveira et al., 2003). Thermally activated bentonite possesses highest adsorptive capacity for U(VI) from aqueous solutions (Aytas et al., 2009). It has been demonstrated that bentonite-polyacrylamide composite effectively adsorb Cu^{2+} ions from the ground and surface waters (Zhao et al., 2010). Magnesium incorporated bentonite (MB) showed enhanced capacity for fluoride removal and therefore serve as an potent adsorbent for defluorinating water (Zhao et al., 2010). Besides magnesium, manganese and lanthanum (La) incorporated bentonite were also explored for fluoride adsorption and both were showed to be potent in deflourinating water, however, the efficiency of La-bentonite was much higher as compared to Mg-bentonite, Mn-bentonite (Kamble et al., 2009).

Another type of clay that gained much popularity in the recent years is allophane nanoclay. Allophane nanoclays are inexpensive, eco-friendly and most importantly, recoverable after use. It contains greater concentration of active aluminum, hence competent in adsorbing and precipitating phosphates. Hence, this clay was reported to be effective in treating eutrophic water, meatworks effluent and sewage water (Yuan and Wu, 2007). In batch experiments carried out by Etci et al. (2010) demonstrated that beidellite, an inexpensive natural mineral, holds excellent adsorption capacity for lead and cadmium ions present in aqueous solutions.

Organically modified clays known as organoclays are often applied in the upstream sector of petrochemical industry for eliminating hydrocarbons from water produced during refining processes (Pereira et al., 2005). Various organoclays have also been explored for treating toxic organic chemicals released by pharmaceuticals and pesticides industries. Carbamazepine is most widely prescribed anticonvulsant and mood stabilizer. Owing to its low biodegradability, it is highly persistent in the water systems. Trimethylphenylammonium (TMPA), and hexadecyltrimethylammonium (HDTMA) smectites showed greater sorption for Carbamazepine from water (Zhang et al., 2010). Also, poly(4-vinylpyridine-co-styrene)-montmorillonite has been reported to remediate atrazine from water (Zadaka et al., 2009).

5.3 NANO-BASED TECHNOLOGIES VS CONVENTIONAL TECHNOLOGIES FOR WATER TREATMENT

Conventional technologies employed for water treatment include filtration, UV radiation, chemical treatment, and desalination. Even though they have been extensively used for many decades, they have certain drawbacks too. For instance, chemical treatment such as chlorination not only destroys disease-causing microbes, but also gives a possibility of producing toxic chlorinated fragments. In addition, chlorination gives the undesirable odor and taste to the treated water. UV radiation is considered as effective means of disinfection; however, it is inefficient in treating organic contaminants of water. Even though, filtration techniques such as reverse osmosis (RO) and ultrafiltration play inevitable part in water treatment, these techniques are not competent in treating water contaminated with chemicals, drugs, personal care products, industrial additives and surfactants, and chemicals. Taken together, the conventional technologies are not efficient enough to address all potential issues of water contamination. Moreover, new microcontaminants released into the aquatic environment also cause water contamination and potential threat to aquatic life.

Nanotechnology has been considered holding tremendous potential in solving problems related to water contamination. Nanotechnological approaches hold promises in treating a wide spectrum of existing and emerging water contaminants. Further, it aids to improve the performance of existing technologies of water treatment. In recent years, various nanomaterials have been extensively studied in eliminating various organic, inorganic pollutants from water. Nanomaterials such as nano zero-valent iron, TiO_2, silver nanoparticles, carbon nanotubes, nanomembranes, and nanoclays act as major contributors for the development of more efficient water treatment strategies. Figure 5.6 displays various conventional and nano-based technologies available to treat contaminated water.

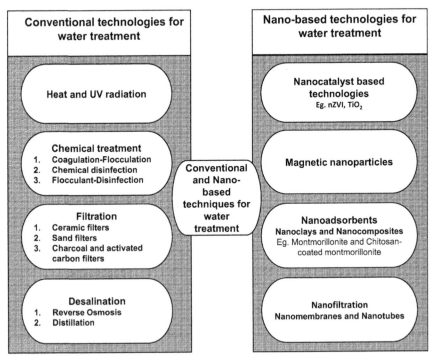

FIGURE 5.6   Conventional and nano-based technologies for water treatment.

5.4 CONCLUSION AND PERSPECTIVES

Nanotechnology is an innovative emerging technology and its potential environmental applications include pollution detection, monitoring, remediation and control. In this chapter, we presented an overview of recent advances in the applications of selected nanomaterials in water remediation. Table 5.1 displays a list of nanomaterials used in the remediation of various inorganic, organic, and microbial contaminants of water. This clearly reveals that nanomaterials have already provided a significant impact on water remediation, which is expected to exponentially grow during the next years. However, most of the nanomaterials have been examined only in the lab-scale so far. Owing to the nanoscale nature and potentially high reactivity resulting from the large surface to volume ratio of nanomaterials, the potential risks associated with all new nanosystems need to be evaluated. However, at this point of time, not much information is perceived about the fate, transport, and toxicity of nanomaterials in the environment. Hence, it is of immense significance to carry out systematic studies to explore the toxicological aspects of nanomaterials with regard to the environment. Nevertheless, the current scenario evidently demonstrates that nanotechnology holds a great promise in the realm of water remediation. Further, the knowledge gained in this field so far, paves the way to establish novel strategies for applying nanomaterials for practical and large-scale water remediation.

TABLE 5.1 Summary of Nanomaterials Used in the Remediation of Various Inorganic, Organic, and Microbial Contaminants of Water.

Nanomaterials	Examples	Target Pollutants
TiO_2 based nanoparticles (NPs) and nanopowders	TiO_2 NPs, mesoporous TiO_2 nanosized powder	Organic pollutants, phenol, rhodamine B dye
Iron based nanoparticles	Nano zero valent iron	chlorinated solvents, polychlorinated biphenyls, arsenic, nitrates, humic acid
Silver nanoparticles	Ag NPs	Antibacterial activity (against *E. coli*)
Metal oxide based nanoparticles	ZnO NPs, CUO NPs	Antimicrobial activity (against *E. coli*, *S. aureus*)
Bimetallic nanoparticles	Pd/Au, Fe/Pd, Cu/Fe	Chlorinated compounds, lindane, atrazine, nitrate
Magnetic nanoparticles	$MnFe_2O_4$, $MgFe_2O_4$, $ZnFe_2O_4$, $CuFe_2O_4$	Cr(VI), As(III), As(VI), organic pollutants

TABLE 5.1   *(Continued)*

Nanomaterials	Examples	Target Pollutants
Nanoclays and nanocomposites	Montmorillonite and composites, bentonite and composites, allophane, organoclays	Nitrates, lead, zinc, Cd, Cu^{2+}, U(VI), phosphates, herbicide such as atrazine
Nanotubes	Carbon nanotubes	Heavy metals such as Pb, Cu, Co, Mn, Zn, lead, fluoride, organic pollutants
Nanomembranes	Nanofiltration membranes with TiO_2	Nitrates, natural organic matters such as fulvic acid and humic acid
Dentrimers	Polyamindoamine, diaminobutane poly dendrimers	Copper, lead, polycyclic aromatic hydrocarbons

KEYWORDS

- Antimicrobial activity
- Heavy metals
- Inorganic ions
- Nano zero-valent iron
- Nanoclays
- Nanomaterials
- Nanomembranes
- Nanotechnology
- Nanotubes
- Organic pollutants
- Silver nanoparticles
- Titanium dioxide
- Water remediation

REFERENCES

Agnihotri, S.; Mukherji, S.; Mukherji, S. Immobilized silver nanoparticles enhance contact killing and show highest efficacy: Elucidation of the mechanism of bactericidal action of silver. *Nanoscale* 2013, 5, 7328–7340.

Ake, C. L.; Mayura, K.; Huebner, H.; Bratton, G. R.; Phillips, T. D. Development of porous clay-based composites for the sorption of lead from water. *J Toxicol Environ Health Sci* 2001, 63, 459–475.

An, J. H.; Dultz, S. Adsorption of tannic acid on chitosan-montmorillonite as a function of pH and surface charge properties. *Appl Clay Sci* 2007, 36, 256–264.

Arnold, W.; Roberts, L. Pathway and kinetics of chlorinated ethylene and chlorinated acetylene reaction with Fe(0) particles. *Environ Sci Technol* 2000, 34(9), 1794–1805.

Asahi, R.; Morikawa, T.; Ohwaki, T.; Aoki, K.; Taga, Y. Visible-light photocatalysis in nitrogen-doped titanium oxides. *Science* 2001, 293, 269–271.

Aytas, S.; Yurtlu, M.; Donat, R. Adsorption characteristic of U(VI) ion onto thermally activated bentonite. *J Hazard Mater* 2009, 172, 667–674.

Bhattacharyya, K. G.; Gupta, S. S. Adsorption of a few heavy metals on natural and modified kaolinite and montmorillonite: A review. *Adv Colloid Interface Sci* 2008, 140, 114–131.

Bleiman, N.; Mishael, Y. G. Selenium removal from drinking water by adsorption to chitosan-clay composites and oxides: Batch and columns tests. *J Hazard Mater* 2010, 183, 590–595.

Breen, C. The characterisation and use of polycation-exchanged bentonites. *App Clay Sci* 1999, 15, 187–219.

Cao, J.; Elliott, D.; Zhang, W. Perchlorate reduction by nanoscale iron particles. *J Nanopart Res* 2005, 7, 499–506.

Chitose, N.; Ueta, S.; Seino, S.; Yamamoto, T. A. Radiolysis of aqueous phenol solutions with nanoparticles. 1. Phenol degradation and TOC removal in solutions containing TiO_2 induced by UV, gamma-ray and electron beams. *Chemosphere* 2003, 50, 1007–1013.

Chong, M. N.; Jin, B.; Chow, C. W. K.; Saint, C. Recent developments in photocatalytic water treatment technology: A review. *Water Res* 2010, 44, 2997–3027.

Chou, W. L.; Yu, D. G.; Yang, M. C. The preparation and characterization of silver loading cellulose acetate hollow fiber membrane for water treatment. *Polym Adv Technol* 2005, 16, 600–607.

Churchman, G. J. Formation of complexes between bentonite and different cationic polyelectrolytes and their use as sorbents for non-ionic and anionic pollutants. *Appl Clay Sci* 2002a, 21, 177–189.

Churchman, G. J. The role of clays in the restoration of perturbed ecosystems. *Dev Soil Sci* 2002b, 28A, 333–350.

Churchman, G. J.; Gates, W. P.; Theng, B. K. G.; Yuan, G. Clays and clay minerals for pollution control. In *Handbook of Clay Science,* Bergaya, F., Theng, B. K. G., Lagaly, G., Eds.; Elsevier Ltd: Amsterdam, 2006, pp 625–675.

Danilczuk, M.; Lund, A.; Sadlo, J.; Yamada, H.; Michalik, J. Conduction electron spin resonance of small silver particles. *Spectrochim Acta A Mol Biomo Spectrosc* 2006, 63, 189–191.

Dhakras, P. A. Nanotechnology applications in water purification and waste water treatment: A review. In *International conference on nanoscience, engineering and technology (ICONSET)*, Chennai, Nov 28–30, 2011, 285–291.

Di, Z. C.; Ding, J.; Peng, X. J.; Li, Y. H.; Luan, Z. K.; Liang, J. Chromium adsorption by aligned carbon nanotubes supported ceria nanoparticles. *Chemosphere* 2006, 62, 861–865.

Domínguez-Espíndola, R. B.; Silva-Martínez, S.; Ortiz-Hernández, M. L.; Román-Zubillaga, J. L.; Guardián-Tapia, R. Photocatalytic disinfection of municipal wastewater using TiO_2 film and Ag/TiO_2 powders under UV and solar light irradiation. *Mexican J Sci Res* 2013, 2(2), 60–68.

Esparza, P.; Borges, M. E.; Díaz, L.; Alvarez-Galván, M. C.; Fierro, J. L. G. Photodegradation of dye pollutants using new nanostructured titania supported on volcanic ashes. *Appl Catal A Gen* 2010, 388, 7–14.

Etci, O.; Bektaş, N.; Öncel, M. S. Single and binary adsorption of lead and cadmium ions from aqueous solution using the clay mineral beidellite. *Environ Earth Sci* 2010, 61, 231–240.

Feng, Q. L.; Wu, J.; Chen, G. Q.; Cui, F. Z.; Kim, T. N.; Kim, J. O. A mechanistic study of the antibacterial effect of silver ions on *Escherichia coli* and *Staphylococcus aureus*. *J Biomed Mater Res* 2000, 52, 662–668.

Gecol, H.; Miakatsindila, P.; Ergican, E.; Hiibel, S. R. Biopolymer coated clay particles for the adsorption of tungsten from water. *Desalination* 2006, 197, 165–178.

Gotovac, S.; Yang, C. M.; Hattori, Y.; Takahashi, K.; Kanoh, H.; Kaneko, K. Adsorption of polyaromatic hydrocarbons on single wall carbon nanotubes of different functionalities and diameters. *J Colloid Interface Sci* 2007, 314, 18–24.

Hara, S. O.; Krug, T.; Quinn, J.; Clausen, C.; Geiger, C. Field and laboratory evaluation of the treatment of DNAPL source zones using emulsified zero-valent iron. *Remediat J* 2006, 16, 35–56.

Hatchett, D. W.; White, H. S. Electrochemistry of sulfur adlayers on the low-index faces of silver. *J Phys Chem* 1996, 100, 9854–9859.

Hedderman, T. G.; Keogh, S. M.; Chambers, G.; Byrne, H. J. In-depth study into the interaction of single walled carbon nanotubes with anthracene and *p*-terphenyl. *J Phys Chem B* 2006, 110, 3895–3901.

Ireland, J. C.; Klostermann, P.; Rice, E. W.; Clark, R. M. Inactivation of *Escherichia coli* by titanium dioxide photocatalytic oxidation. *Appl Environ Microbiol* 1993, 59, 1668–1670.

Jin, L.; Dai, B. TiO_2 activation using acid-treated vermiculite as a support: Characteristics and photoreactivity. *Appl Surf Sci* 2012, 258, 3386–3392.

Kamble, S. P.; Dixit, P.; Rayalu, S. S.; Labhsetwar, N. K. Defluoridation of drinking water using chemically modified bentonite clay. *Desalination* 2009, 249, 687–693.

Kanel, S. R.; Grenèche, J. M.; Choi, H. Arsenic(V) removal from groundwater using nano scale zero-valent iron as a colloidal reactive barrier material. *Environ Sci Technol* 2006, 40(6), 2045–2050.

Kemnetz, S.; Cody, C. A. Oil spill flocculating agent and method of remediating oil spills. US Patent 5558777, 1996.

Kim, J. S.; Kuk, E.; Yu, K. N.; Kim, J. H.; Park, S. J.; Lee, H. J.; Kim, S. H.; Park, Y. K.; Park, Y. H.; Hwang, C. Y.; Kim, Y. K.; Lee, Y. S.; Jeong, D. H.; Cho, M. H. Antimicrobial effects of silver nanoparticles. *Nanomed Nanotech Biol Med* 2007, 3, 95–101.

Kim, Y. C.; Lee, K. H.; Sasaki, S.; Hashimoto, K.; Ikebukuro, K.; Karube, I. Photocatalytic sensor for chemical oxygen demand determination based on oxygen electrode. *Anal Chem* 2000, 72, 3379–3382.

Lee, S. Y.; Kim, J. H.; Patel, R.; Im, S. J.; Kim, J. H.; Min, B. R. Silver nanoparticles immobilized on thin film composite polyamide membrane: Characterization, nanofiltration, antifouling properties. *Polym Adv Technol* 2007, 18, 562–568.

Li, J. M.; Meng, X. G.; Hu, C. W.; Du, J. Adsorption of phenol, *p*-chlorophenol and *p*-nitrophenol onto functional chitosan. *Bioresour Technol* 2009, 100, 1168–1173.

Li, Y. H.; Di, Z.; Ding, J.; Wu, D. H.; Luan, Z. K.; Zhu, Y. O. Adsorption thermodynamic, kinetic and desorption studies of pb^{2+} on carbon nanotubes. *Water Res* 2005, 39, 605–609.

Linsebigler, A. L.; Lu, G.; Yates, J. T. Photocatalysis on TiO_2 surfaces: Principles, mechanisms, and selected results. *Chem Rev* 1995, 95, 735–758.

Liu, Y.; Chen, X.; Li. J.; Burda, C. Photocatalytic degradation of azo dyes by nitrogen-doped TiO$_2$ nanocatalysts. *Chemosphere* 2005, 61, 11–18.

Livraghi, S.; Votta, A.; Paganini, M. C.; Giamello. E. The nature of paramagnetic species in nitrogen doped TiO$_2$ active in visible light photocatalysis. *Chem Commun* 2005, 4, 498–500.

Lo, I. Solidification/stabilization of phenolic waste using organic-clay complex. *J Environ Eng* 1996, 12, 850–855.

Makarova, O. V.; Rajh, T.; Thurnauer, M. C. Surface modification of TiO$_2$ nanoparticles for photochemical reduction of nitrobenzene. *Environ Sci Technol* 2000, 34, 4797–4803.

Makhluf, S.; Dror, R.; Nitzan, Y.; Abramovich, Y.; Jelinek, R.; Gedanken, A. Microwave assisted synthesis of nanocrystalline MgO and its use as a bacteriocide. *Adv Funct Mater* 2005, 15, 1708–1715.

Matsumura, Y.; Yoshikata, K.; Kunisaki, S. I.; Tsuchido, T. Mode of bactericidal action of silver zeolite and its comparison with that of silver nitrate. *Appl Environ Microbiol* 2003, 69, 4278–4281.

Mayo, J. T.; Yavuz, C.; Yean, S.; Cong, L.; Shipley, H.; Yu, W.; Falkner, J.; Kan, A.; Tomson, M.; Colvin, V. L. The effect of nanocrystalline magnetite size on arsenic removal. *Sci Technol Adv Mater* 2007, 8, 71–75.

McNally, B.; Singer, A.; Zhiliang, Y.; Yingjie, S.; Zhipeng, W.; Meller, A. Optical recognition of converted DNA nucleotides for single-molecule DNA sequencing using nanopore arrays. *Nano Lett* 2010, 10, 2237–2244.

Mills, A.; Davies, R. H.; Worsley, D. Water-purification by semiconductor photocatalysis. *Chem Soc Rev* 1993, 22, 417–425.

Mishra, P. C.; Patel, R. K. Removal of lead and zinc ions from water by low cost adsorbents. *J Hazard Mater* 2009, 168, 319–325.

Mostafavi, S. T.; Mehrnia, M. R.; Rashidi, A. M. Preparation of nanofilter from carbon nanotubes for application in virus removal from water. *Desalination* 2009, 238, 271–280.

Nahar, M. S.; Hasegawa, K.; Kagaya, S. Photocatalytic degradation of phenol by visible light-responsive iron-doped TiO$_2$ and spontaneous sedimentation of the TiO$_2$ particles. *Chemosphere* 2006, 65, 1976–1982.

Ni, M.; Leung, M. K. H.; Leung, D. Y. C.; Sumathy, K. A review and recent developments in photocatalytic water-splitting using TiO$_2$ for hydrogen production. *Renew Sust Energ Rev* 2007, 11, 401–425.

Noworyta, K.; Augustynski, J. Spectral photoresponses of carbon-doped TiO$_2$ film electrodes. *Electrochem Solid-State Lett* 2004, 7, E31–E33.

Oliveira, L. C. A.; Rios, R. V. R. A.; Fabris, J. D.; Sapag, K.; Garg, K.; Lago, R. M. Clay–iron oxide magnetic composites for the adsorption of contaminants in water. *Appl Clay Sci* 2003, 22, 169–177.

Otto, M.; Floyd, M.; Bajpai, S. Nanotechnology for site remediation. *Remediat J* 2008, 19, 99–108.

Pal, S.; Tak, Y. K.; Song. J. M. Does the antibacterial activity of silver nanoparticles depend on the shape of the nanoparticle? A study of the gram-negative bacterium *Escherichia coli*. *Appl Environ Microbiol* 2007, 73, 1712–1720.

Peng, X.; Luan, Z.; Ding, J.; Di, Z.; Li, Y.; Tian, B. Ceria nanoparticles supported on carbon nanotubes for the removal of arsenate from water. *Mater Lett* 2005, 59, 399–403.

Pereira, K. O.; Hanna, R. A. Brazilian organoclays as nanostructured sorbents of petroleum-derived hydrocarbons. *Mater Res* 2005, 8, 77–80.

Rao, G. P.; Lu, C.; Su, F. Sorption of divalent metal ions from aqueous solution by carbon nanotubes: A review. *Sep Purif Technol* 2007, 58, 224–231.

Shalini, C.; Pragnesh, N. D.; Shah, N. K. Applications of nano-catalyst in new era. *J Saudi Chem Soc* 2012, 16, 307–325.

Shrivastava, S.; Bera, T.; Roy, A.; Singh, G.; Ramachandrarao, P.; Dash, D. Characterization of enhanced antibacterial effects of novel silver nanoparticles. *Nanotechnology* 2007, 18, 225103.

Sobczyk-Guzenda, A.; Pietrzyk, B.; Szymanowski, H.; Gazicki-Lipman, M.; Jakubowski, W. Photocatalytic activity of thin TiO_2 films deposited using sol–gel and plasma enhanced chemical vapor deposition methods. *Ceram Int* 2013, 39, 2787–2794.

Son, W. K.; Youk, J. H.; Lee, T. S.; Park, W. H. Preparation of antimicrobial ultrafine cellulose acetate fibers with silver nanoparticles. *Macromol Rapid Comm* 2004, 25, 1632–1637.

Sondi, L.; Salopek-Sondi, B. Silver nanoparticles as antimicrobial agent: A case study on *E. coli* as a model for gram-negative bacteria. *J Colloid Interface Sci* 2004, 275, 177–182.

Srinivasan, R. Advances in application of natural clay and its composites in removal of biological, organic, and inorganic contaminants from drinking water. *Adv Mater Sci Eng* 2011, 872531, doi:10.1155/2011/872531

Srivastava, A.; Srivastava, O. N.; Talapatra, S.; Vajtai, R.; Ajayan, P. M. Carbon nanotube filters. *Nat Mater* 2004, 3, 610–614.

Theng, B. K. G.; Churchman, G. J.; Gates, W. P.; Yuan, G. Organically modified clays for pollution uptake and environmental protection. In *Soil mineral microbe-organic interactions*, Qiaoyun, H., PanMing, H., Antonio, V., Eds.; Springer: Heidelberg, 2006, Vol. 12, pp 850–855.

Theron, J.; Walker, J. A.; Cloete, T. E. Nanotechnology and water treatment: Applications and emerging opportunities. *Crit Rev Microbiol* 2008, 34, 43–69.

Upadhyayula, V. K. K.; Deng, S.; Mitchell, M. C.; Smith, G. B. Application of carbon nanotube technology for removal of contaminants in drinking water: A review. *Sci Total Environ* 2009, 408, 1–13.

USEPA. Nanotechnology and radiation clean up, 2013, http://www.epa.gov/radiation/clean-up/nanotechnology.html

Watlington, K. *Emerging nanotechnologies for site remediation and wastewater treatment.* U.S. Environmental Protection Agency: Washington DC, 2005.

Yan, W.; Herzing, A.; Kiely, C.; Zhang, W. Nanoscale zero-valent iron (nZVI): Aspects of the core-shell structure and reactions with inorganic species in water. *J Contam Hydrol* 2010, 118, 96–104.

Yan, X. M.; Shi, B. Y.; Lu, J. J.; Feng, C. H.; Wang, D. S.; Tang, H. X. Adsorption and desorption of atrazine on carbon nanotubes. *J Colloid Interface Sci* 2008, 321, 30–38.

Yang, K.; Zhu, L.; Xing, B. Adsorption of polycyclic aromatic hydrocarbons by carbon nanomaterials. *Environ Sci Technol* 2006, 40, 1855–1861.

Yuan, G. Natural and modified nanomaterials as sorbents of environmental contaminant. *J Environ Sci Health A* 2004, 39, 2661–2670.

Yuan, G.; Wu, L. Allophane nanoclay for the removal of phosphorus in water and wastewater. *Sci Technol Adv Mater* 2007, 8, 60.

Yuan, P.; Fan, M.; Yang, D.; He, H.; Liu, D.; Yuan, A.; Zhu, J.; Chen, T. Montmorillonite-supported magnetite nanoparticles for the removal of hexavalent chromium [Cr(VI)] from aqueous solutions. *J Hazard Mater* 2009, 166, 821–829.

Zadaka, D.; Nir, S.; Radian, A.; Mishael, Y. G. Atrazine removal from water by polycation-clay composites: Effect of dissolved organic matter and comparison to activated carbon. *Water Res* 2009, 43, 677–683.

Zhang, W. X. Nanoscale iron particles for environmental remediation: An overview. *J Nanopart Res* 2003, 5, 323–332.

Zhang, W.; Ding, Y.; Boyd, S. A.; Teppen, B. J.; Li, H. Sorption and desorption of carbamazepine from water by smectite clays. *Chemosphere* 2010, 81, 954–960.

Zhao, G.; Zhang, Q.; Fan, Q.; Ren, X.; Li, J.; Chen, C.; Huang, Y.; Wang, X. Sorption of copper(II) onto super-adsorbent of bentonite–polyacrylamide composites. *J Hazard Mater* 2010, 173, 661–668.

Zhao, Y.; Sugiyama, S.; Miller, T.; Miao, X. Nanoceramics for blood-borne virus removal. *Expert Rev Med Devices* 2008, 5, 395–405.

Zhu, B.; Zou, L. Trapping and decomposing of color compounds from recycled water by TiO_2 coated activated carbon. *J Environ Manage* 2009, 90, 3217–3225.

Zodrow, K.; Brunet, L.; Mahendra, S.; Li, D.; Zhang, A.; Li, Q.; Alvarez, P. J. Polysulfone ultrafiltration membranes impregnated with silver nanoparticles show improved biofouling resistance and virus removal. *Water Res* 2009, 43, 715–723.

CHAPTER 6

FUNGAL DEHALOGENATION: AN OVERVIEW

RAGHUNATH SATPATHY[1], VENKATA SAI BADIREENATH KONKIMALLA[2], and JAGNYESWAR RATHA[1]

[1]*School of Life Sciences, Sambalpur University, Jyoti Vihar, Burla, Odisha 768019, India*

[2]*School of Biological Sciences, National Institute of Science Education and Research (NISER), Bhubaneswar, Odisha 751005, India*

CONTENTS

6.1   INTRODUCTION

Microbes, such as fungi play a vital role in preserving the balance of global bio-geochemical cycling of organic components in our environment. The mechanism includes both synthesis and degradation of organic compounds in conjunction with their derivatives and intermediates (Madsen, 2011). Some of the fungi adapt to these organic chemicals in the environment and utilize them as a substrate for metabolism whereas some secrete extracellular enzymes to degrade the organic chemicals. This mechanism has been established by identification and characterization of specialized enzyme systems and metabolic pathways in different microorganisms of diverse habitats (Van Der Meer, 1997; Göbel et al., 2004). The genetic architecture of these microbes is established based on the gene sequence and their divergence pattern has been studied earlier. The study infers that due to presence of gene clusters, the microbes have developed different metabolic pathways for degradation of varied toxic organo-chemicals (Van Der Meer, 1992; Ding et al., 2012). This mechanism, by which the toxic organic compound is converted to non-toxic metabolite by a biological organism is referred to as bioremediation. One amongst the potential bioremediation technique is dehalogenation, where the microbe has the ability to degrade halogenated toxic organic compounds (Singh et al., 2004). The organo-halide compounds are among the largest group of environmental chemicals and also important intermediate substances of the global halogen cycle occurring in nature. Thus, the dehalogenation phenomena are grabbing attention for its great environmental significance (Gribble, 2003). Many types of microorganisms like algae, fungi, bacteria, and archea have been studied previously for the dehalogenation process and currently attempts are made to replace halogenated toxic chemicals by using the suitable one. Out of about 4000 natural halogenated compounds, the most common biological origin is generally chlorinated phenols and phenolic ethers, halogenated terpenes, chlorinated amino acids and peptides, halogenated alkaloids, bromo- and chloro-substituted pyrroles, chlorinated insole, halogenated thiophenes, chlorinated prostaglandins, and varied antibiotics (Hardman, 1991). However, the toxic effect exhibited by different organo-halide compounds, and their capability for bioaccumulation in the food chain, food web, and the creation of environmental contamination is of great concern in the current scenario. Several type of microorganism posses the dehalogenation potential by diverging their metabolism. Thus, it is all important to study and understand the diverse microbial impact (in molecular level) on the dehalogenation process, which would open the door to understanding more about the carbon–halogen cleaving process (Erable et al., 2006).

6.2 AVAILABLE ENZYME SYSTEMS AND BIOREACTOR CONSIDERATION IN FUNGI

Until now, little is understood about the role of yeasts and fungi in the deha-logenation process of the toxic pesticides as compared especially to bacteria. Basically the white rot fungi grow by hyphal extension through the soil and having an advantage for gaining better access to some of the pollutant chem-icals especially halogenated pesticides and herbicides (Atagana, 2004; D' Baldrian, 2008). Being aerobic in nature in case of fungi, the complete deg-radation of halogenated substance occurs through the Krebs cycle, but the best-known transformations take place through co-metabolism. However, different factors like soil type, pH, organic matter, fungal biomass, moisture, and aeration are also important equally that affect the process. The fungi induce specific dehalogenating enzymes, which is based on the adaptation to the halogenated pollutant of interest. The important biochemical reactions in the fungal degradation of organohalogens are alkylation, dealkylation, am-ide/ester hydrolysis, dehalogenation, dehydrogenation, hydroxylation, ether cleavage, ring cleavage, oxidation, reduction, condensation, and conjugate formation. There are four broad categories of enzymes and their response mechanism available in fungi as presented in Figure 6.1.

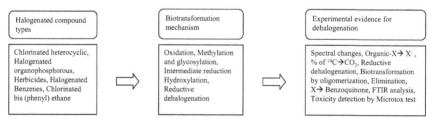

FIGURE 6.1   Different biotransformation strategy and methods used to detect halogenated metabolites during fungal dehalogenation.

6.2.1 PEROXIDASES

Peroxidases enzymes are equally well known as oxido-reductases, which uti-lizes the hydrogen peroxide to catalyze corresponding oxidation reactions. Many numbers of peroxide enzymes have been identified in fungal sources and characterized at the molecular level (Ana et al., 2000). The extracellular peroxidases of white rot fungi include three basic enzymes, such as lignin peroxidase (LiP), manganese peroxidase (MnP), and versatile peroxidase (VP). The general mechanism of the enzyme is given below:

Step 1: Enzyme (LiP, MnP, VP) + $H_2O_2 \rightarrow$
Intermediate Cationic Radical (Oxidized) + H_2O

Step 2: Intermediate Cationic Radical (Oxidized) + Substrate Oxidized $\rightarrow$
Intermediate (neutral) + substrate (reduced)

Step 3: Oxidized intermediate (neutral) + Substrate $\rightarrow$
Enzyme (LiP, MnP, VP) + Oxidized Product

These enzymes have been classified to the Class II fungal heme peroxidases and both LiPs and MnPs belong to a family of multiple isozymes coded by multiple genes (Cameron et al., 2000).

6.2.2 LACCASES

Laccase enzymes are widely distributed in fungi, higher plants, bacteria, and insects. More than 60 fungal strains, belonging to various classes, such as Ascomycetes, Basidiomycetes, and Deuteromycetes have been demonstrated to produce laccase (Higson, 1991). Laccase enzyme catalyzes the substrate by a specialized oxido-reductase activity via a mediator, schematically explained below (Fig. 6.2a).

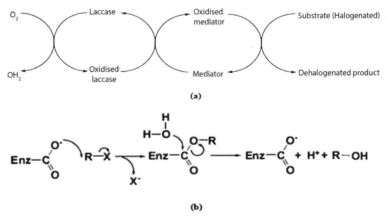

FIGURE 6.2 (a) Showing the general mechanism of laccase enzyme action; (b) Catalytic mechanism of halo acid dehalogenase.

Laccase activity was also noticed in the cultures of a wide range of fungi, from ascomycetes to basidiomycetes, and from wood and litter decomposing fungi to ectomycorrhizal fungi. The white rot fungus *Trametes pubescence* MB 89 is a source of the laccase production at the industrial level (Gianfreda et al., 1999; Muhammad et al., 2012).

6.2.3   HALO ACID DEHALOGENASE FROM FUNGI

The haloacid dehalogenase are the group of enzymes, which removes the halide group from haloacid. The general mechanism of this enzyme is represented in Figure 6.2b. Various works on the characterization and identification of bacterial haloacid dehalogenase enzyme have been performed and from some of their report a possible haloacid dehalogenase activity in fungi has been proposed (Ridder et al., 1997; Loubna et al., 2005; Papajak et al., 2006). Evidence of dehalogenation reaction from the experiment is normally obtained by radioactive labeling procedure of CO_2, O_2 in the microbial metabolism (Liu et al., 1995). As per Expasy data base records (www.expasy.org), currently there are about 228 numbers of predicted sequences of halo acid dehalogenase from different fungal sources.

6.2.4   P450 SYSTEM IN FUNGAL DEHALOGENATION

The cytochrome P450 consists of a large and diverse group of enzymes belongs to the super family of monooxygenases. These enzymes catalyze diverse reactions the organic substances include oxidation of primary and secondary metabolites also in xenobiotic detoxification (Chigu et al., 2010; Nazir et al., 2011). The enzymatic properties and substrate specificity are influenced by their redox partners. Filamentous fungi contain many numbers of cytochrome P450 often possess cytochrome P450 reductases as redox partners (Lah et al., 2011). Many studies have been done to distinguish the potential function of cytochrome P450-catalyzed dehalogenation of organo-halide compounds. Recently the dehalogenation process of *Polaromonas* sp. strain JS666 has been conformed on the substrate *cis*-1,2-dichoroethene (*c*DCE) and in the mechanism for *c*DCE degradation involvement of P450 monooxygenase has been established (Luke et al., 2001). The general mechanism of fungal cytochrome 450 enzyme system includes epoxidation of C=C double bonds followed by hydroxylation of the halo organic compounds.

Many toxic halogenated compounds like perchlorobenzenes and perfluoro-benzenes are converted to their non-toxic phenols and keto-forms.

$$\underset{\text{(Halogenated toxic substance)}}{\text{R-X}} \xrightarrow{\text{Epoxidation}} \underset{\text{(Epoxide derivatives)}}{\text{R-O-OX}} \xrightarrow{\text{Hydroxylation}} \underset{\text{(Hydroxylated non-toxic derivatives)}}{\text{R-O-H}}$$

Much considerable success has been accomplished by the applying the fungal cells (cultures) in several ways, such as immobilization method and utilizing them in different bioreactors. Different types of fungal bioreactors are established for the treatment of various halogenated pollutants (Table 6.1). The potential of certain fungi in the field of bioremediation can be raised by a number of elements, such as media composition, static and agitated culture condition, pH and temperature, C and N sources and salt concentration, initial concentration of halogenated pollutant, and so on.

TABLE 6.1 Showing Some Common Type of Bioreactor Systems Involving Fungal Species for Dehalogenation Purpose.

Bioreactor Types	Fungal Organism	Utility	References
Upflow column bioreactor	*Penicillium camemberti*	Treatment of paper and pulp effluent	Taeli et al. (2004)
Fixed film bioreactor	*P. chrysosporium*	Chlorinated low-molecular-weight phenols	Paszczynski et al. (1985)
(Immobilized) Cell entrapment	*Trametes versicolor*	AOX reduction of pulp and paper wastes	Pallerla and Chambers (1995)
(Immobilized) Adsorption	*Coriolus versicolor*	Monoaromatic chlorophenolics removal	Roy-Arcand and Archibald (1991)
Two-step sequential bioreactor	*Paecilomyces* sp.	AOX reduction	Singh et al. (2005)
Static flask culture	*P. chrysosporium*	Pentachlorophenol dehalogenation	Udayasooriyan et al. (2007)

Unlike fungi, bacteria are often unable to degrade substituted molecules due to low water solubility and availability. Fungi possess a wide range of enzyme systems and most of them being extracellular and non-specific, hence capable for degradation of a large group of chemicals. Fungi are cosmopolitan in nature and they are capable of degrading a list of commonly used halo-organic pollutants individually or in mixtures (Table 6.2).

TABLE 6.2 Example of Fungal Sources that Degrade Halogenated Pollutants with Proposed Mechanism.

Fungal Sources	Act on Halogenated Substances	Proposed Mechanism	References
Fusarium solani AM203, *Botrytis cinerea* AM235, *Beauveria bassiana* AM278	Halolacetones	Hydrolytic dehalogenation	Johnson (2009)
Phanerochaete chrysosporium, Trametes versicolor, Inonotus dryophilus	Pentachlorophenol	Oxidation-reduction	Liu and Zhao (2004), Krupp et al. (2006)
Pycnoporus cinnabarinus	2-hydroxy-5-chloro-biphenyl	Laccase mediated dehalogenation	Bansal (2005)
Candida maltose	Monochlorophenol	Cycloisomerization of the *cis, cis* chloromuconic acid	Moore et al. (2010)
Pleurotus ostreatus, Trametes versicolor	Polychlorinated biphenyls	Laccase mediated dehalogenation	Zhang et al. (2006)
Mucor sp., *Trichoderma* sp.	3-chloropropionic acid (3CP)	Haloacid dehalogenation	Barr and Aust (1994)
Trametes elegans, Phlebia radiata, Panus crinitus, Trametes villosa	Pulp and textile industry waste waters	Partial demethylation	Grabarczyk (2012)
Phanerochaete sordid, Trametes versicolor, Bjerkandera sp. BOS55	Industrial waste water with halogenated substances	Laccase and lignin peroxidases	Mileski et al. (1988)
Mucor plumbeus	pentachlorophenol (PCP)	Oxidation followed by dehalogenation	Alleman et al. (1995)
Pleurotus pulmonarious, Phanerochaete chrysosporium	Atrazine	Hydroxylation	Schultz et al. (2001)
Phelbia tramellosa	Alachlor	Oxidation	Elke et al. (1992)
Dichomitus squalens	2,4-dichlorophenoxyacetic acid	Mn^{2+} mediated hydration	Zeddel et al. (1993)
Stachybotrys sp. strain DABAC 3	2,6-dichloroaniline	Laccase and lignin peroxidises activity	Sepideh et al. (2013)
Penicillium camemberti	Lindane	-	Peralta-Zamora et al. (1999)
Penicillium frequentans Strain Bi 7/2	3,4-dichlorophenol, 2,4-dichlorophenol	Oxidative dehalogenation	Ed de and Jim (1997)
Trametes versicolor	Perchloroethylene (PCE)	Cytochrome P-450 system	Carvalho et al. (2011)

TABLE 6.2   *(Continued)*

Fungal Sources	Act on Halogenated Substances	Proposed Mechanism	References
Phanerochaete chrysosporium	TCE	Lignin peroxidase enzyme	Mougin et al. (1997)
Trichoderma sp. Gc1 and *Penicillium miczynskii* Gc5	DDD	-	Ferrey et al. (1994)
Caldariomyces fumago	4-Fluorophenol	Chloroperoxidases enzyme system	Reddy et al. (1997)
Phlebia brevispora TMIC33929	Coplanar polychlorinated biphenyls (Co-PCBs)	-	Annibale et al. (2006)

6.3   INTERDISCIPLINARY APPROACH TOWARD FUNGAL DEHALOGENATION

Along with traditional molecular level analysis, various modern and novel methodologies have been implemented in the recent studies on the microbial dehalogenation mechanisms and the same can be utilized for in-depth study about fungi. Starting from establishment of fungal culture to characterize it as an dehalogenating agent have provided key knowledge for understanding the catalytic mechanism and engineering of bio-dehalogenation process (Fig. 6.3). Especially, the structural study of the enzymes provides the path to trace out the specificities and activities involved in this type of bio-catalytic mechanism (Hlavica, 2013). There are several aspects, in which the study about dehalogenation couples the approaches including

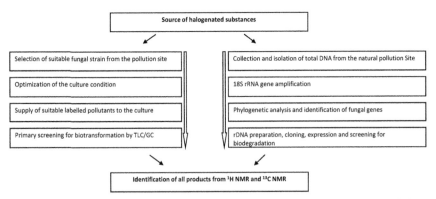

FIGURE 6.3   Approaches from identification to characterization of a known (*left*) or unknown (*right*) fungal species as an acceptable dehalogenating microbe.

metagenomics, which is a culture-independent technique for analysis of the genetic and metabolic potential of natural and model microbial communities that degrade organic pollutants (Nishino et al., 2013). In addition to this, the proteomics has also having significant contribution to understand the individual organisms at the molecular level thereby finding various protein functions with respect to the responses (Alcalde et al., 2006).

During the last few years, the field of dehalogenation has been seen in a rapid progress in the identification of novel fungi as well as gene, sequencing of whole genomes, and the genetic adaptability potential. Hence, combined effort of microbiology, biotechnology, and bioinformatics is required to analyze the fungal dehalogenation systems. Although a large number of fungi as well as their enzymes have been isolated, identified, and studied, however, there is a little knowledge available about the genes encoding these enzymes. From the advance sequencing platforms, there is a substantial amount of sequence information is available in the public database, that include the gene sequences, EST, whole genomes, proteome, metagenome, and so on. This mass of data provides a bigger opportunity in bioinformatics to annotate and evaluate the novel information associated with them (Peralta-Zamora et al., 1999; Pöritz et al., 2013). The development of a large number of databases, software tools to analyze data in the past few years has facilitated the researcher for a wide range of applications to understand the microbial dehalogenation process (Fig. 6.4). Hence the interdisciplinary methodology is truly essential to determine the metabolic potential as well as the diversity in fungal systems to understand better the complex molecular regulation and control in them (Field, 2003; Seifert, 2009; Carvalho et al., 2011).

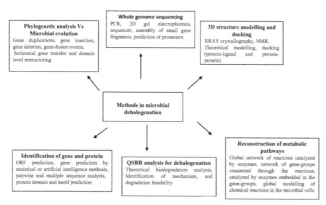

FIGURE 6.4 Various interdisciplinary strategies applied in molecular data to explore dehalogenation process.

6.4 CONCLUSION AND FUTURE PERSPECTIVES

Biological treatment strategies for the bioremediation of organo-halide pollutants are widely utilized as it is eco-friendly and also linked with low-cost treatment methods. Therefore, several researchers are curious about the development of suitable fungal technology for the biodegradation of halogenated pollutants. Filamentous fungi, including zygomycetes, ascomycetes, and basidiomycetes (especially white rot fungi) have been shown to be potentially useful to degrade recalcitrant xenobiotics (Kordon et al., 2010). In the bioremediation process like dehalogenation, the bacterial consortia are mostly come into focus and fungi are much less studied. In actual there is a greater potential of dehalogenation mechanism exists in fungi in comparison to bacteria, due to their rapid growth, capable of more biomass production and aggressive hyphal growth in soil. Hence, more research should be focused on fungal flora for dehalogenation purpose.

The preliminary and advanced knowledge about established molecular structures of a particular enzyme and its products are essential to study the carbon–halogen bond making and bond breaking catalyzing process. To date there is no three-dimensional (3D) experimental structure available for halo acid dehalogenase enzyme from fungal source in protein data bank (PDB, www.rcsb.org/pdb). Hence conducting research in elucidating of structure of novel proteins and understanding their function and mechanism would be of prime importance.

Genetic engineering technology is another approach normally practiced to produce recombinant genes coding for suitable enzymes with a strategy of their large-scale production. This strategy could be accustomed to produce dehalogenating enzymes from fungi to study the synthesis and degradation kinetics of the enzymes (Pieper et al., 2004).

Identification of new fungal strain from microbial community, discovery of specific genetic and metabolic pathways followed by protein and metabolic engineering approach will generate enzymes with altered biological properties for biodegradation of harmful halogenic substances (Schneider et al., 2010; Yin et al., 2011).

Halogenated products are used in the synthesis of many pesticides and pharmaceuticals. These substances also create a risk to the health of the living organism; can be degradable by a process called dehalogenation. The fungi are considered as a suitable organism for degradation of halogenated organic substances. Many new fungal strains with efficient potential are isolated and their metabolic pathways in the process have been elucidated. Starting from the discovery, development of the novel environmental biocatalysis attracts several scientists to pursue research on this. Hence, a thorough interdisciplinary analysis about the genetic, enzyme system, and the mechanism of induction of the enzymes would make fungi as a great chunk of resource material for the future research and industry.

KEYWORDS

- **Bioinformatics**
- **Bioreactor**
- **Bioremediation**
- **Dehalogenase**
- **Dehalogenation**
- **Enzyme system**
- **Fungal source**
- **Halogenated substances**
- **Metabolism**
- **Pollutants**
- **Xenobiotics**

REFERENCES

Alcalde, M.; Ferrer, M.; Plou, F. J.; Ballesteros, A. Environmental biocatalysis: From remediation with enzymes to novel green processes. *Trends Biotechnol* 2006, 24, 281–287.

Alleman, B. C.; Logan, B. E.; Gilbertson, R. L. Degradation of pentachlorophenol by fixed films of white rot fungi in rotating tube bioreactors. *Water Res* 1995, 29, 61–67.

Ana, C.; Cees, A. M. J. J.; Van den, H.; Peter, J. P. Studies on the production of fungal peroxidases in *Aspergillus niger*. *Appl Environ Microbiol* 2000, 66, 3016–3023.

Annibale, D.; Rosetto, F.; Leonardi, V.; Federici, F.; Petruccioli, M. Role of autochthonous filamentous fungi in bioremediation of a soil historically contaminated with aromatic hydrocarbons. *Appl Microbiol Biotechnol* 2006, 72, 28–36.

Atagana, H. J. Biodegradation of phenol, *o*-cresol, *m*-cresol and *p*-cresol by indigenous soil fungi in soil contaminated with cresolate. *World J Microbiol Biotechnol* 2004, 20, 851–858.

Bansal, A. K. Bioinformatics in microbial biotechnology – a mini review. *Microb Cell Fact* 2005, 4, 19.

Barr, D. P.; Aust, S. D. Pollutant degradation by white rot fungi. In *Reviews of environmental contamination and toxicology*, Ware, G. W., Ed.; Springer: New York, 1994, pp 49–72.

Cameron, M. D.; Timofeevski, S.; Aust, S. D. Enzymology of *Phanerochaete chrysosporium* with respect to the degradation of recalcitrant compounds and xenobiotics. *Appl Microbiol Biotechnol* 2000, 54, 751–758.

Carvalho, M. B.; Tavares, S.; Medeiros, J.; Núñez, O.; Gallart-Ayala, H.; Leitão, M. C.; Galceran, M. T.; Hursthouse, A.; Pereira, C. S. Degradation pathway of pentachlorophenol by *Mucor plumbeus* involves phase II conjugation and oxidation-reduction reactions. *J Hazard Mater* 2011, 198, 133–142.

Chigu, N. L.; Hirosue, S.; Nakamura, C.; Teramoto, H.; Ichinose, H.; Wariishi, H. Cytochrome p450 monooxygenases involved in anthracene metabolism by the white-rot basidiomycete *Phanerochaete chrysosporium*. *Appl Microbiol Biotechnol* 2010, 87, 1907–1916.

D' Baldrian, P. Wood-inhabiting ligninolytic basidiomycetes in soils: Ecology and constraints for applicability in bioremediation. *Fungal Ecol* 2008, 1, 4–12.

Ding, C.; He, J. Molecular techniques in the biotechnological fight against halogenated compounds in anoxic environments. *Microb Biotechnol* 2012, 5, 347–367.

Ed de, J.; Jim, A. F. Biosynthesis and biodegradation of organohalogens by basidiomycetes. *Annu Rev Microbiol* 1997, 51, 375–414.

Elke, P.; Helmut, K.; Heiko, F.; Hofmann, K. H. Degradation and dehalogenation of monochlorophenols by the phenol-assimilating yeast *Candida maltosa*. *Biodegradation* 1992, 2, 193–199.

Erable, B.; Goubet, I.; Lamare, S.; Legoy, M. D.; Maugard, T. Bioremediation of halogenated compounds: Comparison of dehalogenating bacteria and improvement of catalyst stability. *Chemosphere* 2006, 65, 1146–1152.

Ferrey, M. L.; Koskinen, W. C.; Blanchette, R. A.; Burnes, T. A. Mineralization of alachlor by lignin-degrading fungi. *Can J Microbiol* 1994, 40, 795–798.

Field, J. A. Biodegradation of chlorinated compounds by white rot fungi. In *Dehalogenation: Microbial processes and environmental applications*, Häggblom, M., Bossert, I. D., Eds.; Springer: New York, 2003, pp 159–204.

Gianfreda, L.; Xu, F.; Bollag, J. M. Laccases: A useful group of oxidoreductive enzymes. *Biorem J* 1999, 3, 1–26.

Göbel, M.; Kranz, O. H.; Kaschabek, S. R.; Schmidt, E.; Pieper, D. H.; Reineke, W. Microorganisms degrading chlorobenzene via a meta-cleavage pathway harbor highly similar chlorocatechol 2,3-dioxygenase-encoding gene clusters. *Arch Microbiol* 2004, 182, 147–156.

Grabarczyk, M. Fungal strains as catalysts for the biotransformation of halolactones by hydrolytic dehalogenation with the dimethylcyclohexane system. *Molecules* 2012, 17, 9741–9753.

Gribble, G. W. The diversity of naturally produced organohalogens. *Chemosphere* 2003, 52, 289–297.

Hardman, D. J. Biotransformation of halogenated compounds. *Crit Rev Biotechnol* 1991, 11, 1–40.

Higson, F. K. Degradation of xenobiotics by white rot fungi. *Rev Environ Contam Toxicol* 1991, 122, 111–152.

Hlavica, P. Evaluation of structural features in fungal cytochromes P450 predicted to rule catalytic diversification. *Biochim Biophys Acta* 2013, 1834, 205–220.

Johnson, A. D. Single-nucleotide polymorphism bioinformatics: a comprehensive review of resources. *Circ Cardiovasc Genet* 2009, 2, 530–536.

Kordon, K.; Mikolasch, A.; Schauer, F. Oxidative dehalogenation of chlorinated hydroxybiphenyls by laccases of white-rot fungi. *Int Biodeter Biodegr* 2010, 64, 203–209.

Krupp, M.; Weinmann, A.; Galle, P. R.; Teufel, A. Actin binding LIM protein 3 (abLIM3). *Int J Mol Med* 2006, 17, 129.

Lah, L.; Podobnik, B.; Novak, M.; Korošec, B.; Berne, S.; Vogelsang, M.; Kraševec, N.; Zupanec, N.; Stojan, J.; Bohlmann, J.; Komel, R. The versatility of the fungal cytochrome p450 monooxygenase system is instrumental in xenobiotic detoxification. *Mol Microbiol* 2011, 81, 1374–1389.

Liu, J. Q.; Kurihara, T.; Miyagi, M.; Esaki, N.; Soda, K. Reaction mechanism of L-2-haloacid dehalogenase of *Pseudomonas* sp. YL. Identification of Asp10 as the active site nucleophile by ^{18}O incorporation experiments. *J Biol Chem* 1995, 270, 18309–18312.

Liu, Y.; Zhao, H. A computational approach for ordering signal transduction pathway components from genomics and proteomics data. *BMC Bioinformatics* 2004, 5, 158.

Loubna, Y.; Thomas, J. S.; Javier, A. The *Neurospora crassa* gene responsible for the cut and ovc phenotypes encodes a protein of the haloacid dehalogenase family. *Mol Microbiol* 2005, 55, 828–838.

Luke, A. K; Burton, S. G. A novel application for *Neurospora crassa*: Progress from batch culture to a membrane bioreactor for the bioremediation of phenols. *Enzyme Microb Technol* 2001, 29, 348–356.

Madsen, E. L. Microorganisms and their roles in fundamental biogeochemical cycles. *Curr Opin Biotechnol* 2011, 22, 456–464.

Mileski, G. J.; Bumpus, J. A.; Jurek, M. A.; Aust, S. D. Biodegradation of pentachlorophenol by the white rot fungus *Phanerochaete chrysosporium. Appl Environ Microbiol* 1988, 54, 2885–2889.

Moore, J. H.; Asselbergs, F. W.; Williams, S. M. Bioinformatics challenges for genome-wide association studies. *Bioinformatics* 2010, 26, 445–455.

Mougin, C.; Laugero, C.; Asther, M.; Chaplain, V. Biotransformation of s-triazine herbicides and related degradation products in liquid cultures by the white rot fungus *Phanerochaete chrysosporium. Pestic Sci* 1997, 49, 169–177.

Muhammad, I.; Asad, M. J.; Hadr, S. H.; Sajid, M. Production and industrial applications of laccase enzyme. *J Cell Mol Biol* 2012, 10, 1–11.

Nazir, K. N. H.; Ichinose, H.; Wariishi, H. Construction and application of a functional library of cytochrome P450 monooxygenases from the filamentous fungus *Aspergillus oryzae. Appl Environ Microbiol* 2011, 77, 3147–3150.

Nishino, S. F.; Shin, K. A.; Gossett, J. M.; Spain, J. C. Cytochrome p450 initiates degradation of *cis*-dichloroethene by *Polaromonas* sp. strain JS666. *Appl Environ Microbiol* 2013, 79, 2263–2272.

Pallerla, S.; Chambers, R. P. Continuous decolorization and AOX reduction of bleach plant effluents by free and immobilized *Trametes versicolor. J Environ Sci Health A* 1995, 30, 423–437.

Papajak, E.; Kwiecień, R. A.; Rudziński, J.; Sicińska, D.; Kamiński, R.; Szatkowski, Ł.; Kurihara, T.; Esaki, N.; Paneth, P. Mechanism of the reaction catalyzed by DL-2-haloacid dehalogenase as determined from kinetic isotope effects. *Biochemistry* 2006, 45, 6012–6017.

Paszczynski, A.; Huynh, V. B.; Crawford, R. L. Enzymatic activities of an extracellular, manganese-dependent peroxidase from *Phanerochaete chrysosporium. FEMS Microbiol Lett* 1985, 9, 37–41.

Peralta-Zamora, P.; Kunz, A.; de Moraes, S. G.; Pelegrini, R.; Moleiro, P. D.; Reyes, J.; Duran, N. Degradation of reactive dyes - I. A comparative study of ozonation, enzymatic and photochemical processes. *Chemosphere* 1999, 38, 835–852.

Peralta-Zamora, P.; Kunz, A.; de Moraes, S. G.; Pelegrini, R.; Moleiro, P. D.; Reyes, J.; Duran, N. Degradation of reactive dyes - I. A comparative study of ozonation, enzymatic and photochemical processes. *Chemosphere* 1999, 38, 835–852.

Pieper, D. H.; Martins dos Santos, V. A.; Golyshin, P. N. Genomic and mechanistic insights into the biodegradation of organic pollutants. *Curr Opin Biotechnol* 2004, 15, 215–224.

Pöritz, M.; Goris, T.; Wubet, T.; Tarkka, M. T.; Buscot, F.; Nijenhuis, I.; Adrian, L. Genome sequences of two dehalogenation specialists – *Dehalococcoides mccartyi* strains BTF08 and DCMB5 enriched from the highly polluted Bitterfeld region. *FEMS Microbiol Lett* 2013, 343, 101–104.

Reddy, G. V. B.; Joshi, D. K.; Gold, M. H. Degradation of chlorophenoxyacetic acids by the lignin-degrading fungus *Dichomitus squalens*. *Microbiology* 1997, 143, 2353–2360.

Ridder, I. S.; Rozeboom, H. J.; Kalk, K. H.; Janssen, D. B.; Dijkstra, B. W. Three-dimensional structure of l-2-haloacid dehalogenase from *Xanthobacter autotrophicus* GJ10 complexed with the substrate-analogue formate. *J Biol Chem* 1997, 272, 33015–33022.

Roy-Arcand, L.; Archibald, F. S. Comparison and combination of ozone and fungal treatments of a kraft bleachery effluent. *Tappi J* 1991, 74, 211–218.

Schneider, T.; Riedel, K. Environmental proteomics: Analysis of structure and function of microbial communities. *Proteomics* 2010, 10, 785–798.

Schultz, A.; Jonas, U.; Hammer, E.; Schauer, F. Dehalogenation of chlorinated hydroxybiphenyls by fungal laccase. *Appl Environ Microbiol* 2001, 67, 4377–4381.

Seifert, K. A. Progress towards DNA barcoding of fungi. *Mol Ecol Resour* 2009, 9, 83–89.

Sepideh, P.; Tengku, H.; Tengku, A. H.; Fahrul, H. Molecular identification and biodegradation of 3-chloropropionic acid (3CP) by filamentous fungi - *Mucor* and *Trichoderma* species isolated from UTM agricultural land. *Malays J Microbiol* 2013, 9, 120–124.

Singh, A.; Ward, O. P. Biotechnology and bioremediation - an overview. In *Biodegradation and bioremediation,* Singh, A., Ward, O. P., Eds.; Springer: Berlin, 2004, pp 1–17.

Singh, P.; Thakur, I. S. Colour removal of anaerobically treated pulp and paper mill effluent by microorganisms in two steps bioreactor. *Bioresour Technol* 2005, 97, 218–223.

Taeli, B. K.; Gökçay, C. F.; Taeli, H. Upflow column reactor design for dechlorination of chlorinated pulping wastes *Penicillium camemberti*. *J Environ Manage* 2004, 72, 175–179.

Udayasooriyan, C.; Prabhu, P. C.; Balasubramanian, G. Degradation of pentachlorophenol by white rot fungus (*Phanerochaete chrysosporium*) TL1 grown in ammonium lingosulphonate media. *Biotech* 2007, 6, 76–80.

Van Der Meer, J. R. Evolution of novel metabolic pathways for the degradation of chloroaromatic compounds. *Antonie van Leeuwenhoek* 1997, 71, 159–178.

Van Der Meer, J. R.; De Vos, W. M.; Harayama, S.; Zehnder, A. J. Molecular mechanisms of genetic adaptation to xenobiotic compounds. *Microbiol Rev* 1992, 56, 677–694.

Yin, X.; Chen, Y.; Zhang, L.; Wang, Y.; Zabriskie, T.M. Enduracidin analogues with altered halogenation patterns produced by genetically engineered strains of *Streptomyces fungicidicus*. *J Nat Prod* 2010, 73, 583–589.

Zeddel, A.; Majcherczyk, A.; Hüttermann, A. Degradation of polychlorinated biphenyls by white-rot fungi *Pleurotus ostreatus* and *Trametes versicolor* in a solid state system. *Toxicol Environ Chem* 1993, 40, 255–266.

Zhang, S.; Golbraikh, A.; Tropsha, A. Development of quantitative structure-binding affinity relationship models based on novel geometrical chemical descriptors of the protein-ligand interfaces. *J Med Chem* 2006, 49, 2713–2724.

CHAPTER 7

INSIGHT OF BIOFUEL PROSPECTS FROM MICROALGAE AS RENEWABLE ENERGY SOURCE FOR ENVIRONMENTAL SUSTAINABILITY

GANAPATHI SIBI

Department of Biotechnology, Centre for Research and Post Graduate Studies, Indian Academy Degree College, Bangalore 560043, Karnataka, India

CONTENTS

7.1 INTRODUCTION

Around 70% of the total global energy requirement is represented by fuels, and finding sufficient supplies of energy for the future is one of the most daunting challenges. Fluctuating oil prices, increasing gaseous emissions and their effect on green house, climatic change, and global warming stresses the urgent need to find new feedstocks for fuels. Biofuels promote environmental sustainability in terms of non-toxic, clean energy with a consequent decrease in greenhouse effect. Microalgae offer great promise to contribute a significant portion of the renewable fuels due to their higher photosynthetic activity, biomass productivity, CO_2 fixation, O_2 production, faster growth rate than higher plants, and ability to grow in non-arable land unsuitable for agricultural purposes. The properties of biodiesel from microalgal oil in terms of density, viscosity, acid value, and heating value are comparable to those of fossil fuels. However, prior to industrial scale application, a series of key challenges have to be resolved. The capital costs to transform microalgal biomass into biofuel are high and face a broad range of grand challenges to become technologically and economically viable. A systematic approach and process integration are critical factors in a successful future for algal bio-refineries. This chapter presents views and opinions on key technical challenges associated with microalgal culture system, cultivation conditions, growth medium, strain selection for highest growth rate, increased biomass productivity integrated with CO_2 capture, lipid accumulation with adequate composition, and extraction methods.

7.2 OPEN AND CLOSED CULTURE SYSTEMS

The selection of culture system for microalgal biofuel production can be in either open or closed systems. Under open system, microalgae are mass cultured in artificial circular, open ponds, and cascades from long time (Richmond, 1986; Becker, 1994). Open ponds can be further categorized as raceway, circular, inclined, and unmixed ponds. Raceway ponds are the most applicable for both the pilot-study level and commercial scale because of their easy and higher productivity (Borowitzka, 2005; Putt et al., 2011). Circular ponds and inclined ponds are capable of achieving algal growth rates as high as 21 g m^{-2} d^{-1} (Benemann and Oswa, 1993) and 31 g m^{-2} d^{-1} (Doucha et al., 2006), respectively. Unmixed open algal ponds are generally avoided due to the low productivities and suitable for only selected algal species.

Higher biomass production and lesser energy requirement are the advantages of open system (Hase et al., 2000; Jorquera et al., 2010). At the same time, open systems require large area of land and are continuously threatened by invading species, such as undesired algae and bacteria. Monocultures of algae under open culture system are achieved by maintaining an extreme culture environment, such as high pH, salinity, and nutritional status (Lee, 1986). However, such approaches do not necessarily exclude bacteria, zooplankton, and other biological contaminants. Another disadvantage of open system is that local weather conditions influence the cultivation conditions, which is further hardly controlled and makes the production seasonal (Perez-Garcia et al., 2011). Further, lower atmospheric CO_2 concentration could slow down the cell growth of microalgae, as there is a poor mass transfer limitation in open systems. Due to these constraints, it is difficult to maintain monocultures of desired algae in open culture systems (Lee, 2001), which have led to the development of enclosed tubular (Gudin and Chaumont, 1983; Pirt et al., 1983; Robinson, 1987; Tredici and Materassi, 1992; Lee et al., 1995; Borowitzka, 1999) and flat bed (Tredici et al., 1991; Pulz, 1994; Hu et al., 1996) photo bioreactors.

The design of a photobioreactor is more complicated compared to an open pond and one should consider the fundamental principles, such as light regime, gas liquid mass transfer, nutrient supply, and oxygen removal system. Irradiance supply influences the cell growth, CO_2 fixation, and biochemical composition of microalgae (Chrismadha and Borowitzka, 1994). Reducing the light path increases the light available to each cell and increases growth rate. Poor mixing of gas liquid mass transfer leads to increased O_2 concentration and CO_2 stripping in the photobioreactor, which inhibits the growth of the microalgae (Hoekema et al., 2002). However, cultivation in closed systems is costlier compared with open ponds, which include light illumination, CO_2 feed, cultivation medium, and circulator system. In return, microalgal productivity in a photobioreactor is higher and has less contamination (Wu et al., 2005).

7.3   BATCH AND CONTINUOUS CULTURE

Batch culture of microalgae is defined as a culture period, where the cultivated microalgal cells are harvested at once and continuous culture is conducted based on the dilution rate. Growth rates and biomass can be regulated and maintained for extended time periods by varying the dilution rate. Higher microalgal production can be achieved in the continuous culture mode compared with batch culture. McGinn et al. (2012) found that the biomass productivity of *Scenedesmus* sp. was two times greater when growing in a continuous chemostat than in batch cultivation. Similar findings were obtained by Wen et al. (2014) by cultivating *Chlorella pyrenoidosa* in the continuous culture under varying nitrate conditions. Further, maximum lipid productivity of 144.93 mg L^{-1} d^{-1} when compared to 96.28 mg L^{-1} d^{-1} in batch culture during the study. Under semi-continuous cultivation with nitrogen limitation and pH regulation, 3.64-fold higher lipid productivity was obtained in *C. pyrenoidosa* (Han et al., 2013). In another study, excess light combined with a growth-limiting nitrogen supply resulted in up to 12.4% w/w triacyl glycerol (TAG) accumulation in turbidostat cultures of *Neochloris oleoabundans* (Klok et al., 2013). In contrast, Tang et al. (2012) reported that varying dilution rates has increased biomass productivity of *Chlorella minutissima* and *Dunaliella tertiolecta* but had no effect on lipid productivity. Sobczuk and Chisti (2010) found that lipid content in *Choricystis minor* did not change significantly with various dilution rates in chemostat culture.

7.4   CULTIVATION STRATEGIES

Under optimal growth conditions, microalgae synthesize fatty acids for esterification into glycerol based membrane lipids, which make up about 5–20% of their dry cell weight. Under stress conditions, algae have been found to produce significantly higher concentrations of neutral lipids. The ability of microalgae to adapt their metabolism to varying cultural conditions provides opportunities to modify or control the formation of targeted compounds. Various systems are used for production of microalgae, which can be autotrophic, mixotrophic, and heterotrophic. Heterotrophic cultivation has been known for decades, as it is regarded as the most practical and promising way to promote productivity of biomass and high levels of lipids (Grima et al., 2003; Olaizola, 2003; Miao and Wu, 2006). Under autotrophic and heterotrophic cultivation, light penetration is inversely proportional to the cell concentration and algal biomass hence, light limitation is the major factor during microalgal cultivation (Chen, 1996). Another method of microalgal cultivation for higher biomass and lipid production is two-stage cultivation. In this strategy, microalgal cells first grow rapidly under growth-optimized conditions, and then are transferred to conditions where light irradiance (Zhang et al., 2009), nutrition (Su et al., 2011), culture pH (Han et al., 2013), as well as other factors (Liu et al., 2008; Das et al., 2011) are adjusted to promote lipid accumulation at the expense of cell growth. It is possible to increase both biomass and lipid production in microalgae by choosing optimum cultural conditions and nutrient composition of the growth medium (Sibi, 2015a).

7.5 SELECTION OF MICROALGAL SPECIES

Microalgae are good option for biodiesel production than terrestrial crops. However, selection of the most adequate species needs to take into account for productivity and economic viability. Factors, such as the ability of microalgae to develop using the nutrients available or under specific environmental conditions, fatty acid composition of the different microalgae species should be considered simultaneously in the selection of the most adequate species or strains for biodiesel production. *Chlorella, Crypthecodinium, Cylindrotheca, Dunaliella, Isochrysis, Nannochloris, Nannochloropsis, Neochloris, Nitzschia, Phaeodactylum, Porphyridium, Scenedesmus, Schizochytriu,* and *Tetraselmis* have oil levels between 20 and 50% and should be considered for biodiesel production (Table 7.1).

TABLE 7.1 Lipid Content of Microalgal Species.

Microalgal Species	Lipid Content (% dry weight biomass)	Lipid Productivity (mg L^{-1} day^{-1})
Ankistrodesmus sp.	24.0–31.0	–
Botryococcus braunii	25.0–75.0	–
Chaetoceros muelleri	33.6	21.8
Chaetoceros calcitrans	14.6–16.4/39.8	17.6
Chlorella emersonii	25.0–63.0	10.3–50.0
Chlorella minutissima	57	
Chlorella protothecoides	14.6–57.8	1214
Chlorella sorokiniana	19.0–22.0	44.7
Chlorella vulgaris	14–40/56	11.2–40.0
Chlorella pyrenoidosa	2.0	–
Chlorococcum sp.	19.3	53.7
Crypthecodinium cohnii	20.0–51.1	–
Dunaliella salina	6.0–25.0	116.0
Dunaliella primolecta	23.1	–
Dunaliella tertiolecta	16.7–71.0	–

TABLE 7.1　*(Continued)*

Microalgal Species	Lipid Content (% dry weight biomass)	Lipid Productivity (mg L^{-1} day^{-1})
Dunaliella bioculata	8.0	
Dunaliella salina	14–20	33.5
Ellipsoidion sp.	27.4	47.3
Euglena gracilis	14.0–20.0	–
Haematococcus pluvialis	25.0	–
Isochrysis galbana	7.0–40.0	–
Monodus subterraneus	16.0	30.4
Monallanthus salina	20.0–22.0	–
Nannochloropsis oculata	22.7–29.7	84.0–142.0
Nannochloropsis sp.	12.0–53.0	37.6–90.0
Neochloris oleoabundans	29.0–65.0	90.0–134.0
Nitzschia sp.	16.0–47.0	
Oocystis pusilla	10.5	–
Pavlova salina	30.9	49.4
Pavlova lutheri	35.5	40.2
Phaeodactylum tricornutum	18.0–57.0	44.8
Porphyridium cruentum	9.0–18.8/60.7	34.8
Scenedesmus dimorphus	6–40	
Scenedesmus obliquus	11.0–55.0	–
Scenedesmus quadricauda	1.9–18.4	35.1
Skeletonema costatum	13.5–51.3	17.4
Spirulina platensis	4.0–16.6	–
Spirulina maxima	4.0–9.0	–
Thalassiosira pseudonana	20.6	17.4
Tetraselmis suecica	8.5–23.0	27.0–36.4

Adapted from Becker (1994), Illman et al. (2000), Miao and Wu (2006), Spolaore et al. (2006), Liu et al. (2008), Natrah et al. (2008), Xiong et al. (2008), and Mata et al. (2010).

7.6　ENVIRONMENTAL PARAMETERS AFFECTING LIPID PRODUCTION

Different nutritional and environmental factors, cultivation conditions, and growth phases may affect the fatty acid composition. Lipids act as a secondary metabolite in microalgae, maintaining specific membrane functions, and cell signaling pathways while responding to the environment changes. The quantity and quality of oils produced by algal cells are directly proportional to the stimulus received from the surroundings. Stressful environmental conditions change the use of carbon uptake by algae for proliferation to energy storage in the form of oil. The lipid content, composition, and the proportions of various fatty acids of microalgae vary according to the environmental or culturing variables, such as light intensity, growth phase, photoperiod, temperature, salinity, CO_2 concentration, nitrogen, and phosphorous concentration. Further, both the quantity and quality of lipids produced will vary with the identity of the algal species.

7.6.1 NUTRIENT CONTENT

The growth and lipid accumulation of microalgae are affected by nutrition concentration of the growth medium. Cellular lipid levels of microalgae will increase under nutrient stress with triacyl glycerols as the dominant proportions (Illman et al., 2000). Under nitrogen limitation or starvation conditions, excess carbon from photosynthesis is channeled into storage molecules, such as triglyceride or starch (Scott et al., 2010). Further transferring microalgal cells from normal nutrient to nitrogen-depleted media will gradually change the lipid composition from free fatty acid-rich lipid to mostly triglyceride-containing lipid (Takagi et al., 2000). The deprivation of nitrogen enhances the lipid production in microalgae (Chen et al., 2008; Li et al., 2008; Li et al., 2013) and produces more favorable triacyl glycerols by inducing changes in fatty acid chain length and saturation for biofuel conversion. Phosphate limitation caused significant changes in the fatty acid and lipid composition of *Monodus subterraneus* (Khozin-Goldberg and Cohen, 2006). However, nitrogen deficiency was more effective than phosphate deficiency in *C. zofingiensis* (Feng et al., 2012). Significant increase in lipid content of *Desmodesmus* sp., *Nannochloropsis oculata*, *C. minutissima* and *Botryococcus* spp. were observed under nitrogen starvation conditions (Liu et al., 2008; Yeesang and Cheirsilp, 2011; Cao et al., 2014; Surendhiran and Vijay, 2014; Rios et al., 2015).

7.6.2 CARBON DIOXIDE

Microalgae are capable of taking zero-energy form of carbon and synthesizing it into a high-density liquid form of energy and are capable of storing carbon in the form of natural oils or as a polymer of carbohydrates. CO_2 is known to influence the lipid content of algae and alterations in the composition of the fatty acids are dependent on the CO_2 concentration during the algal growth. CO_2 concentration in the range of 10–40 ml min^{-1} was reported to increase the lipid productivity of microalgae (Chiu et al., 2009; Ho et al., 2010; Lv et al., 2010; Ortiz Montoya et al., 2014). However, under high concentrations of CO_2, the algal growth was affected as unutilized CO_2 will be converted to H_2CO_3 thereby reducing the pH of the medium. Hence, optimum CO_2 levels are required to obtain maximum biomass and enhanced lipid production using microalgae.

7.6.3 TEMPERATURE

Temperature is known to influence the lipid accumulation in microalgal cells. Many microalgae have displayed increasing growth and lipid content with increasing temperature. At low temperature, microalgae synthesize higher ratio of saturated fatty acids and increased temperature resulted in decreased neutral lipid and polyunsaturated fatty acids. Total lipid content was increased at lower temperature in fresh water and marine microalgae (Hoffmann et al., 2010; Bohnenberger and Crossetti, 2014). At the same time, increasing temperature has resulted in higher lipid production in *Nannochloropsis oculata* and *Ettlia oleoabundans* (Converti et al., 2009; Yang et al., 2013; Subhash et al., 2014).

7.6.4 SALINITY

Salinity influences physiological and biochemical mechanisms of microalgae and can lead to increment in the lipid content due to changes in the fatty acid metabolism. Restoration of turgor pressure, regulation of the uptake and export of ions through the cell membrane, and accumulation of osmo-protecting solutes and stress proteins get activated when cells are exposed to salinity. This leads to stress generation inside the algal cells causing increased total lipid accumulation, which act as a reserve energy material until favorable conditions arise (Alkayal et al., 2011; Talebi et al., 2013). Further, increase in unsaturated fatty acids proportion was observed under salt stress on the other hand higher levels of saturated fatty acids has also been reported under high salt conditions (Kan et al., 2012). Another advantage of cultivating microalgae under high alkaline salt conditions is to limit contaminants and competing microorganisms. Salinity acts as growth-limiting factor within natural biotic communities in open pond systems by inhibiting invasive organisms.

Higher lipid contents were observed in *Scenedesmus* species that were subjected to salt stress (Walsby, 1982; Kirrolia et al., 2011; Kaewkannetra et al., 2012). Duan et al. (2012) have reported a 21.1% increase of lipid yield in *C. vulgaris* under salt induced osmotic stress. Increased salt concentrations resulted in a higher intracellular lipid content to 70% in *Dunaliella* cells (Takagi et al., 2006). Salinity stress triggered both biomass growth and lipid synthesis in microalgae significantly with total and neutral lipid content of 23.4 and 9.2% along with higher amounts of saturated fatty acid methyl esters (Mohan and Devi, 2014). However, higher salt conditions inhibited the cell growth at the same time. Rao et al. (2007) reported increase in the relative proportions of palmitic acid and oleic acid in *Botryococcus braunii* at different levels of salinity. Similar results were obtained by Ruangsombon (2012) along with varying light and nutrient conditions. Lipid accumulation of *Nannochloropsis salina* under salinity stress has significantly increased at a concentration of 36%. Elevated salinity conditions leads to increased non-polar lipid content and decrease in membrane lipid content (Bartley et al., 2013).

7.6.5 METAL IONS

Metal ions influence the algal biomass and lipid production. Heavy metals like cadmium, copper, and zinc are known to increase the total lipid content of *Euglena gracilis* (Einicker-Lamas et al., 2002). The total lipid content and lipid productivity of *Scenedesmus* increased 28.2% and 29.7% in the presence of iron, magnesium, and calcium with the addition of EDTA during cultivation (Ren et al., 2014). Liu et al. (2008) reported the effect of iron on *C. vulgaris* and the total lipid content was raised up to 56.6%. Lipid accumulation in *C. protothecoides* by copper stressed lipid biosynthesis was studied by Li et al. (2013) where optimized biomass and lipid yield were achieved by 6.47 g L^{-1} and 5.78 g L^{-1}. Copper stress has influenced the lipid production in *Chlorella* at qualitative and quantitative manner. Higher concentrations of fatty acids were observed in *C. vulgaris*, *C. protothecoides,* and *C. pyrenoidosa* at copper levels of 4 mg L^{-1} (Sibi et al., 2014). The effect of hexavalent chromium on fatty acid composition and lipid peroxidation was studied in *Euglena gracilis* by Rocchetta et al. (2006).

7.6.6 OXIDATIVE STRESS

Environmental stresses trigger the excessive formation and accumulation of intracellular reactive oxygen species (ROS) in algae, which cause damage through the oxidation of cellular components. However, algal cells are able to mediate anti-oxidative defense and under oxidative stress, the lipid profile of many microalgae is reported to be altered. Kang et al. (2014) have used oxidative stress to induce lipid production in *Chlorella vulgaris*. Yilanciooglu et al. (2014) have determined the oxidative stress-mediated increased cellular lipid content up to 44% by application of exogenous H_2O_2. Nitrogen depletion results in the co-occurrence of ROS and lipid accumulation in diatoms (Liu et al., 2012). Association of increased ROS levels and cellular lipid accumulation under different environmental stress conditions was also shown in green microalgae. Osundeko et al. (2013) have reported that lipid content of *Chlorella luteoviridis* and *Parachlorella hussii* were increased under oxidative stress, which could be used for biofuel feed stock production. However, a mechanistic understanding of the connection between oxidative stress and increased algal lipid accumulation requires further investigation (Hong et al., 2008).

7.7 ALTERNATE SUBSTRATES FOR ALGAL CULTIVATION

Growth medium provides necessary nutrient sources for algal growth and under heterotrophic conditions, the cost of growth medium is high, therefore economic considerations demand much cheaper and easily available resources. Cellulosic biomasses are abundant renewable resources for the production of biofuel. A large amount of valuable compounds are present in crop residues after harvesting, which could be used as nutrient sources. Utilizing lignocellulosic biomass offers the possibility of renewable source of carbon and nitrogen that can be used to cultivated microalgae. Heterotrophic microalgae are capable of converting organic carbon sources to intracellular oil that could be used to produce biodiesel efficiently. Organic carbon sources like starch hydrolysates from Jerusalem artichoke (Cheng et al., 2009), sweet sorghum (Gao et al., 2010), cassava (Lu et al., 2010), waste molasses (Yan et al., 2011), rice straw (Li et al., 2011), wheat bran (El-Sheekh et al., 2012), sweet sorghum, and rice straw (Sibi, 2015b) were utilized to cultivate microalgae as cost-effective approach to displace glucose.

7.8 EXTRACTION METHODS

Lipids due to their high energy density are a very attractive feedstock for biofuel production. The biodiesel product ion process from microalgae consists of two major steps. First step includes biomass cultivation, harvesting, drying, and lipid extraction. This is followed by the conversion of extracted lipids to biodiesel and final purification. One of the main challenges in the biodiesel from microalgae is extraction of lipids from the harvested biomass. Lipid extraction continues to be a significant challenge toward the commercial production of microalgal oil, even though a multitude of extraction methods have been followed. The method of Folch et al. (1957) that uses chloroform and methanol (2:1) was optimized for isolation and purification of lipids from animal tissues. The method of Bligh and Dyer (1959) that uses chloroform and methanol (1:2) followed by chloroform extraction was originally optimized for extraction of phospholipids from fish muscles. It should be noted that microalgae contain unusual lipid classes and fatty acids differing from higher animal and plant organisms (Guschina and Harwood, 2006). Further, following different lipid extraction methods can result in widely varying estimations. Hence, careful choice and validation of analytical methodology in microalgal lipid extraction is needed. The efficient extraction of lipids is highly dependent on the polarity of the organic solvents used (Hamilton et al., 1992; Lewis et al., 2000). In general, solvent mixtures containing a polar and a non-polar solvent could extract a greater amount of lipids.

Various solvent systems were performed for extraction of lipids from microalgae. Li et al. (2014) have compared the extraction methods and found supercritical CO_2 technique for lipid extraction in *Tetraselmis* sp. Ryckebosch et al. (2012) found chloroform-methanol (1:1) was best solvent mixture for total lipid extraction from lyophilized microalgae. Lipid extraction from *Botryococcus braunii* was carried out using chloroform-methanol (2:1), hexane-isopropanol (3:2), dichloroethane-ethanol (1:1), and acetone-dichloromethane (1:1) by Lee et al. (1998) and chloroform-methanol (2:1) has produced highest lipid content. Grima et al. (1994) have used seven solvent mixtures to extract lipids from *Isochrysis galbana* and obtained 93.8% of lipid using chloroform-methanol-water (1:2:0.8).

Chloroform known for its carcinogenicity and its decomposition yields phosgene and hydrochloric acid inflicts chemical modification of lipids (Schmid et al., 1973). Matyash et al. (2008) have used methyl tert-butyl ether (MTBE) extraction to avoid carcinogenic chloroform for lipid recovery. Hexane can be considered for lipid extraction considering its low toxicity and removal of non-polar lipids from crude lipids efficiency (Prommuak et al., 2012). Chen et al. (2011) have produced 88% total lipid recovery by using hexane-ethanol (3:1). Yang et al. (2014) indicated ethanol extraction of lipids from wet biomass of *Picochlorum* sp. and obtained 33.04% yield, whereas, Fajardo et al. (2007) used ethanol and hexane to extract and purify lipids from dried microalga *Phaeodactylum tricornutum*.

7.9   CONCLUSION

All of the elements for the production of lipid-based fuels from algae have been demonstrated in this chapter. It is clear that algal lipids can be extracted and converted to biodiesel or other transportation fuels but the relevant question is not whether biofuels from algae are possible, but rather whether they can be made economically and at a scale sufficient to help contribute to global fuel demand. A number of major technical challenges are needed to achieve this goal. Significant attention and support should be given to both basic and applied research on algae for biofuels. Photosynthetic microalgae are technically viable and attractive alternatives for terrestrial crops and lignocellulosic biomass. The promise of algal biofuels comes with the vision of a novel form of large-scale production with economic viability of alternate fuels for sustainable environment.

KEYWORDS

- **Algae**
- **Autotrophic**
- **Biodiesel**
- **Biofuel**
- **Biomass**
- *Chlorella*
- **Environmental sustainability**
- **Fatty acid**
- **Heterotrophic**
- **Lipid production**

- **Microalgae**
- **Open pond**
- **Photobioreactor**
- **Photosynthetic**
- **Renewable energy**
- *Scenedesmus*
- **Triacyl glycerols**
- **Tubular bioreactor**

REFERENCES

Alkayal, F.; Albion, R. L.; Tillett, R. L.; Hathwaik, L. T.; Lemos, M. S.; Cushman, J. C. Expressed sequence tag (EST) profiling in hyper saline shocked *Dunaliella salina* reveals high expression of protein synthetic apparatus components. *Plant Sci* 2011, 179, 437–449.

Bartley, M. L.; Boeing, W. J.; Corcoran, A. A.; Holguin, F. O.; Schaub, T. Effects of salinity on growth and lipid accumulation of biofuel microalga *Nannochloropsis salina* and invading organisms. *Biomass Bioenerg* 2013, 54, 83–88.

Becker, E. W. *Microalgae: Biotechnology and microbiology.* Cambridge University: Cambridge, 1994.

Benemann, J. R.; Oswald, W. J. *Systems and economic analysis of microalgae ponds for conversion of carbon dioxide to biomass.* Pittsburgh Energy Technology Center: Pittsburgh, PA, 1993.

Bligh, E. G.; Dyer, W. J. A rapid method of total lipid extraction and purification. *Can J Biochem Physiol* 1959, 37(8), 911–917.

Bohnenberger, J. E.; Crossetti, L. O. Influence of temperature and nutrient content on lipid production in fresh water microalgae cultures. *Ann Brazilian Acad Sci* 2014, 86, 1239–1248.

Borowitzka, M. A. Culturing microalgae in outdoor ponds. In *Algal culturing techniques,* Andersen, R. A., Ed.; Academic Press: London, UK, 2005; pp 205–218.

Borowitzka, M. A. Commercial production of microalgae: Ponds, tanks, tubes and fermentors. *J Biotechnol* 1999, 70, 313–321.

Cao, J.; Yuan, H.; Li, B.; Yang, J. Significance evaluation of the effects of environmental factors on the lipid accumulation of *Chlorella minutissima* UTEX 2341 under low-nutrition heterotrophic condition. *Bioresour Technol* 2014, 152, 177–184.

Chen, F. High cell density culture of microalgae in heterotrophic growth. *Trends Biotechnol* 1996, 14, 421–426.

Chen, M.; Chen, X. L.; Liu, T. Z.; Zhang, W. Subcritical ethanol extraction of lipid from wet microalgae paste of *Nannochloropsis* sp. *J Biobased Mater Bio* 2011, 5, 385–389.

Chen, M.; Tang, H.; Ma, H.; Holland, T. C.; Ng, K. Y. S.; Salley, S. O. Effect of nutrients on growth and lipid accumulation in the green algae *Dunaliella tertiolecta. Bioresour Technol* 2008, 102, 1649–1655.

Cheng, Y.; Zhou, W.; Gao, C.; Lan, K.; Gaw, Y.; Wu, Q. Biodiesel production from jerusalem artichoke (*Helianthus tuberosus* L.) tuber by heterotrophic microalgae *Chlorella protothecoides*. *J Chem Technol Biotechnol* 2009, 84, 777–781.

Chiu, S. Y.; Kao, C. Y.; Tsai, M. T.; Ong, S. C.; Chen, C. H.; Lin, C. S. Lipid accumulation and CO_2 utilization of *Nannochloropsis oculata* in response to CO_2 aeration. *Bioresour Technol* 2009, 100, 833–838.

Chrismadha, T.; Borowitzka, M. A. Effect of cell-density and irradiance on growth, proximate composition and eicosapentaenoic acid production of *Phaeodactylum-Tricornutum* grown in a tubular photobioreactor. *J Appl Phycol* 1994, 6, 67–74.

Converti, A.; Casazza, A. A.; Ortiz, E. Y.; Perego, P.; Del Borghi, M. Effect of temperature and nitrogen concentration on the growth and lipid content of *Nannochloropsis oculata* and *Chlorella vulgaris* for biodiesel production. *Chem Eng Process* 2009, 48, 1146–1151.

Das, P.; Aziz, S. S.; Obbard, J. P. Two phase microalgae growth in the open system for enhanced lipid productivity. *Renew Energy* 2011, 36, 2524–2528.

Doucha, J.; Livansky, K. Productivity, CO_2/O_2 exchange and hydraulics in outdoor open high density microalgal (*Chlorella* sp.) photobioreactors operated in a Middle and Southern European climate. *J Appl Phycol* 2006, 18, 811–826.

Duan, X.; Ren, G. Y.; Liu, L. L.; Zhu, W. X. Salt-induced osmotic stress for lipid overproduction in batch culture of *Chlorella vulgaris*. *Afr J Biotechnol* 2012, 11, 7072–7078.

Einicker-Lamas, M.; Mezian, G. A.; Fernandes, T. B.; Silva, F. L. S.; Guerra, F.; Miranda, K.; Attias, M.; Oliveira, M. M. *Euglena gracilis* as a model for the study of Cu^{2+} and Zn^{2+} toxicity and accumulation in eukaryotic cells. *Environ Pollut* 2002, 120, 779–786.

El-Sheekh, M. M.; Bedaiwy, M. Y.; Osman, M. E.; Ismail, M. M. Mixotrophic and heterotrophic growth of some microalgae using extract of fungal-treated wheat bran. *Int J Recycl Org Waste Agric* 2012, 1, 12.

Fajardo, A. R.; Cerdan, L. E.; Medina, A. R.; Fernandez, F. G. A.; Moreno, P. A. G.; Grima, E. M. Lipid extractionfrom the microalga *Phaeodactylum tricornutum*. *Eur J Lipid Sci Technol* 2007, 2, 120–126.

Feng, P.; Deng, Z.; Fan, L.; Hu, Z. Lipid accumulation and growth characteristics of *Chlorella zofingiensis* under different nitrate and phosphate concentrations. *J Biosci Bioeng* 2012, 114, 405–410.

Folch, J.; Lees, M. Sloane Stanley, G. H. A simple method for the isolation and purification of total lipids from animal tissues. *J Biol Chem* 1957, 226, 497–509.

Gao, C.; Zhai, Y.; Ding, Y.; Wu, Q. Application of sweet sorghum for biodiesel production by heterotrophic microalga *Chlorella protothecoides*. *Appl Energy* 2010, 87, 756–761.

Grima, E. M.; Belarbi, E. H.; Fernandez, F. G. A.; Medina, A. R.; Chisti, Y. Recovery of microalgal biomass and metabolites, process options and economics. *Biotechnol Adv* 2003, 20, 491–515.

Grima, E. M.; Medina, A. R.; Gimenez, A. G.; Perez, J. A. S.; Camacho, F. G.; Sanchez, J. L. G. Comparison between extraction of lipids and fatty acids from microalgal biomass. *JAOCS* 1994, 71, 955–959.

Gudin, C.; Chaumont, D. Solar biotechnology study and development of tubular solar receptors. In *Energy from biomass*, Palz, W., Pirruitz, D., Eds.; Reidel: Dordrecht, 1983, pp 184–193.

Guschina, I. A.; Harwood, J. L. Lipids and lipid metabolism in eukaryotic algae. *Prog Lipid Res* 2006, 45, 160–186.

Hamilton, S.; Hamilton, R. J.; Sewell, P. A. Extraction of lipids and derivative formation. In *Lipid analysis: a practical approach*, Hamilton, R. J., Hamilton, S., Eds.; IRL Press: Oxford, 1992; pp 13–64.

Han, F.; Huang, J.; Li, Y.; Wang, W.; Wan, M.; Shen, G.; Wang, J. Enhanced lipid productivity of *Chlorella pyrenoidosa* through the culture strategy of semicontinuous cultivation with nitrogen limitation and pH control by CO_2. *Bioresour Technol* 2013, 136, 418–424.

Hase, R.; Oikawa, H.; Sasao, C.; Morita, M.; Watanabe, Y. Photosynthetic production of microalgal biomass in a raceway system under greenhouse conditions in Sendai city. *J Biosci Bioeng* 2000, 89, 157–163.

Ho, S. H.; Chen, W. M.; Chang, J. S. *Scenedesmus Obliquus* CNW-N as a potential candidate for CO_2 mitigation and biodiesel production. *Bioresour Technol* 2010, 101, 8725–8730.

Hoekema, S.; Bijmans, M.; Janssen, M.; Tramper, J.; Wijffles, R. H. A pneumatically agitated flat-panel photobioreactor with gas re-circulation: Anaerobic photoheterotrophic cultivation of a purple non-sulfur bacterium. *Int J Hydrogen Energy* 2002, 27, 1331–1338.

Hoffmann, M.; Marxen, K.; Schulz, R.; Vanselow, K. H. TFA and EPA productivities of *Nannochloropsis salina* influenced by temperature and nitrate stimuli in turbidostatic controlled experiments. *Mar Drugs* 2010, 8, 2526–2545.

Hong, Y.; Hu, H. Y.; Li, F. M. Physiological and biochemical effects of allelochemical ethyl 2-methyl acetoacetate (EMA) on cyanobacterium *Microcystis aeruginosa*. *Ecotoxicol Environ Saf* 2008, 71, 527–534.

Hu, Q.; Guterman, H.; Richmond, A. A flat inclined modular photobioreactor (FIMP) for outdoor mass cultivation of photoautotrophs. *Biotechnol Bioeng* 1996, 51, 51–60.

Illman, A. M.; Scragg, A. H.; Shales, S. W. Increase in *Chlorella* strains calorific values when grown in low nitrogen medium. *Enzyme Microb Technol* 2000, 27, 631–635.

Jorquera, O.; Kiperstok, A.; Sales, E. A.; Embirucu, M.; Ghirardi, M. L. Comparative energy life-cycle analyses of microalgal biomass production in open ponds and photobioreactors. *Bioresour Technol* 2010, 101, 1406–1413.

Kaewkannetra, P.; Enmak, P.; Chiu, T. Y. The effect of CO_2 and salinity on the cultivation of *Scenedesmus obliquus* for biodiesel production. *Biotechnol Bioprocess Eng* 2012, 17, 591–597.

Kan, G.; Shi, C.; Wang, X.; Xie, Q.; Wang, M.; Wang, X.; Miao, J. Acclimatory responses to high-salt stress in *Chlamydomonas* (chlorophyta, chlorophyceae) from Antarctica. *Acta Oceanol Sin* 2012, 31, 116–124.

Kang, N. K.; Lee, B.; Choi, G. G.; Moon, M.; Park, M. S.; Lim, J. K.; Yang, J. W. Enhancing lipid productivity of *Chlorella vulgaris* using oxidative stress by TiO_2 nanoparticles. *Korean J Chem Eng* 2014, 31, 861–867.

Khozin-Goldberg, I.; Cohen, Z. The effect of phosphate starvation on the lipid and fatty acid composition of the fresh water eustigmatophyte *Monodus subterraneus*. *Phytochem* 2006, 67, 696–701.

Kirroliaa, A.; Bishnoia, N. R.; Singh, N. Salinity as a factor affecting the physiological and biochemical traits of *Scenedesmus quadricauda*. *J Algal Biomass Utln* 2011, 2, 28–34.

Klok, A. J.; Martens, D. E.; Wijffels, R. H.; Lamers, P. P. Simultaneous growth and neutral lipid accumulation in microalgae. *Bioresour Technol* 2013, 134, 233–243.

Lee, S. J.; Yoon, B. D.; Oh, H. M. Rapid method for the determination of lipid from the green alga *Botryococcus braunii*. *Biotechnol Tech* 1998, 12, 553–556.

Lee, Y. K. Enclosed bioreactors for the mass cultivation of photosynthetic microorganisms: The future trend. *Trends Biotechnol* 1986, 4, 186–189.

Lee, Y. K. Microalgal mass culture systems and methods: Their limitation and potential. *J Appl Phycol* 2001, 13, 307–315.

Lee, Y. K.; Ding, S. Y.; Low, C. S.; Chang, Y. C.; Forday, W. L.; Chew, P. C. Design and performance of an α type tubular photobioreactor for mass cultivation of microalgae. *J Appl Phycol* 1995, 7, 47–51.

Lewis, T.; Nichols, P. D.; Mcmeekin, T. A. Evaluation of extraction methods for recovery of fatty acids from lipid-producing microheterotrophs. *J Microbiol Methods* 2000, 43, 107–116.

Li, P.; Miao, X.; Li, R.; Zhong, J. *In situ* biodiesel production from fast-growing and high oil content *Chlorella pyrenoidosa* in rice straw hydrolysate. *J Biomed Biotechnol* 2011, 141207, doi:10.1155/2011/141207.

Li, Y.; Horsman, M.; Wang, B.; Wu, N.; Lan, C. Q. Effects of nitrogen sources on cell growth and lipid accumulation of green alga *Neochloris oleoabundans*. *Appl Microbiol Biotechnol* 2008, 81, 629–636.

Li, Y.; Mu, J.; Chen, D.; Han, F.; Xu, H.; Kong, F.; Xie, F.; Feng, B. Production of biomass and lipid by the microalgae *Chlorella protothecoides* with heterotrophic-Cu(II) stressed (Hcus) coupling cultivation. *Bioresour Technol* 2013, 148, 283–292.

Li, Y.; Naghdi, F. G.; Garg, S.; Adarme-Vega, T. C.; Thurecht, K. J.; Ghafor, W. A.; Tannock, S.; Schenk, P. M. A comparative study: The impact of different lipid extraction methods on current microalgal lipid research. *Microb Cell Fact* 2014, 13, 14.

Liu, W. H.; Huang, Z. W.; Li, P.; Xia, J. F.; Chen, B. Formation of triacylglycerol in *Nitzschia closterium* F. *minutissima* under nitrogen limitation and possible physiological and biochemical mechanisms. *J Exper Marine Biol Ecol* 2012, 418, 24–29.

Liu, Z. Y.; Wang, G. C.; Zhou, B. C. Effect of iron on growth and lipid accumulation in *Chlorella vulgaris*. *Bioresour Technol* 2008, 99, 4717–4722.

Lu, Y.; Zhai, Y.; Liu, M.; Wu, Q. Biodiesel production from algal oil using cassava (*Manihot esculenta* Crantz) as feedstock. *J Appl Phycol* 2010, 22, 573–578.

Lv, J. M.; Cheng, L. H.; Xu, X. H.; Zhang, L.; Chen, H. L. Enhanced lipid production of *Chlorella vulgaris* by adjustment of cultivation conditions. *Bioresour Technol* 2010, 101, 6797–6804.

Mata, T. M.; Martins, A. A.; Caetano. N. S. Microalgae for biodiesel production and other applications: A review. *Renew Sustain Energy Rev* 2010, 14, 217–232.

Matyash, V.; Liebisch, G.; Kurzchalia, T. V.; Shevchenko, A.; Schwudke, D. Lipid extraction by methyl-tert-butyl ether for high-throughput lipidomics. *J Lipid Res* 2008, 49, 1137–1146.

Mcginn, P. J.; Dickinson, K. E.; Park, K. C.; Whitney, C. G.; Macquarrie, S. P.; Black, F. J.; Frigon, J.; Guiot, S. R.; O'Leary, S. J. B. Assessment of the bioenergy and bioremediation potentials of the microalga *Scenedesmus* sp. AMDD cultivated in municipal wastewater effluent in batch and continuous mode. *Algal Res* 2012, 1, 155–165.

Miao, X. L.; Wu, Q. Biodiesel production from heterotrophic microalgal oil. *Bioresour Technol* 2006, 97, 841–846.

Mohan, S. V.; Devi, M. P. Salinity stress induced lipid synthesis to harness biodiesel during dual mode cultivation of mixotrophic microalgae. *Bioresour Technol* 2014, 165, 288–294.

Natrah, F.; Yousoff, F. M.; Shariff, M.; Abas, F.; Mariana, N. S. Screening of malaysian indigenous microalgae for antioxidant properties and nutritional value. *J Appl Phycol* 2008, 19, 711–718.

Olaizola, M. Commercial development of microalgal biotechnology: From the test tube to the market place. *Biomol Eng* 2003, 20, 459–466.

Ortiz Montoya, E. Y.; Casazza, A. A.; Aliakbarian, B.; Perego, P.; Converti, A.; De Carvalho, J. C. M. Production of *Chlorella vulgaris* as a source of essential fatty acids in a tubular photobioreactor continuously fed with air enriched with CO_2 at different concentrations. *Biotechnol Prog* 2014, 30, 916–922.

Osundeko, O.; Davies, H.; Pittman, J. K. Oxidative stress-tolerant microalgae strains are highly efficient for biofuel feedstock production on wastewater. *Biomass Bioenerg* 2013, 56, 284–294.

Perez-Garcia, O.; Escalante, F. M. E.; De-Bashan, L. E.; Bashan, Y. Heterotrophic cultures of microalgae: Metabolism and potential products. *Water Res* 2011, 45, 11–36.

Pirt, S. J.; Lee, Y. K.; Walach, M. R.; Pirt, M. W.; Balyuzi, H. H. M.; Bazin, M. J. A tubular bioreactor for photosynthetic production of biomass from carbon dioxide: Design and performance. *J Chem Tech Biotechnol* 1983, 33, 35–58.

Prommuak, C.; Pavasant, P.; Quitain, A. T.; Goto, M.; Shotipruk, A. Microalgal lipid extraction and evaluation of single-step biodiesel production. *J Eng* 2012, 5, 157–166.

Pulz, O. Open air and semi closed cultivation systems for the mass cultivation of microalgae. In *Algal biotechnology in the Asia Pacific region*, Phang, S. M., Lee, Y. K., Borowitzka, M. A., Whitton, B. A., Eds.; University of Malaya: Kuala Lumpur, 1994, pp 113–117.

Putt, R.; Singh, M.; Chinnasamy, S.; Das, K. C. An efficient system for carbonation of high-rate algae pond water to enhance CO_2 mass transfer. *Bioresour Technol* 2011, 102, 3240.

Rao, A. R.; Dayananda, C.; Sarada, R.; Shamala, T. R.; Ravishankar, G. A. Effect of salinity on growth of green alga *Botryococcus braunii* and its constituents. *Bioresour Technol* 2007, 98, 560–564.

Ren, H. Y.; Liu, B. F.; Kong, F.; Zhao, L.; Xie, G. J.; Ren, N. Q. Enhanced lipid accumulation of green microalga *Scenedesmus* sp. by metal ions and EDTA addition. *Bioresour Technol* 2014, 169, 763–767.

Richmond, A. *Handbook of microalgal mass culture*. CRC Press, Boca Raton: Florida, 1986.

Rios, L. F.; Klein, B. C.; Luz, L. F.; Maciel Filho, R.; Wolf Maciel, M. R. Nitrogen starvation for lipid accumulation in the microalga species *Desmodesmus* sp. *Appl Biochem Biotechnol* 2015, 175, 469–476.

Robinson, L. F. Improvements relating to biomass production. European Patent 0239272, 1987.

Rocchetta, I.; Mazzuca, M.; Conforti, V.; Ruiz, L.; Balzaretti, V.; Molina, M. C. R. Effect of chromium on the fatty acid composition of two strains of *Euglena gracilis*. *Environ Pollution* 2006, 141, 353–358.

Ruangsombon, S. Effect of light, nutrient, cultivation time and salinity on lipid production of newly isolated strain of the green microalga, *Botryococcus Braunii* KMITL 2. *Bioresour Technol* 2012, 109, 261–265.

Ryckebosch, E.; Muylaert, K.; Foubert, I. Optimization of an analytical procedure for extraction of lipids from microalgae. *J Am Oil Chem Soc* 2012, 89, 189–198.

Schmid, P.; Hunter, E.; Calvert, J. Extraction and purification of lipids. III. Serious limitations of chloroform and chloroform-methanol in lipid investigations. *Physiol Chem Phys Med* 1973, 5, 151–155.

Scott, S. A.; Davey, M. P.; Dennis, J. S.; Horst, I.; Howe, C. J.; Lea-Smith, D. J.; Smith, A. G. Biodiesel from algae: Challenges and prospects. *Curr Opin Biotechnol* 2010, 21, 277–286.

Sibi, G. Cultural conditions and nutrient composition as effective inducers for biomass and lipid production in fresh water microalgae. *Res J Environ Toxicol* 2015a, 9, 168–178.

Sibi, G. Low cost carbon and nitrogen sources for higher microalgal biomass and lipid production using agricultural wastes. *J Environ Sci Technol* 2015b, 8, 113–121.

Sibi, G.; Anuraag, T. S.; Bafila, G. Copper stress on cellular contents and fatty acid profiles *in Chlorella* species. *Online J Biol Sci* 2014, 14, 209–217.

Sobczuk, T.; Chisti, Y. Potential fuel oils from the microalga *Choricystis minor*. *J Chem Technol Biotechnol* 2010, 85, 100–108.

Spolaore, P.; Joannis-Cassan, C.; Duran, E.; Isambert, A. Commercial applications of microalgae - review. *J Biosci Bioeng* 2006, 101, 87–96.

Su, C. H.; Chien, L. J.; Gomes, J.; Lin, Y. S.; Yu, Y. K.; Liou, J. S.; Syu, R. J. Factors affecting lipid accumulation by *Nannochloropsis oculata* in a two-stage cultivation process. *J Appl Phycol* 2011, 23, 903–908.

Subhash, G. V.; Rohit, M. V.; Devi, M. P.; Swamy, Y. V.; Mohan, S. V. Temperature induced stress influence on biodiesel productivity during mixotrophic microalgae cultivation with wastewater. *Bioresour Technol* 2014, 169, 789–793.

Surendhiran, D.; Vijay, M. Effect of various pre-treatment for extracting intracellular lipid from *Nannochloropsis oculata* under nitrogen replete and depleted conditions. *ISRN Chem Eng* 2014, 536310, dx.doi.org/10.1155/2014/536310.

Takagi, M.; Karseno.; Yoshida, T. Effect of salt concentration on intracellular accumulation of lipids and triacylglyceride in marine microalgae *Dunaliella* cells. *J Biosci Bioeng* 2006, 101, 223–226.

Takagi, M.; Watanabe, K.; Yamaberi, K.; Yoshida, T. Limited feeding of potassium nitrate for intracellular lipid and triglyceride accumulation of *Nannochloris* sp. UTEX LB1999. *Appl Microbiol Biotechnol* 2000, 54, 112–117.

Talebi, A. F.; Tabatabaei, M.; Mohtashami, S. K.; Tohidfar, M.; Moradi, F. Comparative salt stress study on intracellular ion concentration in marine and salt-adapted freshwater strains of microalgae. *Not Sci Biol* 2013, 5, 309–315.

Tang, H.; Chen, M.; Ng, K. Y. S.; Salley, S. O. Continuous microalgae cultivation in a photobioreactor. *Biotechnol Bioeng* 2012, 109, 2468–2474.

Tredici, M. R.; Carlozzi, P.; Zittelli, C. G.; Materassi, R. A vertical aveolar panel (vap) for outdoor mass cultivation of microalgae and cyanobacteria. *Bioresour Technol* 1991, 38, 153–159.

Tredici, M. R.; Materassi, R. From open ponds to vertical alveolar panels: The Italian experience in the development of reactors for the mass cultivation of phototrophic microorganisms. *J Appl Phycol* 1992, 4, 221–231.

Walsby, A. Cell-water and cell-solute relations. In *The biology of cyanobacteria*, Carr, N., Whitton, B., Eds.; University of California Press: California, 1982, pp 237–262.

Wen, X.; Geng, Y.; Li, Y. Enhanced lipid production in *Chlorella pyrenoidosa* by continuous culture. *Bioresour Technol* 2014, 161, 297–303.

Wu, S. T.; Yu, S. T.; Lin, L. P. Effect of culture conditions on docosahexaenoic acid production by *Schizochytrium* sp. S31. *Process Biochem* 2005, 40, 3103–3108.

Xiong, W.; Li, X.; Xiang, J.; Wu, Q. High-density fermentation of microalga *Chlorella protothecoides* in bioreactor for microbiodiesel production. *Appl Microbiol Biotechnol* 2008, 78, 29–36.

Yan, D.; Lu, Y.; Chen, Y. F.; Wu, Q. Waste molasses alone displaces glucose-based medium for microalgal fermentation towards cost-saving biodiesel production. *Bioresour Technol* 2011, 102, 6487–6493.

Yang, F.; Xiang, W.; Sun, X.; Wu, H.; Li, T.; Long, L. A novel lipid extraction method from wet microalga *Picochlorum* sp. at room temperature. *Mar Drugs* 2014, 12, 1258–1270.

Yang, Y.; Mininberg, B.; Tarbet, A.; Weathers, P. At high temperature lipid production in *Ettlia oleoabundans* occurs before nitrate depletion. *Appl Microbiol Biotechnol* 2013, 97, 2263–2273.

Yeesang, C.; Cheirsilp, B. Effect of nitrogen, salt and iron content in the growth medium and light intensity on lipid production by microalgae isolated from freshwater sources in Thailand. *Bioresour Technol* 2011, 102, 3034–3040.

Yilancioooglu, K.; Cokol, M.; Pastirmaci, I.; Erman, B.; Cetiner, S. Oxidative stress is a mediator for increased lipid accumulation in a newly isolated *Dunaliella salina* strain. *PLoS One* 2014, 9, E91957.

Zhang, B. Y.; Geng, Y. H.; Li, Z. K.; Hu, H. J.; Li, Y. G. Production of astaxanthin from *Haematococcus* in open pond by two-stage growth one-step process. *Aquaculture* 2009, 295, 275–281.

CHAPTER 8

INTEGRATED ALGAL INDUSTRIAL WASTE TREATMENT AND BIOENERGY CO-GENERATION

MOHD AZMUDDIN ABDULLAH[1] and ASHFAQ AHMAD[2]

[1]*Institute of Marine Biotechnology, Universiti Malaysia Terengganu, 21030, Kuala Terengganu, Terengganu, Malaysia*

[2]*Department of Chemical Engineering, Universiti Teknologi PETRONAS, 32610, Seri Iskandar, Perak, Malaysia*

CONTENTS

8.1 INTRODUCTION

Algae are a group of photosynthetic prokaryotes and eukaryotes, divided into several divisions and kingdoms (Borowitzka et al., 2012). These are look-like unicellular or simple multicellular organisms, living as distinct individuals, in pairs, in clusters, colonies, or in sheets of individuals. Microalgae do not form roots, stems, or leaves, but with high surface area to volume ratio, may reach sizes visible to the naked eye as minute green particles. They differ from macroalgae in size and exist not only in aquatic but also in terrestrial eco-system and in wide ranging environmental conditions, from icy North Pole to the humid tropics and volcanic area. Prokaryotic algae include cyanobacteria (*Cyanophyceae*), and eukaryotic microalgae are green algae (*Chlorophyta*) and diatoms (*Bacillariophyta*) (Li et al., 2008; Lim et al., 2010). With an esti-mate of more than 50,000 species, only a limited number of around 3000, have been studied and identified (Richmond, 2004). The most plentiful microalgae are single-cell drifters in plankton called phytoplankton which are competent for rapid uptake of nutrients and carbon dioxide, have fast cell growth and much higher photosynthetic efficiency than the land-based plants, although the photosynthetic process is similar (Bajhaiya et al., 2010). As the energy de-mand increases, the solution may lie in the application of microorganisms and algae as the source of bioenergy (Rittmann et al., 2008; Laurens et al., 2012). Among issues to be addressed in developing energy-based crops for biofuels is the competition between fuels and food production, the effect of which has been an increase in food prices (Somerville, 2007; Rude and Schimar, 2009). The photosynthetic and heterotrophic natures of algae and the reported higher oil productivity than the best producing oil crops make them highly potential as alternatives to energy crops (Converti et al., 2009; Bajhaiya et al., 2010).

Algae are the main synthesizers of organic matter in aquatic habitats. Since early time, microalgae have been used in human health food prod-ucts, feeds for fish and livestock, and cultured for high-value of oils (Molina et al., 1999; Spolaore et al., 2006), high-value chemicals for pharma- and nutraceuticals and pigments such as carotenoids (Spolaore et al., 2006; Borowitzka, 2010), and long-chain polyunsaturated fatty acids (PUFAs) and phycobilins (Mendes et al., 2009). Industrial, municipal, and agricul-tural wastewater treatments by algal culture systems enhance degradation and improve CO_2 balance with lower energy demand for oxygen supply in aerobic treatment stages (Samori et al., 2013; Zhang et al., 2013). Among the species for biofuel production includes *Nannochloropsis oculata* and *Tetraselmis suecica* (Fig. 8.1) and those suitable for wastewater treatment include *Scenedesmus* sp., *Chlorella* sp., and *Chlamydomonas reinhardtii*

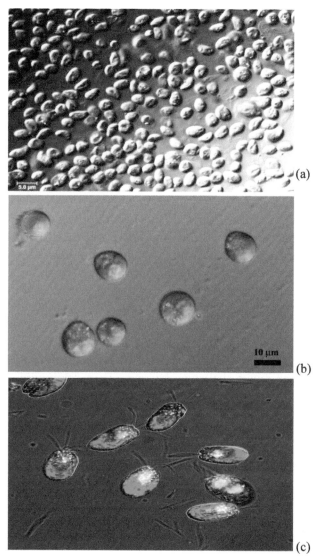

FIGURE 8.1 Productive microalgae species for biofuel production: a) *Nannochloropsis salina,* b) *Dunaliella salina,* c) *Tetraselmis suecica* (Burton et al., 2009; Greenwell et al., 2009).

(Oswald, 2003; Leadbeater, 2006). *Nannochloropsis* belongs to the class of *Eustigmatophyceae* and is the most widely studied species owing to its relatively high growth rate and lipid content with high nutritional values and resistance to mixing and contamination, which fit the needs of the biofuel industry and aquaculture (Roncarati et al., 2004; Bentley et al., 2008). *Tetraselmis suecica*, a marine green flagellate belonging to *Chlorophyceae*, has good nutritional properties and contains C16:0 and C18:1 as predominant fatty acids for biofuel and as feeds for bivalve molluscs, penaeid shrimp larvae, and rotifers (Harwood et al., 2009). *Isochrysis galbana*, a *Haptophyceae* also has good nutritive values for aquaculture, principally to feed mollusk larvae, as well as fish and crustaceans in the early stages of growth (Wikfors, 1994). *Cyanobacteria* and *Chlorella* sp. are efficient for the treatment of organic pollutants (Kirkwood et al., 2003). *Chlorella* sp. can be found in freshwater-bodies and grows well in municipal and agricultural wastewaters and sludge (Kong et al., 2010).

In many Chinese lakes such as Lake Chaohu, Lake Taihu, and Lake Dianchi, algal blooms have caused major water pollution, resulting in the death of fish and illnesses. At the same time, these actually provide large amount of biomass for value-added utilization (Miao et al., 2004). The role of algae in waste treatment is both to incorporate nutrients and to provide oxygen to bacteria, which in turn involve in the bacterial degradation of organic materials in the wastewater, the same process utilized in activated sludge. The presence of microalgae infact reduces the chemical oxygen demand (COD) and biological oxygen demand (BOD), phosphorus, nitrogen, and pathogens in a more cost effective way than activated sludge (Singh and Dhar, 2011). The major challenge, however, is in determining a way that allows downstream processing of algae to make it suitable for producing biofuel and other bioproducts (Christenson et al., 2011). The integrated processes that combine algae cultivation and wastewater treatment system for biomethane production can reduce the cost and especially when combined with CO_2 mitigation. This versatility makes algae among the most interesting organisms under research to solve global problem related to greenhouse gases (GHG) emission, waste remediation, and bioenergy generation.

8.2   INDUSTRIAL WASTE REMEDIATION

8.2.1   LIQUID WASTES

Organic and inorganic substances which are freed into the environment as a result of domestic, agricultural, and industrial wastewater can lead to pollution. Most are hazardous and must be treated prior to disposal into the waterways and land surfaces. Conventional primary and secondary treatments of the wastewaters remove the easily settled materials and oxidize the organic materials. However, secondary treatment still releases large amount of phosphorus and nitrogen which are directly responsible for eutrophication of rivers, lakes, and seas (Lau et al., 1997; Trepanier et al., 2002). Secondary effluents are also loaded with refractory organics and heavy metals. This is where algal may become elegant solution in the tertiary and quaternary treatments harnessing the ability of algae to use inorganic nitrogen, phosphorus, and amino acids for their growth and the capacity to remove heavy metals, as well as toxic organic compounds to prevent secondary pollution (Abdel-Raouf, 2012; Guiry and Guiry, 2014). Algae cultivation actually requires high amount of nutrients rendering the process economically and environmentally not competitive (Halleux et al., 2008; Sialve et al., 2009). Wastewaters can be a cost-effective alternative to synthetic culture media, especially those derived from agro-industrial facilities which usually contain high nutrient concentration (Markou and Georgakakis, 2011). Algal–bacterial systems for agro-industrial wastewater treatment have gained attention where algae assimilate nutrients and through photosynthesis, produce dissolved oxygen that is immediately available to bacteria for the oxidization of wastes whilst releasing CO_2 needed for algal growth (De-Bashan et al., 2004; Shilton et al., 2008). These avoid external oxygen supplementations as required in the conventional processes, allowing nutrients recovery into the biomass and reducing CO_2 emissions (Molinuevo-Salces et al., 2010). The tertiary algal biotreatment can then be coupled with the production of bioenergy and extraction of lipids or biocompounds.

Heavy metals are other major pollutants as a result of industrial and mining activities, use of fertilizers and pesticides, and release from fuels and microelectronic products. Although toxic level of heavy metals may hinder photosynthesis and kill the cells algae readily take up heavy metals from the environment and induce heavy metal stress responses in the form of binding factors and proteins. The removal rate of metal ions such as aluminum, calcium, ferum, manganese, and magnesium from plastic manufacturing and electroplating waste water vary among the algae species, ranging from 50 to 99% (Wang et al., 2009a; Woertz et al., 2011). Macroalgal species such as *Lamiaria, Sargassum, Macrocystis, Ecklonia, Ulva, Lessonia*, and *Durvillaea* have been reportedly efficient for binding copper, nickel, lead, zinc, and cadmium (McHugh, 2003). Combination of these algal nutrient uptake, elevated pH, and high dissolved oxygen concentration avoids any chemical additives and eutrophication leading to improved, safer, less expensive and more efficient approach to wastewater treatment and heavy metals removal (Munoz and Guieysse, 2006).

8.2.2 HYDROCARBONS

Leaks and accidental spills of petroleum-based products occur frequently from oil exploration, production, refining, transport, and storage. The amount of natural crude oil seepage estimated at 600,000 metric tons per year (Kvenvolden and Cooper, 2003). These have become major contributors to water and soil pollution (Holliger et al., 1997) which cause serious damage on eco-system with the accumulation in animals and plant tissues, resulting in deaths or mutations (Alvarez and Vogel, 1991). Bioremediation by microbial consortia can detoxify or remove pollutants owing to their diverse metabolic capabilities. It is an evolving, non-invasive and relatively inexpensive method for the removal and degradation of many environmental pollutants (Leahy and Colwell, 1990; April et al., 2000; Ulrici, 2000; Medina-Bellver, 2005). Algae and protozoa are the essential members of the microbial community in both aquatic and terrestrial eco-systems, but reports are scanty on their capabilities for hydrocarbon biodegradation. Isolated alga *Prototheca zopfi* could utilize crude oil, a mixed hydrocarbon substrate and exhibit extensive degradation of *n*-alkanes and isoalkanes as well as aromatic hydrocarbons (Walker et al., 1957). Nine cyanobacteria, five green algae, one red alga, one brown alga, and two diatoms have been reportedly capable of oxidizing naphthalene (Cerniglia et al., 1980), but protozoa has not been shown to utilize hydrocarbons. Diatoms species *Skeletonema costatum* and *Nitzschia* sp. could degrade hydrocarbons simultaneously though the latter is more efficient. Microalgae species show comparable or greater efficiency in removing the mixture of hydrocarbons suggesting that the presence of polyaromatic hydrocarbon may stimulate the degradation of the other hydrocarbon (Hong et al., 2008).

Phenolic compounds such as nitrophenols and chlorophenols are the toxic industrial pollutants (Khan et al., 1981a; Shigeoka et al., 1988; Wang et al., 2001b; Aravindhan et al., 2009). The halophenols can be found in pulp mill effluents (Xie et al., 1986), agricultural and residential runoff (Ahlborg and Thunberg, 1980), and sewage and wastewater discharges (Stasinakis et al., 2008) from wood pulp bleaching, water chlorination, textile dyes, oil refineries, chemical, agro-chemical, and pharmaceutical industries (Rodriguez et al., 1996; Perez et al., 1997; Fahr et al., 1999). Some of the phenolic compounds are suspected to be endocrine disruptors and have deleterious effects on humans and other organisms in the natural eco-system at concentrations lower than the discharge standard for phenols (Schafer et al., 1999). Their abatement in STPs is often insufficient because phenolics are highly toxic to anaerobic and aerobic bacteria (Capasso et al., 1995). The detoxification potential of phenol by different microrganisms, mainly bacteria and fungi, has been extensively studied (Kahru et al., 1998). Microalgae that are capable of phenol degradation include *Chlorella* sp., *Scenedesmus obliquus* and *S. maxima* (Kleckner and Kosaric, 1992). Four species of freshwater microalgae have been shown to mineralise phenol (Ellis, 1977) and green *Ankistrodesmus* degrade various phenols (Pinto et al., 2002). After three months of selective enrichment with ρ-chlorophenol and ρ-nitrophenol, two microalgal species, *Chlorella vulgaris* and *Coenochloris pyrenoidosa*, have been isolated from the aquatic communities recovered from the waste discharge container fed with several aromatic pollutants. As an axenic culture, the consortia grown under 24 h light regime are capable of biodegrading 50 mgL^{-1} of ρ-chlorophenol within 5 days. Addition of zeolite as an adsorbing material does not improve ρ-chlorophenol removal. However, with ρ-chlorophenol at 150 mg L^{-1} introduced to the culture supplemented with zeolite, the growth rate of the consortia improves, but with longer lag phase (16 against 14 days without zeolite) (Lima et al., 2004). A golden brown unicellular chrysophyte *Ochromonas danica*, investigated for the degradation of phenols in the dark and in aerobic conditions, has been found to affect the metacleavage of exogenous phenol by utilizing the reactions for its energetic requirements (Semple et al., 1999). *O. danica* metabolize phenol and shows heterotrophic growth at 96 mgL^{-1} phenol (Semple and Cain, 1997). However, *Ochromonas* may not be completely suitable for wastewater treatment as far as mass cultivation is concerned. As with other chrysophytes, it possesses endogenous mechanism of regulation, limiting the number of vegetative cells within a population (Van Den Hoek et al., 1995).

8.2.3   GASES

Algae assimilate inorganic carbon during photosynthesis in two steps: (1) during light reaction, solar energy or other sources are converted to chemical energy with oxygen as by-product; and (2) during dark reaction, the chemical energy is used to assimilate carbon dioxide and converted into sugars (Fig. 8.2) (Larsdotter et al., 2006). This can be simplified as follows:

$$2H_2O + 2NADP^+ + 3ADP + 3Pi + light \rightarrow 2NADPH + $$
$$2H^+ + 3ATP + O_2 \tag{8.1}$$

$$3CO_2 + 9ATP + 6NADPH + 6H^+ \rightarrow C_3H_6O_3\text{–phosphate} + $$
$$9ADP + 8Pi + 6NADP + 3H_2O \tag{8.2}$$

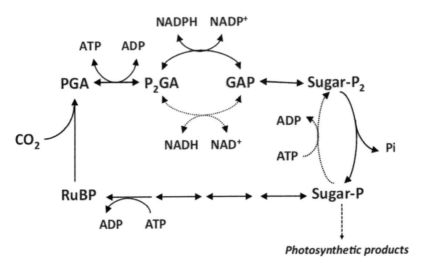

FIGURE 8.2 A generalized structure of the Calvin cycle (Igamberdiev and Kleczkowski, 2011).

The concept of fixating the carbon into biomass through photosynthesis is a sustainable way of sequestering CO_2. Important considerations are the growth rates and photosynthetic efficiency, resource requirements such as land, nutrients, and water, resistance to environmental stress, as well as the possibility of acquiring useful end products from cultivation. Compared to conventional forest, agricultural and aquatic plants, algal growth rate is approximately 50 times higher with better CO_2 fixation and more efficient carbon capture, utilizing small growth areas and lower energy consumption and costs. However, the use of algae is only considered feasible if they are used as biofuel feedstock rather than merely as a carbon sequester. Turning CO_2 emissions into a fuel via photosynthesis ensures recycling of carbon and reducing the demand for virgin resources (Packer et al., 2009). With simple metabolism and reproduction system, algal species could grow under harsh and varied conditions, with no necessity for land or freshwater making CO_2 biofixation economically viable (Packer et al., 2009).

In addition to pure CO_2, flue gases from industrial plants can also be used as feeds. For coal-fired thermal power plants, algal conversion of photosynthetic function and solar energy has been explored in two ways: (1) to use CO_2 gas separated from the flue gas; and (2) to use the flue gas directly. Direct use of flue gas in the cultivation system does not adversely affect algal growth and the production has been shown to be reliable, possible (Suali and Sarbatly, 2010), and advantageous due to energy saving, but may encounter problems such as high temperature, CO_2 concentration up to 15%, and the presence of SO_x and NO_x. Since flue gases from industries such as steel-making plants and thermal power stations contain about 500 times higher concentration of CO_2 [10–20% (v/v)] than that in the air, there may be inhibition of algal growth with requirement for large amount of nutrients such as nitrogen and phosphorus but low CO_2 conversion due to short gas retention time. Screening and selection of suitable algal strains having tolerance to high CO_2 concentration have been extensively carried out (Mata et al., 2010). The focus is to identify suitable algal strains that can grow under high CO_2 concentration while producing lipid. The desired strains should have the following characteristics: (1) high growth rate and biomass productivity, (2) high tolerance to trace amount of acidic components from NO_x and SOx, and (3) the ability to sustain growth even under extreme culture conditions (e.g. high temperature of water due to direct introduction of flue gases) (Chisti, 2007). Moderate SO_x and NO_x contents (a few tens of ppm) can be tolerated by algae (Brown, 1996; Lee et al., 2000; Lee et al., 2002), but higher concentrations may have moderate (Negoro et al., 1993) to strong inhibitory effects, depending on culture conditions and species (Yanagi et al., 1995). The causes of toxicity are not clear as these molecules can act directly on the cell physiology, or indirectly by changing the properties of culture medium. The deleterious effects of SO_2 (Matsumoto et al., 1997) and NO (Lee et al., 2002; Jin et al., 2005) can be significantly altered, if the pH of the media is regulated within the physiologically acceptable ranges.

Microalgal species suitable for CO_2 fixation include *Chlorella, Spirulina platensis, Emiliania huxley, Phaeodactylum*, and *Nannochloropsis* sp. (Negoro et al., 1993). The challenge of limited accessibility of land for large scale CO_2-capturing from industrial or power plants by microalgae can be overcome by sophisticated area-efficient techniques to recycle CO_2 (Suali and Sarbatly, 2010). A pilot scale system has been successfully developed to look into CO_2 recycling where *Scenedesmus obliquus* is shown to tolerate high concentration of CO_2 up to 12% (v/v) with optimal removal efficiency of 67% (Li et al., 2011). CO_2 tolerance of *Chlorella vulgaris* is enhanced by gradual increase of CO_2 concentration while *S. obliquus, Chlorella kessleri*, and *Spirulina* sp. have exhibited good tolerance (up to 18% CO_2) indicating great potentials for CO_2 sequestration from CO_2-rich streams (Ho et al., 2010). CO_2 consumption rate of 549.9 $mgL^{-1}d^{-1}$ for maximum *S. obliquus* biomass productivity of 292.5 mg/L and lipid productivity of 78.73 $mgL^{-1}d^{-1}$ (38.9% lipid content per dry weight of biomass), has been reported in two-stage system with 10% CO_2 (Ho et al., 2010). A study on *Dunaliella tertiolecta, Chlorella vulgaris, Thalassiosira weissflogii*, and *Isochrysis galbana*, representing four different phyla, grown with CO_2-enriched air or with a mixture of gases mimicking the composition of a typical cement flue gas (CFG) has suggested no effect of CFG. Dusts added at realistic concentrations also do not show any impact on growth. In the second stage, the culture exposed to the CFG receives an increasing concentration of dust characteristic of cement industry at concentration two ranges of magnitude higher and microalgal growth is inhibited (Amélie et al., 2013). Dust in flue gases may contain toxic compounds such as soot (Matsumoto et al., 1997) or trace metals (Borkenstein et al., 2011) and at inhibitory level can kill the cells.

8.2.4 SOLID WASTES

Nutrient-rich wastes from animal farms can pose great environmental challenges with regards to regional eutrophication and global warming. Excess runoff and discharge of nutrients such as 18% of N from farms and 25% of P from animal wastes have partly caused water quality deficiency in the Chesapeake Bay (the largest estuary in the United States) (Chesapeake Bay Foundation, 2004). Animal waste is a rising source of GHG emissions, including methane and nitrous oxide. The U.S. Environmental Protection Agency (EPA; http://www.epa.gov/methane/) estimates that GHG emissions from animal wastes have inceased by almost 60% between 1990 and 2009 (EPA, 2010). There is therefore a considerable interest in leveraging possible synergies between algae-derived energy production and the animal waste management initiative for environmental sustainability and economic consideration. Co-digestion of *Spirulina platensis* with waste activated sludge (WAS) could improve volatile solid (VS) reduction, but *Chlorella* sp. has a slight negative effect on dewaterability of the digestate as compared to WAS alone (Yuan et al., 2012). The major aspect in co-utilizing the solid waste with microalgae is to achieve optimized C/N ratio for biogas production. Olive mill solid waste (OMSW) is a pollutant coming from olive oil extraction by the two-phase centrifugation system and contains high organic matter content and unbalanced C/N ratio (31:1), resulting in reduced methane yields in the anaerobic digester. *Dunaliella salina* used as co-substrate with the OMSW in anaerobic digestion has enhanced substrate biodegradability when *D. salina* is increased from 25 to 50%. Maximum methane production of 330 mLg^{-1} VS is achieved at co-digestion mixture of 75% OMSW-25% *D. salina*, keeping C/N ratio at 26.7/1 (Fernández-Rodríguez et al., 2014). In another study, co-digestion of algae biomass residue and fats, oil, and grease

(FOG), each at 50% of the loading, allowing organic loading rates (OLR) up to 3 g^{-1} VSL^{-1} d^{-1} has resulted in a specific methane rate of 0.54 Lg^{-1} VS d^{-1} and a volumetric reactor productivity of 1.62 Lg^{-1} VS d^{-1}. Lipid-rich FOG is the key factor to achieve high methane yields, accounting for 68–83% of the total methane produced (Park and Li, 2012).

8.3 BIOENERGY

The world's primary energy need is projected to grow by 55% between 2005 and 2030, where fossil fuels will remain as the major source (International Energy Agency, 2009). Coal reserves are predicted to last over 200 years (Khan et al., 2009b) and will most likely account for half of the world's baseline electricity generation by 2015, before oil production reaching its peak by 2020 (Almeida et al., 2009). The main concerns are related to economic, ecological, and environmental impact of carbon fossil fuel and whether or not the conventional fuels should be left underground unmined and replaced with greener options. Alternative renewable energy such as biodiesel (Naik et al., 2010) and the bio-refinery set-up utilizing nutrient-rich wastes and capturing and utilizing CO_2 from flue gases can partly address the issue on CO_2 mitigation (Stewart et al., 2005). For biodiesel production, the required amount of biomass can be huge and the production cost should fall below $400/tonne of biomass to be economically feasible. This is still far from the price currently reported in a full-scale plant, where the cost for even a medium-scale plant still 173 times more expensive (Chisti, 2010; Acién et al., 2012). The concept of using algal biomass as a potential source of biofuel is therefore promising and now being taken seriously because of the fluctuating petroleum prices and the concerns about global warming and climate change (Gavrilescu and Chisti, 2005). It has become pertinent to improve the economics of bioenergy generation from algae by understanding and improving the algal biology through genetic and metabolic engineering, the use of hybrid bioreactors for a more controlled environment (Chisti, 2007; Abdullah et al., 2015b) and efficient downstream processes. A conceptual model for integrated microalgal biomass and biofuel production is shown in Figure 8.3. Renewable biofuels that can be developed based on algae include biodiesel (from algal oil), biomethane (by anaerobic digestion of the algal biomass), photobiologically produced biohydrogen, and bio-ethanol (by fermentation of the algal carbohydrates) (Gavrilescu and Chisti, 2005; Spolaore et al., 2006; Park et al., 2012).

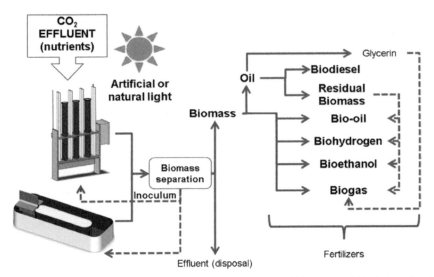

FIGURE 8.3 A conceptual model for integrated microalgal biomass and biofuel production (Demirbas, 2011).

8.3.1 BIODIESEL

Biodiesel is a promising alternative to petro-diesel for vehicles and internal combustion engines (Sgroi et al., 2005). Lipids, presently extracted for bio-diesel production, are mainly derived from oily seeds such as soybean, palm, castor bean, peanut, sunflower, corn, rapeseed, and cotton. Cultivation of these oil crops does not actually replace the natural energy resources as there may be a need for large available areas, suitable soil, high-quality water, seasonality and in some cases causing demineralization, salinization, desert-ification, soil erosion, reduction of water sources, and extensive use of pesti-cides (Gerpen, 2005; Knothe, 2005; Cheng et al., 2014). In a tropical climate region, the mean annual productivity of microalgal biomass is 1.53 $kgm^{-3}d^{-1}$, with a mean lipid content of 30%. The concentration per hectare of the total area is around 123 m^3 for 90% of the year, since the remaining 10% may be for perpetuation and cleansing of the reactors, for the yield of biodiesel from microalgae around 98.4 m^3 ha^{-1}. For the production of 5.4 billion m^3 of bio-diesel, an area of approximately 5.4 Mha must therefore be cultivated, which represents only 3% of the area currently used for cultivation of plants for biodiesel production. This is a possible scenario even with algal lipid content of 15% dry weight (Chisti, 2008). To meet the required demand and for 5% biodiesel (B5) addition to mineral diesel oil, the production of vegetable oils may be boosted by 50–100%, which is a difficult goal to achieve as it represents a proportional increase in arable land with oil crops, and the current agricultural productivity makes it harder to increase. As the concentration of fatty acids and productivity of algae are much higher than that of plants, the effort to increase oil production with energy crop cultivation would not be so great (Maa and Hanna, 1999). Biodiesel output per required land area is estimated at: corn – 145 kg oil ha^{-1}, soybeans – 375 kg oil ha^{-1}, palm oil – 5000 kg oil ha^{-1}, and algae – 80000 kg oil ha^{-1} (Skjanes et al., 2007). The amount of land needed for the corresponding production using microalgae would be around 100–200 times less (Costa et al., 2011).

Algal cultivation can be an exceptional source of lipids and fatty acids (Colla et al., 2004; Johnson and Wen, 2009; Cheng et al., 2014). Among microalgae species, the levels of 20–50% lipids are quite common. *Chlorella* has been reported with 50% lipids and *Botryococcus* achieves 80% lipids (Powell and Hill, 2009). The variations can be attributed to different growing conditions especially the CO_2 levels and methods of extraction of lipids and fatty acids. In areas where algae are grown for biodiesel production alongside fossil-fuel power stations, CO_2 from flue gases can be utilized and the lipid content has been reportedly increased (Brown and Zeiler, 1993; Sawayama et al., 1995). The alkane chain distribution of microalgae is very similar to that of mineral diesel (Miao et al., 2004). Analyses of the saturated fraction of biofuel from *Chlorella protothecoides* demonstrate that the alkane chain reaches 10–30 carbons, while the alkane chain of the saturated fraction of biofuel from *Microcystis aeruginosa* records 10–28 carbons. Microalgal biofuel also has lower O content and a higher H/C ratio than biofuel from plants, sunflower and cotton although the content in the former may be lower. The H/C and O/C mean molar ratios of microalgal biofuel are 1.72 and 0.26, while the plant biofuel is 1.38 and 0.37, respectively (Miao et al., 2004). High O content is not attractive for the production of transportation fuels. Biofuels from *C. protothecoides* and *Microcystis aeruginosa* have high calorific values of 30 and 29 MJkg^{-1}, respectively, due to the high C and H content and low O content, conferring higher stability, lower viscosity, and lower density than biofuel from plants. This high H content of algal biofuel is due to the chlorophyll and proteins (Miao et al., 2004).

8.3.2   BIOMETHANE

Microbial activities annually generate some 590–880 million tonnes of methane released into the atmosphere worldwide with about 90% coming from the biogenic sources. Methane is over 20 times more effective in trapping heat in the atmosphere than CO_2 over a 100-year period (EPA, 2010). Methane can be used as fuel gas and converted to produce electricity (Vergara-Fernandez et al., 2008) and CO_2 can be removed should pure methane is to be used (Hankamer et al., 2007). Anaerobic digestion of plant-based lignocelluloses and organic waste materials such as cow dung, pig slurry, effluent from slaughter houses, palm oil mill effluent (POME), and landfill produce biomethane and biohydrogen (Ahmad et al., 2003; Basri et al., 2009). There are four sequential stages in anaerobic digestion process: hydrolysis, acidogenesis, acetogenesis, and methanogenesis. During hydrolysis, complex organic biopolymers (e.g. carbohydrates, lipids, and proteins) are hydrolyzed and broken down into soluble sugars. Fermentation carried out by bacteria converts sugars into alcohols, acetic acid, volatile fatty acids and gases containing H_2 and CO_2. The acids are primarily metabolized by methanogenesis into CH_4 (60–70%) and CO_2 (30–40%). Anaerobic production based on algae similarly produces a mixture of methane (55–75%) and CO_2 (25–45%). The main advantages of algae-based over the conventionally used plant biomass are that algae are grown in a liquid medium where the space available for cultivation is three-dimensional and the absence of lignin and lower cellulose content which lead to good process stability and high conversion efficiencies (Harun et al., 2010a). Furthermore, anaerobic digestion is suitable for high moisture content (80–90%) organic wastes which are the characteristic of wet algal biomass, and the remaining biomass can be reprocessed into fertilizers for sustainable agricultural practices, thus economizing the production costs. The conversion of algal biomass into biogas even recovers energy through the extraction of lipids for biodiesel production (Prasertsan, 1996; Li et al., 2008), or high-value biocompounds (Khan et al., 2005; Abdullah et al., 2015b).

8.3.3 BIOHYDROGEN

Hydrogen has wide applications in fuel cells, liquefaction of coal, and upgrading of heavy oils. The technological viability is dependent on the development of cost-effective, large-scale, sustainable production systems capable of replacing classical steam reforming of natural gas, petroleum refining, and coal gasification (Rupprecht et al., 2006). Efficient containers and absorbers/adsorbers for liquid hydrogen are needed to minimize leakage and risk of explosion. Hydrogen can be produced by steam reformation of bio-oils, dark and photo fermentation of organic materials, and photolysis of water catalyzed by special algal species (Kapdan and Kargi, 2006; Ran et al., 2006; Wang et al., 2009a). Biohydrogen production by photosynthetic microorganisms requires a simple solar reactor such as a transparent closed box, with low energy requirements as opposed to thermochemical or electrochemical technique via solar battery-based water splitting which need high energy requirements. Photosynthetic production of H_2 from water via biological process converts sunlight into useful chemical energy and the underlying phenomenon is discovered long time ago (Gaffron and Rubin, 1942), but the biotechnology progress is slow. Cyanobacteria are able to diverge the electrons emerging from the two primary reactions of oxygenic photosynthesis directly into the production of H_2 based on the use of solar energy and water (Tamagnini et al., 2007). Immobilized cells of *Clostridium acetobutylicum* are used to ferment various microalgae (*Arthrospira platensis*, *Nannochloropsis* sp., *Dunaliella tertiolecta*, *Galdieria partita*, *Chlorella vulgaris*, *Cosmarium* sp., *Nostoc* sp.) where the highest productivity of 0.35 mmol $H_2 L^{-1} h^{-1}$ has been reported with *Nannochloropsis* sp. (Efremenko et al., 2012). The use of 2.5 g *Nannochloropsis* with short thermal treatment and without acid addition elevates productivity to 1.07 mmol $H_2 L^{-1} h^{-1}$ (Nobre et al., 2013). Biohydrogen production from *C. vulgaris* and *D. tertiolecta* yields of 10.8 mL $H_2 g^{-1}$ VS and 12.6 mL $H_2 g^{-1}$ VS, respectively, using untreated anaerobic digested-sludge as inoculum (Lakaniemi et al., 2011). High H_2 production of 81 mLg^{-1} alga is obtained after 65 h fermentation with *C. vulgaris* (acid hydrolyzed) by *C. butyricum* (Liu et al., 2012).

In recent years, efforts have been directed toward decreasing the costs related to microalgae culture systems for the production of biofuels. Dark fermentation for biohydrogen generates CO_2 emissions and soluble metabolites (e.g. volatile fatty acids) with high COD as the by-products, which necessitate further treatments. Mixotrophic culture of an isolated *Chlorella vulgaris* ESP6 could be utilized to simultaneously consume CO_2 and CODs from dark fermentation, converting them to valuable biomass. Light intensity is adjusted to 150 mmolm^{-2}s^{-1} and food to microorganism (F/M) ratio of 4.5 to improve the efficiency of assimilating the soluble metabolites. The mixotrophic microalgae culture reduces the CO_2 content of dark fermentation effluent from 34 to 5% with nearly 100% consumption of soluble metabolites (mainly butyrate and acetate) in 9 days. The obtained microalgal biomass is hydrolyzed with 1.5% HCl and subsequently used as the substrate for H_2 production with *Clostridium butyricum* CGS5, giving a cumulative H_2 production of 1276 mlL^{-1}, a H_2 production rate of 240 mlL^{-1} h^{-1}, and a H_2 yield of 0.94 mol mol^{-1} sugar (Liu et al., 2013). Simple and inexpensive strategy to bio-prospect and cultivate mixed indigenous chlorophytes with high carbohydrate content for biomethane and biohydrogen production have been developed. Mixed microalgae from four different water-bodies in Queretaro, Mexico are grown in bold basal mineral medium and secondary effluent from a wastewater treatment plant using inexpensive photo-bioreactors. Large variations in microalgal genera diversity are observed based on different culture media and nitrogen sources. In secondary effluent, *Golenkinia* sp. and *Scenedesmus* sp. proliferate, while the carbohydrate content varies between 12 and 57%, with the highest volumetric productivity achieved at 61 mgL^{-1}d^1 and 4.6 gm^{-2}d^{-1}, respectively. The results indicate that mixed microalgae are a good feedstock for biomethane and biohydrogen production (Glenda et al., 2014).

8.3.4   BIO-ETHANOL

Ethanol has been globally considered as an alternative to petrol fuel as it reduces the levels of CO_2 and CO emission, lead, sulfur, and particulates (Willke and Vorlop, 2004). The largest bioenergy program that has taken place is in Brazil for sugar cane ethanol (Proálcool), which begins in 1976. Since the 1980s, ethanol has been an established alternative to fossil fuels in Brazil and produced mainly from sugar and starch (sugar cane, corn). The use of ethanol as a gasoline substitute avoids the emission of 9.56×10^6 tonnes of carbon per annum (about 15% of Brazil's total emissions). The government encourages ethanol production from sugar cane and the adaptation of motors to the Otto cycle to run on "pure" ethanol (hydrated alcohol with 96% ethanol and 4% water) or gasohol (78% gasoline and 22% anhydrous ethanol). Alcohol addition increases the gasoline's octane and removes the highly toxic tetraethyl lead additive. Ethanol has a calorific value of 22 MJL^{-1}, while gasoline is 33 MJL^{-1}, but the higher octane rating of ethanol and the compatibility to the engines and injection systems mean that the technical equivalence of ethanol per liter of gasoline is about 1.15 (Rupprecht, 2009). There are, however, issues related to environmental sustainability of bioenergy after the growth of ethanol in the world market especially with regards to the competition for agricultural land for food. The value of agricultural commodities reaches an unprecedented high in 2006, mainly those of grains. In 2008, the United States produces 600 million litres of alcohol from cereals for beverages where the conversion utilizes more energy than the sugar conversion into alcohol. This can even result in negative yields, which precludes the use of cereals from an energy point of view.

Thus, identifying alternative sources of bio-ethanol feedstock is of high priority (Xuan et al., 2009).

The common methods for bio-ethanol production are fermentation (biochemical process) or gasification (thermo-chemical process) (Singh and Gu, 2010). Traditionally, ethanol is produced by fermentation of biomass such as from energy crops and organic wastes (Xuan et al., 2009). The biomass preparation can be carried out with mechanical press or enzymatic cell wall break down to make carbohydrates more accessible, as well as breaking down the large molecules. When cells are disrupted, *Saccharomyces cerevisiae* is added for fermentation, where sugar is converted into ethanol and purified by distillation (Amin, 2009). The energetic yield of converting sugar into ethanol is estimated at 40%. Alcoholic fermentation is one of the end products for pyruvate at the end of the glycolytic pathway. It consists of anaerobic conversion to ethanol and CO_2 in two steps. In the first step, pyruvate is decarboxylated by pyruvate decarboxylase, releasing CO_2 and forming acetaldehyde, which is then reduced to ethanol by alcohol dehydrogenase (Lehninger et al., 2004; Park et al., 2012). CO_2 can be recycled during fermentation for residual biomass in anaerobic digestion for biomethane production, such that in essence all the organic matter is accounted for (Harun et al., 2010b; Harun et al., 2010c; Singh and Gu, 2010).

Algal feedstock can be advantagous as they are less resistant to conversion into simple sugars than plant biomass. Algal species such as *Chamydomonas* sp., *Chlorella* sp., *Oscillatoria* sp., *Cyanothece* sp., *S. platensis*, *Chlorella vulgaris* (Ueno et al., 1998; Branyikova et al., 2011), and *Chlamydomonas reinhardtii* UTEX 90 (Choi et al., 2010) build-up their energy reserves in starch, which is an efficient carbohydrate feedstock. After oil extraction from the biomass, fermentation ensues utilizing gluco–amylase, α–amylase, and yeast, bacteria or fungi to convert sugars to ethanol and CO_2 with used water recycled (Dismukes et al., 2008). Enzymatic hydrolysis of *C. vulgaris* FSP-E biomass (containing 51% carbohydrate per dry weight) gives a glucose yield of 90.4% (or 0.461 g g^{-1} biomass). The separate hydrolysis and fermentation (SHF) and simultaneous saccharification and fermentation (SSF) processes convert the algal hydrolysate into ethanol with 79.9% and 92.3% theoretical yield, respectively. Hydrolysis with 1% sulfuric acid is efficient in saccharifying *C. vulgaris* FSP-E biomass, achieving a glucose yield of nearly 93.6% with starting microalgae biomass concentration of 50 gL^{-1}. Using acidic hydrolysate of *C. vulgaris* FSP-E as feedstock, the SHF produces 11.7 gL^{-1} ethanol at 87.6% theoretical yield. *Chlorococcum* sp. achieves bio-ethanol production of around 38% dry weight (Harun et al., 2010c), while *Chlorella vulgaris* can achieve conversion efficiency above

65% (Ueno et al., 1998). These show the feasibility of using carbohydrate-producing microalgae as feedstock for bio-ethanol fermentation (Ho et al., 2013), attributable to their high carbon composition and direct availability for fermentation or after pre-treatment.

8.4   LARGE-SCALE APPLICATION

8.4.1   FACTORS FOR OPTIMAL PRODUCTIVITY

Successful cultivation of microalgae needs favorable environmental conditions, which may differ from species to species. The main parameters influencing biomass productivity include nutrients, light exposure, intensity and wavelength, temperature, CO_2 concentration, pH, salinities, and mixing.

8.4.1.1   NUTRIENTS

In large scale algal cultivation, providing optimum nutrient balance is esstential for optimum growth rate and maximum lipid productivities. Inorganic elements, macronutrients, vitamins, and trace elements build-up algal cells. The composition variations and different environmental conditions should affect the performance of anaerobic digestion based on their digestion possibility. The mineral composition must meet the nutrient necessities of the anaerobic microflora. Major components in algal composition are carbon, nitrogen, phosphorus, and metals such as iron, cobalt, and zinc (Grobbelaar and Richmond, 2004). The required macronutrients are nitrogen and phosphorus (16N:1P ratio) and silicon (Richmond, 2004). The composition of proteins (6–52%), lipids (7–23%), and carbohydrates (5–23%) is strongly dependant on species (Brown et al., 1997). The normal C/N ratio for freshwater microalgae is 10.2 (Christenson and Sims, 2011). For some species, the high proportion in proteins is characterized by a lower C/N ratio as compared to terrestrial plants. The combination of different substrates is a strategy to increase the performance of a digester by ensuring optimal influent composition and enhanced biogas productivity. When C/N is lower than 20, there is an inequality between carbon and nitrogen necessary for anaerobic microflora, leading to nitrogen release and may become inhibiting from the accumulation of volatile fatty acids (VFAs) (Leadbeater, 2006; Yen and Brune, 2007).

At 120–150 gL^{-1} KNO_3, 12–15 gL^{-1} Na_2HPO_4, and 4–5 gL^{-1} $FeCl_3$ in conway media formulation in seawater, *N. oculata* achieves the highest cell density of 72 × 10^6 cells mL^{-1} with maximum biomass of 0.8 gL^{-1} and 35% lipid, but *T. suecica* only attains 46.5 × 10^6 cells mL^{-1} although almost comparable 0.7 gL^{-1} biomass and 27% lipids. The maximum specific growth rates are generally 0.16–0.18 d^{-1} with doubling time of 3.78–4.62 days in 250 mL shake flask and 5–300 L tanks. Both *I. galbana* and *P. lutheri* achieve 17 × 10^6 cells mL^{-1} with 0.7 gL^{-1} dry weight and 24–37% lipid. The lower cell density suggests that cell division may be lower, but the cellular composition making up the dry weight can be higher. Under nutrient deficiency conditions at 10–65 g L^{-1} KNO_3, 3–7.5 g L^{-1} Na_2HPO_4, and 2.5 gL^{-1} $FeCl_3$, the cell growth (g L^{-1}) of *N. oculata* (0.64), *T. suecica* (0.49), *I. galbana* (0.54), and *P. lutheri* (0.38) is much reduced, but the lipid accumulation remains in the range 23.6–37.3% (Shah, 2014).

8.4.1.2 LIGHT INTENSITY

Microalgae carry out photosynthesis and cellular division in the presence of light that accessibility and light intensity may affect the success or failure of the cultures. The light-harvesting antennae of algal cells are efficient and can absorb all the lights striking them, even if not used for photosynthesis (Richmond, 2004). The net growth may become zero at low intensities, but photosynthesis increases with increasing intensities until a point with maximum growth rate (saturation point). Most algae get light saturated at about 20% of solar light intensities. Increasing light intensity beyond this point may not affect the growth rate and instead can lead to photo-oxidation, damaging the light receptors, and thereby reducing the photosynthetic rate and productivity (photo-inhibition) (Molina et al., 1999). A study on cyanobacterium *Aphanothece nageli* in a bubble column closed photobioreactors (PBRs) shows that a linear reduction is found in the CO_2 fixation rate and biomass production with reduction in light regime, while the 12:12 (night:day) cycle results in average biomass production and carbon fixation, suggesting the pre-adaptation of microalgae to light regime charges (Jacobe-Lopes et al., 2009; Kesaano et al., 2014). Light intensity as high as 400 µmol photons $m^{-2}s^{-1}$ has resulted in the highest biomass yield of *Scenedesmus* sp. (3.88 gL^{-1}) with equally high lipid (41%), neutral lipid content (32.9%), oleic acid (43–52%), palmitic acid (24–27%), and linoleic acid of 7–11% (Liu et al., 2012). Although this may be good in terms of productivity, utilizing high light intensity or prolonged photoperiod may defeat the purpose of developing green-alternatives with reduced additional energy operating cost.

8.4.1.3   TEMPERATURE

Generally, increase in temperature leads to exponential algal growth until an optimum level is reached, after which cell growth starts declining. Ambient temperature fluctuations can result in diurnal temperature differences as much as 20 °C which can affect productivity (Molina et al., 1999). This affects photosynthesis and creates such condition under which photo-inhibition may occur during low level of light intensity and sub-optimal temperatures. Generally, constant temperatures above the optimal range may cause cell death, but temperatures below the favourable range, excluding freezing conditions, may not. Low biomass during dark periods can therefore be enhanced by increasing the temperatures. Maximum formation of bio-ethanol from *Chlorella* sp. cultivated at 30 °C is 448 molg^{-1} dry weight, but is much reduced to 196 molg^{-1} dry weight at 20 °C. Ethanol production decreases when temperature is increased to 35 °C and is completely inhibited at 45 °C. Enzyme activity at 35 °C is lower than at 25 °C, indicating that enzymes may be denatured at too high a temperature but activated within acceptable temperature range (Ueno et al., 1998). Anaerobic digestion can operate in both mesophilic (35 °C) and thermophilic (55 °C) conditions but the difference is in the reaction rate constant where mesophilic digestion of *Ulva* sp. results in 180 mLg^{-1} VS of methane, but with slower breakdown of organic compounds (Otsuka and Yoshino, 2004; Chynoweth, 2005).

8.4.1.4   GAS EXCHANGE

Roughly, 45–50% of algal biomass is made up of carbon. If oxygen concentrations increase above the saturation level, photo-oxidative damage may occur to the chlorophyll reaction centers which inhibit photosynthesis and decrease productivity (Molina et al., 1999; Pulz and Gross, 2004). With low percentage of CO_2 in the air (0.033%), algal growth rate can be limited if additional carbon is not supplied. Generally, CO_2 is blended with air in aerated cultures or injected into the cultures through gas exchange systems in PBRs or sumps in open raceways. To reduce the losses of CO_2 in open system, bubbling can be affected through air stones, with perforated pipes and plastic dome exchangers, injection into deep sumps, trapping the CO_2 under floating gas exchangers, and maintaining high alkalinities in the culture water (Richmond, 2004).

8.4.1.5   SALINITY AND pH

Although microalgae can tolerate wide range of salinities, changes due to evaporation and rainfall are the major factors affecting the growth of marine microalgae in open system (Richmond, 2004). During hot conditions, salinity will increase due to evaporation and decrease during rainfall. Microalgae are also sensitive to pH changes that control of pH may be essential to maintain high growth rate. Fresh and marine algal species grow at pH 7.6–10.6. Some freshwater species grow well up to 10.6, while some are badly affected by higher pH. The optimum initial pH 8 and salinity of 35 ppt in conway media and seawater achieve the cell dry weight of 0.82, 0.72, and 0.58 gL^{-1} and lipid content of 35.7, 33.5, and 37.3%, respectively, for *N. oculata*, *T. suecica*, and *P. lutheri*. At 25 ppt NaCl, irrespective of pH tested (pH 6–9), the biomass remains below 0.5 gL^{-1} and the lipid content below 25%. This may suggest the role of osmotic stress in regulating influx or efflux of nutrients and ions for improved cell growth and lipid accumulation (Shah et al., 2014b).

8.4.1.6   MIXING

In large scale cultivation, almost all available light may be absorbed only by a thin top layer of algal cells. This can be avoided by having proper mixing to keep the cells in motion, keeping consistent exposure to light and nutrients. While light reduction inside the reactor is not influenced by mixing, there is a complex interaction between culture mixing and the light attenuation as each single algal cell passes through dark and light zones of the reactor in a more or less statistical manner (Barbosa et al., 2003). Mixing is necessary to prevent algae sedimentation, avoiding cell attachment to the reactor wall for equal cell exposure to the light and nutrients, improving gas exchange between the culture medium and the air, and reducing the boundary layer around the cells, facilitating increased uptake, and exudation of metabolic products (Lou and Al-Dahhan, 2004; Richmond, 2004; Kommareddy and Anderson, 2005; Carvalho et al., 2006).

8.4.2   REACTOR ENGINEERING

8.4.2.1   OPEN AND CLOSED SYSTEM

For CO_2 sequestration, issues that need to be addressed include how efficiently the microalgae could use CO_2 in order to avoid excess release into the atmosphere, the design of effective reactor system, and irrigation, planting, fertilization, and harvesting. Since the 1950s, open ponds consisting of natural waters (lakes, lagoons, and ponds) and artificial ponds or containers have been used for algae cultivation (Borowitzka, 1999). The major advantage of open pond is it does not compete for land with accessible agricultural crops, but can be implemented in areas with marginal crop production potential (Chisti, 2007). It also has lower energy input requirement (Rodolfi et al., 2008), and easy regular maintenance and cleaning (Ugwu et al., 2008) with greater potential to return large net energy production (Rodolfi et al., 2008). Raceway ponds, typically closed loop, with mixing and oval shaped recirculation channels, between 0.2 and 0.5 m deep to stabilize algae growth and productivity, are the most frequently used artificial system (Fig. 8.4). Raceways ponds are usually built in concrete, but compacted earth-lined ponds with white plastic have been reported (Jiménez et al., 2003). In a continuous production cycle, algal broth and nutrients are introduced at the front of the paddlewheel and circulated through the loop to the harvest extraction point. The paddlewheel is in continuous operation to provide mixing and

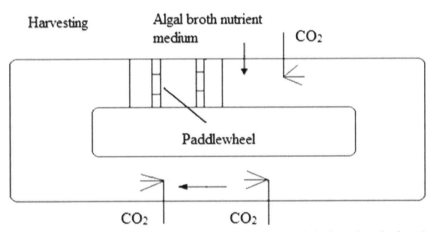

FIGURE 8.4   Plan view of a raceway pond. Algae broth is introduced after the paddlewheel, and completes a cycle while being mechanically aerated with CO_2. It is harvested before the paddlewheel to start the cycle again (Chisti, 2007).

avoid sedimentation. The CO_2 requirement is usually from the surface air, but submerged aerators can be installed to improve distribution (Razzak et al., 2013). For open and outdoor culture, the ability to control temperatures, however, is limited due to atmospheric temperature, solar irradiance, and humidity. To sequester industrial CO_2 outputs, it has to take into account that during night time and cloudy days, algal reproduction rates slowdown, thus taking up less CO_2. Installation of gas storage facilities to cope with the influx of CO_2 at night may be necessary.

Pollution and contamination risks may preclude open pond for the preparation of high-value products in the cosmetics and pharmaceutical industries (Ugwu et al., 2008). Various species in open system with different utilization efficiencies may also directly release excess CO_2 into the atmosphere, while closed systems allow for easy control of cultivation and CO_2 release (Brennan, 2010). A closed-loop system with innovative uses of resources can lead to both more sustainable and cheaper production. PBR technology can overcome some of the major problems associated with open-pond systems. PBRs such as tubular, flat-plate, and column reactors allow culture of single-species for prolonged periods with lesser risk of contamination (Chisti, 2007). Comparison of different closed PBR systems and biomass productivities is shown in Table 8.1. Flat-plate PBRs especially have received much attention for mass cultures due to the large surface area exposed

TABLE 8.1 Comparison of Different Closed Photobioreactor Systems.

Photobioreactor Type	Light Source	Capacity (L)	Algal Strain	Biomass (g/L) [a]mg/(L.d)[b]	References
Tubular	Solar radiation	500	*Scenedesmus obliquus*	284[b]	Hulatt and Thomas (2011a)
	Artificial	1.4	*Dunaliella tertiolecta*	830[b]	Hulatt and Thomas (2011b)
	Artificial	0.26	*Anabaena variabilis*	750[b]	Yoon et al. (2011)
	Artificial	2	*Chlorella* sp.	111.8[b]	Rasoul-Amini et al. (2011)
Airlift	Artificial	3	*Haematococcus pluvialis*	4.09[a]	Kaewpintong et al. (2007)
	Artificial	170	*Chaetoceros*	0.80[a]	Krichnavaruk et al. (2007)
	Artificial	3.7	*Chlorella vulgaris*	750[b]	
	Artificial	3.2	*Scenedesmus* sp.	2.27[a]	Sarah and Jones Susan (2014)

TABLE 8.1 *(Continued)*

Photobioreactor Type Source	Light	Capacity (L)	Algal Strain	Biomass (g/L) [a]mg/(L.d)[b]	References
Bubble column	Artificial	200	*Chlorella ellipsoidea*	31.55[b]	Wang et al. (2014)
	Artificial	1.8	*Cyanobium* sp.	0.071[a]	Henrard et al. (2011)
	Artificial	3.5	*Spirulina*	4.13[a]	De Morais and Costa (2007)
	Artificial	50	*Fistulifera* sp.	0.50[a]	Reiko et al. (2014)
	Artificial	1.8	*Sc. obliquus*	2.12[a]	De Morais et al. (2007)
	Artificial	1.5	*Spirulina platensis*	1.83[a]	Ankita et al. (2014)
Flat plate	Articial	3.4	*Dunaliella*	1.5[a]	Barbosa et al. (2005)
.	Solar radiation	60	*Chlorella zofingiensis*	41.3[b]	Feng et al. (2011)
	Articial	15	*Chlorella pyrenoidosa*	1.3[a]	Huang et al. (2014)
	Articial	440	*Chlorella pyrenoidosa*	124[b]	Qiang et al. (2014)

to illumination (Samson et al., 1985; Ugwu et al., 2008), high densities of photoautotrophic cells (> 80 gL[-1]) (Hu et al., 1998), low accumulation of dissolved oxygen, and high photosynthetic efficiency as compared to tubular PBRs (Richmond et al., 2000). The reactors are made of transparent materials for maximum solar energy capture, and a thin layer of dense culture flows across the flat plate, allowing radiation absorbance in the first few millimetre thickness (Hu et al., 1998; Richmond et al., 2008). Closed flat panels mixed by bubbling air can achieve overall ground–aerial productivity as high as 0.27 gL[-1]d[-1] using 500 L flat-plate glass PBR with 440 L culture volume (Richmond and Cheng, 2001).

Tubular PBRs consist of an array of straight transparent tubes that are usually made of plastic or glass. This tubular array or the solar collector captures the sunlight for photosynthesis (Fig. 8.5). The solar collector tubes are generally less than 0.1 m in diameter to enable light to penetrate into a significant volume of the suspended cells. Algal broth is circulated from a reservoir (such as the degassing column) to the solar collector and back to the reservoir (Chisti, 2007, 2008). Tubular PBR is typically operated as a continuous culture during daylight and cannot be scaled-up indefinitely. There is also a design limitation on the length of the tubes, which affects the potential O_2 accumulation, CO_2 depletion, and the pH difference (Eriksen et al., 2008). Large closed tubular PBRs include the 700 m^3 plant in Klotze, Germany (Pulz et al., 2001) and the 25 m^3 plant at Mera Pharmaceuticals, Hawaii (Olaizola, 2000). The performance of column PBRs compares favourably with tubular PBRs (Sa et al., 2002) as they offer efficient mixing, high volumetric mass transfer rates, low cost, compacted and controllable growth conditions (Eriksen et al., 2008). The vertical column is aerated from the bottom, and illuminated through transparent walls (Eriksen et al., 2008), or internally (Suh et al., 2003).

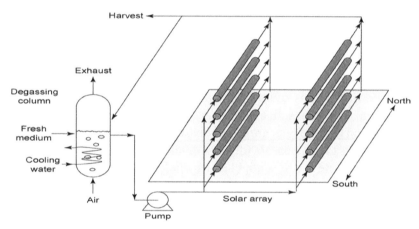

FIGURE 8.5   A tubular photobioreactor with fence-like solar collectors (Chisti, 2007, 2008).

8.4.2.2   LIGHT PENETRATION

Artificial production should attempt to replicate and improve the optimum natural growth conditions where phototrophic algae take up sunlight, absorb CO_2 from the air, and nutrients from aquatic habitat. Algae need light within the photosynthetically active radiation (PAR) to obtain energy by photosynthesis (Fernandes et al., 2010). The wavelength of the PAR ranges from 400 to 700 nm, which is equal to the visible light (Kommareddy and Anderson, 2003). The attenuation of light intensity in the PBR is dependent on its wavelength, cell concentration, reactor geometry, the light penetration distance and the path length, and the algal light absorption (Grima et al., 1994;

Fernandes et al., 2010; Bitog et al., 2011). In a dense culture, the gradient of light varies along the radius of the PBR (Grima et al., 1994; Bitog et al., 2011). The use of natural conditions for commercial algae production should take advantage of using sunlight (Janssen et al., 2003). For outdoor production systems, light is generally the limiting factor (Pulz and Scheinbenbogan, 1998) as this will depend on sunlight availibility which varies with diurnal cycles and seasons; thereby limiting the viability of commercial production to areas with high solar radiation. To address the limitation, artificial means of employing fluorescent lamps are almost exclusively used for cultivation at pilot scale stages (Janssen et al., 2003). It becomes essential to know the absorption spectra of main algal accessory pigments present in different quantities in different algal groups. Diatoms generally have photosynthetic pigments that contain chlorophylls a and c, and fucoxanthin whereas green algae contain chlorophylls a and b, and zeaxanthin (Brennan et al., 2010; Bitog et al., 2011). Artificial lighting allows for constant production, but at considerably higher energy input. Frequently the electricity supply for artificial lighting is derived from fossil fuels thus negating the primary aim of developing a price-competitive fuel and increasing the systems carbon footprint. Because of this, although tubular PBRs are deemed more suitable for outdoor mass cultures since they have larger surface area to sunlight, large-scale production plants are ideally based on the combination of multiple reactor units to compensate for any limitation of individual unit such as the possible alternate use of sunlight and artificial lights.

8.4.2.3   GAS INJECTION

Aeration of CO_2-rich gas through PBR provides CO_2 to the algae, aids in deoxygenating the suspension, and provides mixing to increase the cycle frequency. For very dense cultures, CO_2 originating from the air can be limiting for algae that pure CO_2 supplemented to the air supply is a necessity. CO_2 addition buffers the water against pH changes as a result of the CO_2/ HCO_3 balance (Chen and Durbin, 1994). From economic point of view, a high aeration rate will lead to higher running costs and therefore not always recommended for large-scale application (Zhang et al., 2002; Bitog et al., 2011). Air enriched with 5 or 10% (v/v) CO_2 at 0.025 vvm (volume of air/ medium volume/time) may be cost effective for mass culture (Zhang et al., 2002). In a flat panel PBR, optimum aeration rate of 0.05 v/v has been proposed as sufficient to improve the mixing and mass transfer (Sierra et al., 2008).

8.4.2.4 HYDRODYNAMIC CONDITIONS

Hydrodynamic conditions play major role in ensuring light intensity delivery, sufficient CO_2 transfer, maintaining uniform pH (Kommareddy and Anderson, 2005), and enhancing biomass productivity (Lou and Al-Dahhan, 2004). Poor hydrodynamic condition leads to cell aggregation and three phase system (solid–liquid–gas) that is prone to reduce mass transfer (Panda et al., 1989). The productivity of *S. dimorphus* in a plate PBR is reportedly three times lower than the determined highest productivity of 0.54 $gL^{-1}d^{-1}$, attributable to the lower air volume in the plate PBR, and consequently poor mixing regime. However, too high a mixing rate may lead to shear-induced injury on cells and affects cell viability (Panda et al., 1989; Thomas and Gibson, 1990; Gudin and Chaumont, 1991; Carvalho et al., 2006). In a bubble column and airlift reactor, hydrodynamic condition is characterized by axial dispersion coefficient, mixing and circulation time, and Bodenstein number (Miron et al., 2004). The longer duration for air to reside in the medium with smaller bubbles in the bubble column reactors should result in higher mass transfer. A vertical glass tube with 5 cm diameter and 2.3 m height (4.5 L) established as a bubble column reactor with good light penetration achieves *Monoraphidium* productivity as high as 23 $gm^{-3}d^{-1}$ (Miyamoto et al., 1988). Bubble column has shorter mixing time than airlift reactor, but an airlift reactor is favoured because of its efficiency and biomass productivity (Fan et al., 2007; Oncel and Sukan, 2008; Ranjbar et al., 2008). Airlift reactor has more defined fluid flow and relatively higher gas–liquid mass transfer rates, while a bubble column is likely to cause uneven cell density along the length of the reactor, which may induce cell starvation and death (Fan et al., 2007). The light penetration path is also wider for bubble column than airlift PBR because of the longer path to the center of the column, and the cloud effect caused by chaotic rising bubbles in the column (Oncel and Sukan, 2008). The presence of a draft tube in airlift reactor results in a more effective mixing because of the internal loop for the culture to circulate through the draft, and down through the annulus between the housing column and outside the draft tube (Oncel and Sukan, 2008). Bubbles rising inside the draft tube provide less turbid zone in the annulus region, enabling better exposure to light. The growth and the productivity in a translucent vertical airlift PBR is reported at 109–264 $gm^{-3}d^{-1}$ for *Nannochloropsis* and between 32.5–95.3 $gm^{-3}d^{-1}$ for the *Chlorella* strain (James and Al-Khars, 1990).

8.4.2.5 HYDRAULIC RETENTION TIME (HRT) AND OLR

Retention time is the time required to degrade the organic matter completely depending on the process temperature and batch composition, and it is related to the microbial growth rate. The average retention time for waste treated in a mesophilic plant is 15–30 days and a bit shorter for thermophilic plant (Ekama et al., 2008). There are two types: the solid retention time (SRT) for the average time the bacteria (solids) are in the anaerobic digester, and the HRT. These should be sufficient for the active populations, especially the methanogens, and not limiting the hydrolysis which is usually the limiting-step of the overall conversion of complex substrates to methane. HRT and OLR are the main parameters in anaerobic digestion processes. OLR is the amount of VSs to be fed into the digester each day in a continuous process. As the OLR increases, the biogas yield increases to some extent. Above the optimal OLR, the VS degradation and biogas yield decrease due to overloading (Babaee et al., 2011). For slow biodegradability, HRT is an important deciding factor (Yen and Brune, 2007). When operated at high OLR and HRT, the methane yield is stable and maximal, but at lower OLR or minimum HRT, the methane yield is reduced. Optimal OLR and HRT depend on the type or composition of the algal substrate. When the cells are directly inoculated into the anaerobic process, accessibility of the intracellular content to the anaerobic microflora is limited by the resistance of the algal cell wall to hydrolysis. Thus, characteristics of the species make the difference for a given OLR or HRT (Yen and Brune, 2007; Sialve et al., 2009). For efficient conversion of organic matter, OLR and HRT must be chosen depending on the type or composition of the algal substrate.

8.4.2.6 OPERATING CONDITIONS

Table 8.2 summarizes the conditions and the corresponding methane conversion yield. Two main approaches can be developed: (1) a multi-specific biomass harvested from waste water treatment pond (Leadbeater, 2006); and (2) a mono-specific biomass grown in the laboratory (Munozn and Guieysse, 2006). The methane yield varies from 0.09 to 0.45 Lg VS^{-1} depending on the species and culture conditions, where the CH_4 proportion is 69–75% regardless of species and operating conditions. The most significant factor impacting CH_4 proportion is pH, which controls the speciation of the carbonate system and the release of CO_2. If the pH is high from NH_3 release during digestion, the gas content will shift more to CH_4. The oxidation state of the

TABLE 8.2 Thermal Pretreatment for Microalgae Biogas Production.

Microalgae Species	Reactor Conditions	Methane Yield (L CH$_4$ g VS^{-1})	Pretreatment Conditions	Solubilization Increase	Methane Yield Increase (%)	References
Thermal pretreatment Microalgal biomass[a]	CSTR, 28 days HRT	0.270[b]	100 °C; 8 h	n.d.	33	Chen and Oswald (1998)
Scenedesmus biomass	BMP	0.076[c]	70, 90 °C; 3 h	7, 11-fold	12, 220	González-Fernández et al. (2012)
Microalgal biomass[a]	CSTR, 20 days HRT	0.180	75, 95 °C; 10 h	n.d.	67, 72	Passos and Ferrer (2014)
Chlorella sp. and Scenedesmus sp.	BMP	0.336	120 °C; 30 min	29%	20	Cho et al. (2013)
Nannochloropsis salina	CSTR	0.130	100–120 °C; 2 h	n.d.	108	Schwede et al. (2013)
Acutodesmus obliquus and Oocystis sp.	BMP	35	33	0.97	0.26	Alzate et al. (2012)
Scenedesmus biomass	BMP	0.180	170 °C; 8 bar; 30 min	10-fold	81	Keymar et al. (2013)
Chlorella–Scenesmus	BMP	0.156	160 °C; 20 min	4.5-fold[d]	65	Mendez et al. (2014)

[a]Microalgal biomass grown in wastewater treatment open ponds; [b]Data expressed in L biogas/g VS; [c]Data expressed in L CH$_4$/g COD; [d]Data from solubilization of carbohydrates.

biomass which drives the proportion of methane released also influences the biogas quality. Microalgae have somehow received less consideration as compared to macroalgae as substrates for anaerobic digestion, although they contain sulphurated amino acids that digestion releases lower amount of hydrogen sulphide than the organic substrates. There is, however, the potential presence of ammonia in the biogas due to the high protein content (~50–60%) that hydrolysis may lead to high ammonium concentration which can be toxic to methanogens (Munozn and Guieysse, 2006). The increase in temperature from 15 to 52 °C improves methane production by *Spirulina maxima*, but the energetic balance becomes negative if the energy supplied for heating is considered. The productivity, mutually with the VS reduction, is enhanced with increment upto 35 °C. For multi-specific algae, a temperature increase from 35 to 50 °C improves the rate of algal biodegradability from 5 to 10%. Maximal methane productivity is achieved at 40 °C suggesting mesophilic temperatures as optimal for anaerobic digestion of microalgae (Chen, 1987; Sialve et al., 2009).

8.4.2.7   REACTOR SCALE AND CONFIGURATION

The effects of photoperiod, salinity and pH on growth, and lipid content of *Pavlova lutheri* microalgae for biodiesel production in small-scale and large-scale open-pond tanks have been reported (Shah et al., 2014b). The cultures grow well under 24 h illumination in 250 mL flask with the highest cell density of 13.3–14.1 × 10^6 cells mL^{-1} and dry weight of 0.45 gL^{-1}, achieving μ_{max} of 0.12 d^{-1} and lipid content of 35% as compared to 0.1 d^{-1} and 15% lipid in the dark. The salinity is optimum for the cell growth at 30–35 ppt, but the lipid content of 34–36% is higher at 35–40 ppt. Algal growth and lipid accumulation is optimum at pH 8–9. Cultivation in 5 L and 30 L tanks achieve μ_{max} of 0.13–0.14 d^{-1} as compared to 0.12 d^{-1} in small-scale and 300 L cultures. Comparing between 5 L PBR and 300 L tank under 19.3–24 h illumination and light intensity of 162–198 μmol photons $m^{-2}s^{-1}$, the highest cell density and biomass are shown by *N. oculata* at 82.6 × 10^6 and 63.7 × 10^6 cells mL^{-1} density with 0.96 and 0.72 gL^{-1} biomass, respectively, followed by *T. suecica* (59 × 10^6, 42.7 × 10^6 cells mL^{-1} density with 0.73, 0.58 gL^{-1} biomass, respectively), and *I. galbana* and *P. lutheri* (19.6–21.2 × 10^6, 15.1–15.9 × 10^6 cells mL^{-1} density with 0.52–0.66 gL^{-1} biomass, respectively). The lipid contents, however, are higher in 5 L PBR at 40.1 and 41.8% as compared to 30.7 and 32.1% in 300 L open tank for *N. oculata* and *P. lutheri*, respectively. Fatty acid profile for *N. oculata* suggests that heptadecanoic acid C17:0 (13.7%) and oleic acid C18:1 (7.4%) are enhanced in PBR, but palmitic acid C16:0 (22.1%) and palmitoleic acid C16:1 (9.9%) are reduced. Although the total saturated fatty acids (SFA) (57.0%) and monounsaturated fatty acids (MUFA) (17.7%) are comparable to previous study on optimum pH and salinity, PUFA (22.3%) is enhanced. For *P. lutheri* in PBR, palmitic acid C16:0 (34.4%) is higher than in 300 L tank, while both eicosapentaenoic acid (EPA) C20:5 (8.4%) and docosahexaenoic acid (DHA) C22:6 (6.9%) are slightly increased with the total SFA (47.9%) and MUFA (30.9%) comparable, but PUFA (18.9%) is elevated (Shah, 2014).

8.4.3   INTEGRATED PROCESS ENGINEERING

8.4.3.1   HARVESTING AND EXTRACTION

The reason algal fuels have yet to replace the fossil fuels is due to the cost of production and the major bottleneck for algal-based bulk commodities lies in efficient cell harvesting method. The feasibility for commercialization is hampered by high energy input requirement especially in the upstream processes. Life cycle assessment (LCA) on microalgae biodiesel production shows negative energy balance due to high energy input to harvest and dry the biomass (Sander et al., 2010). The economics can be improved with efficient cultivation technique and simple, low energy, and cost-effective downstream processing. Harvesting from either open pond or PBR also need to separate out the media and algae in the quickest way. Different physical-, chemical-, and biological-based methods can be applied depending on the type of algae, the requirements of the cell sizes and biomass concentrations (Salim et al., 2012). Among conventional methods are centrifugation, filtration, gravity sedimentation, flotation, and flocculation. Continuous centrifugation is currently the preferred process as it is rapid and efficient (Rawat et al., 2011), but requires high energy and a primary concentration step. Filtration making use of packed bed filters can be used with or without additional pressure and works best at low algal cell concentrations (Greenwell et al., 2010). Membrane processes have long been applied in different stages of algal cultivation and processing. These include cross-flow microfiltration, ultra-filtration, dialysis, forward osmosis, membrane contactors, and membrane spargers. Cross-flow, micro, and ultra-filtrations may have the same efficiency as centrifugation, but possibly at a much lower cost (Brennan and Owende, 2010; Greenwell et al., 2010; Mata et al., 2010). It can be implemented both as a standalone and as a coupled system [in membrane biomass retention PBRs (BR-MPBRs) or membrane carbonation PBRs] (C-MPBRs) (Bilad et al., 2014a). For smaller suspended algae, tangential flow filtration is more sufficient than dead-end filtration, but the major drawbacks are membrane fouling, huge costs of replacement (Uduman et al., 2010), and high power requirements (Danquah et al., 2009). Filtration by microstrainers is another commonly used solid–liquid separation technique but the problems encountered include incomplete solids removal and membrane fouling by bacterial biofilms. Although the first problem may be solved by using flocculation, regular cleaning or membrane replacement generate sizable costs (Rawat et al., 2011). The main benefit is in having no chemical additions, but the high power requirements do not make centrifugation and filtration attractive for large-scale applications (Uduman et al., 2010).

Sedimentation is a low cost harvesting option that can typically give concentrations of 1.5% solids (Uduman et al., 2010). Because of the fluctuating density of algal cells, reliability is also low (Shen et al., 2009). At settling rates of 0.1–2.6 cmh^{-1}, sedimentation is relatively slow, and much of the biomass may deteriorate during the settling time (Greenwell et al., 2010). Gravity sedimentation is simple and highly energy-efficient (Rawat et al., 2011), but only work for a relatively large size and that grow to high densities such as *Arthrospira* sp., or when the pH is increased and/or chemical flocculants are added to the water (Knuckey et al., 2006; Chen et al., 2011). The latter, however, incur additional cost. To induce flocculation, organic and inorganic flocculants can be applied (Vandamme et al., 2013). Another option is to induce auto-flocculation by interrupting or limiting CO_2 supply (Demirbas, 2010). Induced flocculation may only be successful for freshwater species (Vandamme et al., 2013), while flocculation of marine microalgae has been suggested as not-feasible (Vandamme et al., 2013). A study has actually shown that cationic polymeric flocculants are viable options to pre-concentrate marine cultivated microalgae before further dewatering (Lam et al., 2014). Different cationic polymeric organic flocculants have been tested on *Phaeodactylum tricornutum* and *Neochloris oleoabundans* grown under marine conditions. The effects of 10 ppm of the commercially available Zetag 7557 and Synthofloc 5080H flocculants on *P. tricornutum*, followed by 2 h sedimentation, show a recovery of 98 and 94%, respectively. The same flocculants and dosage for harvesting *N. oleoabundans* only achieve 52 and 36% recovery, respectively. Use of bioflocculant in combination with sedimentation can reduce energy demand based on centrifugation from 13.8 to 1.34 MJkg^{-1} W (Salim et al., 2012). Chitosan is an example of promising flocculant due to its high efficacy, low dose requirements, and short settling time (Naim et al., 2013). The recovery efficiency of *C. vulgaris* at 120 mgL^{-1} chitosan achieves the highest harvesting efficiency (HE) of 92% within 3 min, with the highest HE (99%) at pH 6.

Direct air flotation (DAF) is often used as an efficient clarification step, notably when treating water containing hydrophobic matter and algae (Demirbas, 2010; Sturm and Lamer, 2011). It involves injecting air at the bottom of a water column to form an upward stream of bubbles. Tiny air bubbles may attach to the algal surface and carry them to the surface, forming a concentrated layer of foam which can be separated by skimming. The main cost is related to the power required for air injection but chemical flocculation is often necessary prior to DAF which further increases the total harvesting costs (Christenson and Sims, 2011). Flotation under vacuum using a vacuum gas lift can be carried out at different airflow rates, bubble sizes, salinities, and harvest volumes. HE and concentration factor (CF) increase by around 50% when the airflow rate of the vacuum gas lift is reduced from 20 to 10 Lmin^{-1}. Reduced bubble size enhances HE and CF by 10 times when specific microbubble diffusers are used or when the water salinity is increased from 0 to 40%. The reduction in harvest volume from 100 to 1 L increases the CF from 10 to 130. An optimized vacuum gas lift allows partial microalgae harvesting using less than 0.2 kWhkg^{-1}, thus reducing the energy costs by 10–100 times as compared to complete harvesting processes, albeit at the expense of a less concentrated biomass (Bertrand et al., 2013). Ozoflotation is used to recover microalgae grown in treated wastewater where 79.6% as TSS of biomass can be harvested with ozone dose of 0.23 mg mg^{-1} of dried biomass. The amount of lipid extracted and fatty acids methyl esters (FAME) recovered doubles with ozone doses of 0.12–0.23 mg mg^{-1} of dried biomass as compared to when using centrifugation. The oxidative stability of biodiesel can also be enhanced by the effect of ozone on the degree of FAME saturation (Velasquez-Orta et al., 2014). Separation using ozoflotation is advantageous over dissolved air flotation and gas bubble flotation because it does not require flocculant or lower pH (Ya-Ling et al., 2010; Nguyen et al., 2013).

Magnetic separation can be applied for quick, simple, efficient, and reliable capture of cells and biomolecules from liquid solution by the functional magnetic particles driven by an external magnetic field (Yang et al., 2009). It can save time and energy associated with harvesting (Ling et al., 2011). A simple and rapid harvesting method by in $situ$ magnetic separation has been developed where Fe_3O_2 nanoparticles are added to culture both leading to the cells of $Botryococcus$ $braunii$ and $Chlorella$ $ellipsoidea$ adsorbed and then separated by an external magnetic field. Maximal recovery efficiency of more than 98% is achieved at 120 rpm stirring within 1 min. Maximal adsorption capacity on Fe_3O_4 nanoparticles are 55.9 mg dry biomass mg^{-1} particles for $B.$ $braunii$ and 5.83 mg dry biomass mg^{-1} particles for $C.$ $ellipsoidea$. Appropriate pH and high nanoparticle dose are favorable for good cell recovery. The functional nanocomposites provide a base for efficient microalgae harvesting with advantages such as rapid execution, low energy consumption, and improves water-use in the algal harvesting process. Fe_3O_4 nanoparticles functionally coated with polyethylenimine (PEI) contain high concentration of ANH_2 groups, for efficient harvesting of microalgae. The functional magnetic nanocomposites are 12 nm in diameter with 69.77 emu/g of saturation magnetization. For harvesting $Chlorella$ $ellipsoidea$ cells, the nanocomposite dosage of 20 mgL^{-1} achieves 97% HE within 2 min, and increasing the temperature could increase the HE. The adsorption capacity of the Fe_3O_4–PEI nanocomposites for the microalgal cells reaches 93.46 g DCW g^{-1} nanocomposites. The adsorption mechanism between naked Fe_3O_4 nanoparticles and the microalgal cells is through electrostatic attraction and nanoscale interactions (Yi-Ru et al., 2014). For simultaneous algal biomass harvesting and cell disruption, cationic surfactant-based harvesting and cell disruption (CSHD) method may be effective. With CSHD, the HE is more than 91% in less than 5 min and 97% in 90 min. Moreover, CSHD exhibits powerful ability to disrupt the cells with lipid recovery from the cells increases by 133% allowing the extraction of up to 100% of the total lipids from wet microalgal biomass with 80% water content (Wen-Can and Jong-Duk, 2013).

8.4.3.2 IMMOBILIZATION AND RECYCLING

Immobilizing microorganisms in a variety of matrices has been applied to enhance nutrient and heavy metal removal and to treat hazardous contaminants (Lebeau and Robert, 2006; Muñoz and Guieysse, 2006; Moreno-Garrido, 2008). Similar to biofiltration, there is a physical barrier between microorganisms and the surrounding environment. The most common way is through gel entrapment and encapsulation in polymers, in which natural polysaccharides such as agars, carrageenans, and alginates are used such as for microbial inoculants in agriculture (Bashan, 1998) due to their low toxicity and high transparency (Moreira et al., 2006; Ignacio et al., 2008; Moreno-Garrido, 2008; Cao et al., 2010). The microorganisms can be immobilized alive within the polymer because its pores are smaller than the microorganisms, while the fluid flows through and sustain cell metabolism and eventual growth (Cohen, 2001). Comparable heavy metal uptake to the free, non-immobilized biomass (Alejandro et al., 2010) has been reported for biomass immobilized in Ca alginate (Sag et al., 1995) and polyacrylamide (Nakajima et al., 1982; Wong et al., 1993). Most immobilization techniques can be easily modified and applied to algae, adding a design factor for the requirement of light. The polymer is mixed with microalgae cells and consequently stabilized with divalent ions to form immobilized microalgal beads through a nozzle. Freshwater *C. vulgaris* has been immobilized in alginate beads and found suitable for simplifying the overall separation process (Lan and Lee, 2012a). Since the immobilized beads are relatively large in size as compared to the free cells, a simple filtration method (e.g. sieving) would be sufficient to separate the beads from water without significant amount of energy input. Handling of microalgae biomass will become easier and feasible to be implemented in commercial scale. Applications are diverse including for organic pollutants and heavy metal removal and biosensor for toxicity measurement (Moreira et al., 2006; Ertúgrul et al., 2008; Ruiz-Marin et al., 2008; Vijayaraghavan et al., 2011).

Co-immobilization of microalgae and nutrients such as plant-growth-promoting bacteria is a solution for the low growth rate of immobilized microalgae (Gonzalez and Bashan, 2000). An entrapment matrix in which algal cells are embedded and grown can be employed at the beginning of the cultivation. For entrapment matrix using filamentous fungi, pelletization with *Aspergillus niger* under photoautotrophic and heterotrophic growth conditions is used to immobilize and grow the freshwater *C. vulgaris* (Zhang and Hu, 2012) where 63 and 24% of *C. vulgaris* are harvested, respectively. Pelletization of oleaginous filamentous fungi with microalgae may contribute to the enhancement of the total oil yield and the fatty acid quality. For cell recycle, a new and effective microalgae cultivation and pre-harvesting have been developed using a membrane PBR (MPBR) in which the bioreactor is coupled to membrane filtration. The membrane completely retains *C. vulgaris* and the biomass can be partly recycled into the bioreactor to maintain a high biomass concentration, enhancing the flexibility and robustness of the system. MPBR can operate at both higher dilution and growth rates, resulting in 9-fold increase at 0.42 gL^{-1} biomass. Pre-harvesting is achieved by applying variable CFs in the filtration stage. The permeate is recycled to the reactor as feed medium without affecting the algal growth, offering a substantial 77% reduction in the water footprint (Bilad et al., 2014b). Owing to the higher biomass productivities, the harvesting costs can be considerably reduced.

8.4.3.3 WASTE TREATMENT AND BIOENERGY CO-GENERATION

The integrated microalgae–bacterial system can improve the feasibility of microalgae biomass production for its further valorization (González-Fernández et al., 2011). Two 5 L PBRs used to treat wastewater from potato processing industry (RPP) and from a treated liquid fraction of pig manure (RTE) inoculated with *Chlorella sorokiniana* and aerobic bacteria at 24 ± 2.7 °C and 6000 lux for 12 h per day of light supply achieve the highest biomass of 26.3 mgL^{-1}d^{-1} with RTE wastewater but with lipid content reaching 30.2% in RPP and only 4.3% in RTE. Methane yield is highly influenced by the lipid content and the substrate/inoculum ratio where maximum methane yield of 518 mLg^{-1} COD is achieved at 30% lipid and the substrate/inoculum ratio of 0.5 (Hernández et al., 2013). In a study where auto/mixotrophic growth is evaluated by using domestic wastewater (WW) amended with glycerol, *C. vulgaris* and *Botryococcus terribilis* show biomass productivities of 118 and 282 mgL^{-1}d^{-1}, producing 18 and 35 mgL^{-1}d^{-1} of lipids, respectively, at the highest glycerol supplements (50 mM). If scaled-up to 200 m^{3}d^{-1} of WW for 240 days per year, the estimated biomass and lipid yields turn out to be 5.6 tonnes y^{-1} and 0.9 tonne y^{-1}, for *C. vulgaris*, or 13.5 tonnes y^{-1} and 1.6 tonnes y^{-1} for *B. terribilis*, respectively. The mixotrophic production of lipids can generate high-quality biodiesel based on the estimation fatty acids profiles and the whole process can be combined with the production of methane and bio-ethanol in a bio-refinery set-up (Cabanelas et al., 2013).

Co-digestion leads to the dilution of toxic compounds by maintaining reaction under toxic threshold level. Some co-substrates can stimulate enzymatic synthesis for improved anaerobic digestion yield. Methane production rate of 1.61 L L^{-1}d^{-1} has been reported under mesophilic condition when algal sludge is mixed with 60% waste paper with the OLR of 5 gVS L^{-1}d^{-1} (Yen and Brune, 2007). Digestion of algal cell walls with waste paper has a positive effect on the anaerobic digestion, an indication of the increased cellulase activity stimulated by specific nature of the waste paper. Increasing the C/N ratio from 4.2 to 6.2 using co-digestion of *S. maxima* with sewage sludge, enhance methane yield to 60%. Co-digestion of algae with effluent from canning facility and protein-extracted algae achieve optimal methane production of 62% at C/N ratio between 25 and 35. The optimal C/N ratio is between 20 and 35, close to the prescribed range known to have a positive effect on the methane yield. Lower ratios lead to potential inhibition due to the presence of excessive ammonia released whereas higher ratios may lead to nitrogen limitations (Golueke et al., 1957). Algal biomass can be fermented directly with co-fermentation of residual algae biomass after lipid extraction. Produced methane is comparable to the pig manure substrate co-cultivation with *Chlorella* sp. producing 0.45 m^3 methane kg^{-1} dry weight and the reported yield methane is typically around 0.3 L g^{-1} VS, which is about half of the theoretical maximum based upon biochemical composition of biomass (Kelly and Dwarjanyn, 2008). The low yield can be attributed to the recalcitrance of a few algal species to biodegradation and inhibition of microbiological conversion by ammonia released from the biomass. Mesophilic reactors have shown concentrations below ammonia toxicity with HRT of 20–30 days and OLR of 1–2 kg VS m^{-3}d^{-1} (Schwede et al., 2013; Passos and Ferrer, 2014). This is more critical under thermophilic conditions with higher possibility of NH_3 release. The tough algal cell wall may prevent high methane production, since organic matter retained in the cytoplasm is not easily accessible to anaerobic bacteria. Many organic substrates such as activated sludge and lignocellulosic biomass similarly consist of a complex structure, which reduces hydrolysis rate in the anaerobic digestion process (Carrère et al., 2010; De La Rubia et al., 2013; Monlau et al., 2013). Suitable pre-treatment must be developed to solubilize particulate biomass and improve anaerobic digestion rate and extent. Thermochemical and mechanical pre-treatments can solubilize biomass by breaking down cell walls which are resistant to biodegradation. These result in significant increase in methane production rates (ca. 30%) for sewage ponds treated with microalgal biomass (Chen and Oswald, 1998).

The cost of closed system is substantially higher than the open-pond system (Carvalho et al., 2006). The unit cost of producing *Dunaliella salina*, one of the commonly cultivated algae strains in an open-pond system is about $2.55 kg^{-1} of dry biomass, which is considerably too high to justify the production for biofuels. LCA and life cycle costing (LCC) have examined the integrated algae bioenergy production and nutrient management on small dairy farms. The most important challenge is in the procurement of low cost and low energy-intensive nutrients, most notably nitrogen and phosphorus. Nutrient procurement can represent up to 50% of energy consumption during algae cultivation when fertilizers are used (Clarens et al., 2010; Stephenson et al., 2010). Four cases are being considered: a reference land application scenario (REF), anaerobic digestion with land-application of liquid digestate (AD), and anaerobic digestion with recycling of liquid digestate to either an open-pond algae cultivation system (OPS) or an algae turf scrubber (ATS). LCA indicates that all the three improved scenarios (AD, OPS, and ATS) are environmentally good as compared to REF, exhibiting increases in net energy output up to 854 GJ y^{-1}, reductions in net eutrophication potential up to 2700 kg PO_4-eq y^{-1} and reduction in global warming potential up to 196 Mg CO_2-eq y^{-1}. LCC reveals that the integrated algae systems are much more financially attractive than either AD or REF, where the net present values (NPV) are estimated at $853,250 (OPS), $790,280 (ATS), $211,126 (AD) and $62,279 (REF). However, these results are highly dependent on the sale price for nutrient credits. Comparison of LCA and LCC results indicates that robust nutrient credit markets or other policy tools are required to align financial and environmental management, preferably with the ability for energy production systems and foster widespread adoption of sustainable nutrient management systems (Zhang et al., 2013).

8.5   CASE STUDY – PALM OIL MILL WASTES

8.5.1   OIL PALM WASTES AND UTILIZATION

It is estimated that more than 50 million tonnes of biomass are generated from the palm oil industry in Malaysia alone, mostly from the oil extraction process. This is about 4 kg of dry biomass from every kg of palm oil production, which comes in the form of the mesocarp fiber, empty fruit bunch (EFB) palm oil mill sludge (POMS), palm kernel cake (PKC), decanter cake, and POME. Most have not been extensively commercially re-used by the industry (Habib, 1979; Wu, 2009). Other solid residues generated during harvesting and replanting are shells, fronds, and trunks from the plantation area. In 2010, out of 88.74 Mt of fresh fruit bunches (FFB) processed, approximately 87 Mt of biomass being produced, excluding the fronds and trunks (Wu et al., 2007). With high moisture content of 60–70%, EFBs are difficult to use as fuel for power boilers but the shells, mesocarp and fibers can be used as boiler fuel to produce steam and electricity for the mills and for power generation. Partial EFB and decanter cake are utilized as fertilizers and soil cover materials in plantation areas. In Thailand, about 60 crude palm oil mills produce approximately 1.24 million tonnes of crude palm oil from 6.4 million tonnes of FFBs (Paepatung et al., 2006), resulting in the million tonnes per year of shell (0.13), decanter cake (0.27), fibers (0.894), and EFB (1.53). Shells are sold as solid fuel to other industry such as cement factories (Paepatung et al., 2006). For old mills, the EFBs applied in the plantation as fertilizer by mulching or disposed off in the landfill or burned in the incinerator to produce potash (Chavalparit et al., 2006; Ludin et al., 2009). PKC is rich in carbohydrate (48%) and protein (19%) and is used as cattle feed or processed into chicken feed. As PKC is nitrogen deficient, additional nitrogen addition such as using poultry manure, with goat manure as supplement to convert it into compost (Ismail, 2004).

8.5.2   POME TREATMENT

POME is a viscous brown liquid with fine suspended solids, with pH ranging between 4 and 5 (Poh and Chong, 2009). It is composed of high organic content mainly oil and fatty acids, carbohydrates (29.55%), proteins (12.75%), nitrogenous compounds, lipids, and a considerable amount of cellulose and non-toxic minerals which can be a good source for microbial fermentation (Wu et al., 2007). The percolation of POME into the waterways and eco-systems remain a major concern (Foo and Hameed, 2010). The high COD and BOD can cause considerable environmental problems and destruction of aquatic biota if discharged without proper treatment (Cheng et al., 2010; Singh and Dhar, 2011). The characteristics and the parameter limits for POME discharge into water courses in Malaysia are summarized in Table 8.3. Figure 8.6 shows different treatment systems used for POME treatment (Yi et al., 2012) which include: (a) anaerobic/facultative ponds (Wong, 1980; Rahim and Raj, 1982); (b) tank digestion and mechanical aeration; (c) tank digestion and facultative ponds; (d) decanter and facultative ponds; (e) physico–chemical and biological treatment (Andreasen, 1982); (f) evaporation (Ma, 1993) and clarification pond coupled with filtration and aeration (UNEP, 1994).

TABLE 8.3   Characteristics and Parameter Limits for POME Discharge into Water Courses in Malaysia.

Parameter[a]	Mean[b]	Range[c]	Parameter Limits for Watercourse Discharge[d]	Major Constituents[e]	Quantity (g/g dry sample)[f]
pH	3.5	3.4–5.2	5.0–9.0	Moisture	6.75
				Crude protein	9.07
Temperature	80	80–90	45	Crude lipid	13.21
BOD 3 days 30°C[b]	24117	10,250–43,750	100	Ash	32.12
COD	65272	15,000–1,00,000	–	Carbohydrate	20.55
Total solids	11,500	11,500–79,000	–	Nitrogen-free extract	19.47
Suspended Solids	68367	5,000–54,000	400	Total carotene	20.07
Volatile Solids	32743	9,000–72,000	–		
Oil and Grease	3546	130–18,000	50		
Ammoniacal Nitrogen	–	4–80	150[c]		
Total Nitrogen	385	180–1400	200[c]		

Units in mg/L except pH and Temperature (°C); The sample for BOD analysis is incubated at 30°C for 3 days; Value of filtered sample; [a,b]Source: Ma (1999) and Ahmad et al. (2014a, 2014b, 2014c); [c,d]Source: EQA (2001); [e,f]Source: Habib et al. (1979).

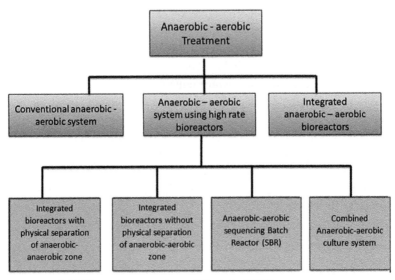

FIGURE 8.6   POME treatment systems (Yi et al., 2012).

At present, 85% of POME treatment is based on anaerobic and faculta-tive pond system, followed by open tank digester attached with extended aeration in a pond (Vijayaraghavan et al., 2007; Subramaniam et al., 2008; Fang et al., 2011). Although require relatively little energy to operate, con-ventional POME treatment has several disadvantages such as extensive land area, long retention time, low treatment effectiveness where only a fraction of nitrogen and phosphorus in the wastewater removed, more sludge produc-tion, and release of GHG in form of CO_2 and CH_4 (Órpez et al., 2009; Wu et al., 2010; Lam and Lee, 2011). The under-sized treatment systems in most mills are also unable to cope with increasing POME volume (Cheng et al., 2010; Singh and Dhar, 2011). New technologies that have been explored in-clude membrane technology, up-flow anaerobic filtration, up-flow anaerobic sludge blanket (UASB), and up-flow anaerobic sludge fixed film (UASFF) bioreactor. The main challenge is to balance environmental protection with economic viability and sustainable development (Ahmad et al., 2003), and meeting the tough environmental regulations.

8.5.2.1 PONDING (LAGOON) SYSTEM

Ponding system consists of the sequence of anaerobic, facultative, and aero-bic ponds with low investment cost (Tong and Bakar, 2004; Yacob et al.,

2006). There is less energy to manage due to limited mechanical mixing, operational control and monitoring (Yacob et al., 2006), but requires large land area to accommodate the ponds. Facultative and aerobic ponds are necessary to decrease the organic content in the wastewater before being discharged into rivers (Yacob et al., 2006; Poh and Chong, 2009). For anaerobic ponds, the optimum depth ranges from 5 to 7 m with 30–45 days HRT; for facultative ponds, 1–1.5 m depth with 15–20 days HRT (Tong and Bakar, 2004; Yacob et al., 2006) and a shallower depth of 0.5–1 m for aerobic ponds with 24 days HRT (Yacob et al., 2006). The raw POME feeding to the anaerobic digestion and aerobic/facultative treatment system normally have BOD and COD concentrations in a narrow range of 20,000–25,000 mgL^{-1} and 45,000–50,000 mgL^{-1}, respectively. Overtime, there will be scums developed on the POME surface and the building-up of solid at the bottom. The sludge and scums will clump together, thus reducing the treatment efficiency. These need continuous de-sludging by using submersible pumps or excavators to maintain efficiency.

8.5.2.2 AEROBIC TREATMENT

Aerobic treatment systems may differ in the type of oxygen (aeration system) and the designed loading rates in the form of facultative ponds (maturation ponds), oxidation ponds, aerated lagoons, and polishing ponds. The common method is the aerobic pond, and a few palm oil mills use the more advanced activated sludge system (Environmental Management Guideline for the Palm Oil Industry, 1997). Facultative ponds, oxidation ponds, and polishing ponds establish oxygen supply by photosynthetic activities of algae and plants by absorbing CO_2 from the atmosphere and releasing O_2 for bacteria or vice versa. Aerated lagoons are artificially aerated, where high temperature of the pond enhances the biochemical reactions, resulting in increased substrate removal even at low oxygen solubility in water (Environmental Management Guideline for the Palm Oil Industry, 1997). Large quantities of GHGs including methane and carbon dioxide are produced from open ponds and tanks raising environmental concern. The total emission can be substantially reduced if CH_4 is completely recovered in sealed anaerobic reactors or simply burnt to convert into CO_2 (Yacob et al., 2006).

8.5.2.3   ANAEROBIC TREATMENT

Anaerobic treatment systems have more advantages over aerobic treatment or other methods as they are almost energy free with low capital cost. There is a high organic load, excluding the acidification step and low formation of surplus sludge. No secondary treatment is required if the treated effluent is used for irrigation. However, if the treated effluent is discharged into the river, secondary treatments are essential in the form of aerobic treatment. During anaerobic process, dissolved organic substrates are mostly converted into biogas (a mixture of around 60% CH_4 and 30% CO_2). Only very little substrate is converted into biomass. The biomass should be separated as biological excess sludge or volatile suspended solids (VSS). Under usual operating conditions, less than 0.3 kg VSS kg^{-1} BOD is removed (Shilton et al., 2008).

8.5.2.4   OPEN DIGESTING TANK

Open digesting tank is used when limited land area is available for ponding system (Poh and Chong, 2009). An open digester tank is made up of mild steel with variety of volumetric capacities ranging from 600 to 3600 m^3 (Yacob et al., 2006). Shorter HRT is required ranging from 20 to 25 days as compared to 30–45 days for ponding system. However, the treated POME from the digester will still have to go through the facultative pond, followed by aerobic ponds for more degradation of organic material. Similar to ponding, no mechanical mixing is installed and the biomethane is released directly to the atmosphere (Tong and Bakar, 2004; Poh and Chong, 2009). Methane emission rate of 518.9 kgd^{-1} has been registered in open digesting tank as compared to 1043 kgd^1 in anaerobic pond, with 54.4% methane (Yacob et al., 2006). Open-pond and open digesting tank treatment systems therefore are not certified for Carbon Emission Reduction (CER) trading. One of the advantages of using open digester tank is in the removal of solids build-up at the bottom continuously, thus maintaining the desired treatment efficiency (Tong and Bakar, 2004). The dewatered sludge can be used as fertilizer. The major drawback is corrosion of the steel structures due to long contact with hydrogen sulfide. Incidents such as bursting and collapsed digesters have been reported (Yacob et al., 2006).

8.5.2.5 CLOSED DIGESTING TANK

Clean development mechanism (CDM) promotes closed anaerobic digesting tanks where biogas is captured and straightly used for flaring, boiler fuel or power generation (Tong and Bakar, 2004). A closed digesting tank has comparable design with an open digesting tank, except that a fixed or floating cover is included and set with other facilities like gas collector, safety valves, and monitoring devices (Sulaiman et al., 2009). Methane production rate from a large scale closed digesting tank has been reported at 5019 kgd^{-1} with 62.5% methane which is much higher than from the open-pond and open digesting tank (Tong and Bakar, 2004). The system contains three permanent stirred-tank reactors, in which two reactors are installed with floating roof and another reactor with a fixed cover. The total capacity of the three reactors is 7500 m^3 with HRT of 18 days. The system is equipped with dual-function mixing mechanism: the pump-aided circulation and the gas lifting mixing. The system has been operating continuously for over 20 years. The high efficiency of the design allows the highest recovery of biomethane potential from the optimally controlled POME digestion. The total biomethane captured and utilized as boiler fuel has been estimated to be 1407 tonnes y^{-1}, which is equivalent to 29,547 tonnes y^{-1} of CO_2-equivalent reduction (Hassan et al., 2004).

8.5.2.6 CONTINUOUS STIRRED-TANK REACTOR

Continuous stirred-tank reactor (CSTR) is the same as a closed-tank digester but with a mixer. The mechanical agitators mix the biomass thus improving the biogas production. It has been in use since early 1980s for POME treatment by Keck Seng Oil Palm Mill (Malaysia) Berhad in Masai, Johor (Tong and Bakar, 2004). CSTR has also been applied for coke wastewater treatment in aerobic conditions (Vazquez et al., 2006) and for dilute dairy wastewater treatment (Chen and Shyu, 1996). The CSTR in Keck Seng's mill has approximately 83% COD removal efficiency with 63% methane, while the CSTR treating dairy wastewater has 60% COD removal efficiency with 23–77% methane. High COD removal from POME between 94 and 98% has been achieved using CSTR at lab scale (Ugoji, 1997). The lower COD removal at the plant scale can be assumed due to the large volume of feed resulting in inefficient mixing. The existing CSTR has been improved by incorporating a biofilm support system (BSS) within a low-density nylon mesh, rolled into cylinders and inserted into the CSTR. The BSS functions as a support media for biomass growth and the efficiency of CSTR is improved without the need for biomass recycling (Ramasamy and Abbasi, 2000).

8.5.2.7 UASFF REACTOR

The drawbacks of anaerobic treatment are long HRT and long start-up period. The problem with long HRT can be rectified by using high-rate anaerobic bioreactors. The long start-up period can be shortened by using granulated seed sludge (Wu, 2010; Lam and Lee, 2011) or utilizing seed sludge from the same process or maintaining suitable pH and temperature in the high-rate anaerobic bioreactor for growth of bacteria consortia (Lorestani, 2006; Wu, 2010). Table 8.4 shows methane emission rate from anaerobic digestion of POME. UASB reactor and anaerobic filter have been integrated to form a hybrid bioreactor called up-flow UASFF. While eliminating the respective drawbacks, the hybrid reactor has the advantages of both reactors. UASFF is better in terms of biomass retention, reactor stability at the shock loadings, and operation at high OLRs, whilst overcoming the problems of clogging

TABLE 8.4 Methane Emission Rate from Anaerobic Digestion of POME.

Reactor Type	Conditions	HRT (days)	CH_4 Emission Rate (L g^{-1} COD removed)	COD Removal Efficiency (%)	References
Single stage stirred digester	–	6.2	0.325	96	Borja and Banks (1994)
Modified anaerobic baffled bioreactor	–	6	0.42	95	Faisal and Unno (2001)
Up-flow anaerobic sludge-fixed film (UASFF)	Mesophilic 38 °C	3	0.346	97	Najafpour et al. (2006)
Single stage	Mesophilic 37–42 °C	17	0.15[a]	97	Yacob et al. (2006)
Continuous stirred tank reactor (CSTR)	Mesophilic 37 °C	7	0.30[a]	71	Choorit and Wisarnwan (2007)
Continuous stirred tank reactor (CSTR)	Thermo-philic 55 °C	5	0.27[a]	70	Choorit and Wisarnwan (2007)

[a]Self-estimation.

and biomass washout in anaerobic filter and UASB (Borja and Banks, 1994; Ayati and Ganjidoust, 2006). Investigations using UASFF include the treatment of wood fiber wastewater, sugar wastewater, dairy wastewater, wash waters from virgin olive oil purification, and POME treatment (Najafpour et al., 2006; Wu, 2010). Anaerobic treatment with UASFF reactor can reduce up to 95% COD at an average OLR of 15 g COD $L^{-1}d^{-1}$. A 96% COD removal has been obtained at an OLR of 10.6 g COD $L^{-1}d^{-1}$ at COD concentration of 42,500 mgL^{-1} and HRT of four days (Lorestani, 2006; Siew, 2006; Wu, 2006). COD removal efficiency has been over 70% except for wood fiber wastewater as wood fiber is harder to degrade although the methane production is satisfactory at 0.346 Lg^{-1} COD. The internal packing and high ratio of effluent recycle are the main factors controlling the constancy of the UASFF reactor in POME treatment. Internal packing effectively retains biomass in the column while effluent recycle produces internal dilution to reduce effects of high OLR (Borja and Banks, 1994; Najafpour et al., 2006).

8.5.2.8 MEMBRANE TREATMENT

Membrane technologies in combination with coagulation/flocculation as pre-treatment achieve 78% water recovery from POME (Ahmad et al., 2003; Ahmad et al., 2011). The analyses of the reclaimed water show that the water qualities fulfill the drinking water standard set by the U.S. EPA. The study shows that membrane fouling is reversible and primarily due to cake formation. The treated effluent has a high-quality and crystal clear water that can be recycled or used as the boiler feed water or as the source of drinking water (Ahmad et al., 2003). A pilot plant has been developed for POME treatment where the first stage involves coagulation, sedimentation, and adsorption and second stage involves combination of ultra-filtration and reverse osmosis. Results show a reduction in turbidity, COD, and BOD upto 100, 99, and 99.4%, respectively, with a final pH of 7. Membrane ultra-filtration is also used as the tertiary treatment method. Combination of filtration–ultra-filtration treatment is the best overall treatment efficiency, with a reduction of 93.4% for total nitrogen, suspended solids turbidity, and colour content. For combination of centrifugation–ultra-filtration, the average removal efficiency is only 86.4%, while coagulation–ultra-filtration treatment only achieves an average of 67% removal (Wong et al., 2002).

8.5.2.9 FIBROUS-BED COLUMN

Adsorption is attractive because of its simplicity, convenience, and efficiency. Biosorbents from renewable sources are economical and biodegradable, and have potential for industrial applications (Li et al., 2011; McManamon et al., 2012). Methods for heavy metal removal such as chemical precipitation, chemical oxidation/reduction, reverse osmosis, electrodialysis, and ultra-filtration from aqueous solutions have low efficiency, high cost, and generate toxic wastes. The performance of raw *Ceiba pentandra* (L.) gaertn. [raw kapok fibers (RKF)] for oil sorption and POME treatment have been compared with structurally modified kapok [NaOH-treated (SKF), surface-modified kapok fiber (SMKF)], raw bentonite clay, and HCl-treated bentonite. For filtration under gravity at 0.08 gcm^{-3}, SKF shows high POME sorption of 82 gg^{-1}, but lower diesel sorption of 23 gg^{-1}. With HCl-treated bentonite, POME sorption at 69 gg^{-1} is only slightly higher than diesel sorption of 60 gg^{-1}. Both RKF and raw bentonite achieve higher removal efficiency of BOD, COD, YOC, and TN at 74–98% and 72–94%, respectively, than with SKF at 66–80%, and HCl-treated bentonite at 64–80% (Abdullah et al., 2015a). Fibrous bed filtration column studies have been carried out to evaluate the performance of RKF in removing heavy metals (Fe, Mn and Zn) from POME and residual oil from Crude Palm Oil–water emulsion at varying flow rate (5–10 mL min^{-1}) and packing density (0.04–0.08 gcm^{-3}). The best result is obtained at low flow rate (5 $mLmin^{-1}$) and high packing density (0.08 gcm^{-3}) of RKF achieving the removal percentage for Fe, Mn, and Zn of 98.8, 99.4, and 98.6%, respectively. For all packing densities and flow rates, the COD reductions exceed 99% whilst turbidity reduction was 90.7–95.6%. RKF is therefore a promising adsorbent for POME remediation and the removal of metallic ions from POME with excellent capability to reduce COD and turbidity of effluent contaminated with residual oil (Afzaal, 2014; Afzaal et al., 2014). Eco-friendly extraction of cellulose from EFB has been reported (Nazir et al., 2013), and the extracted cellulose has been successfully surface-modified with acetic acid and EDTA to be used as biosobent for heavy metal removal from water. The highest Pb (II) removal is achieved with EDTA-surface modification at 236.7 mgg^{-1} (Nazir, 2013). Higher sorption capacity is also achieved for the removal of Mn (II), Ni (II), and Cu (II) ions with EDTA-surface modification (39.3–54.2 mgg^{-1}) than acetic acid (36.4–45.3 mgg^{-1}) (Daneshfozoun et al., 2014).

8.5.3 ALGAL BIOMASS CO-UTILIZATION

8.5.3.1 AEROBIC AND ANAEROBIC TREATMENT

Sustainable energy management in palm oil mill has entered a new dynamic era with the opportunity of culturing microalgae using POME (Khan et al., 2005c; Khan et al., 2009b; Procedia, 2009). Algal treatment replacing conventional tertiary POME treatment can offer an oxygenated effluent and an ecologically safe, less expensive, and more efficient mean to remove nutrients and metals. Microalgae as a tertiary treatment of nitrogen and phosphorus not removed during anaerobic digestion can reduce eutrophication at point sources better than can be achieved by conventional treatment (Órpez et al., 2009; De-Bashan et al., 2010). During digestion, bacteria consume the oxygen released by microalgae to decompose organic matter, giving out carbon dioxide, ammonia, and phosphates, which are assimilated by microalgae and methane released as energy. Sludge from wastewater treatment plant can be co-cultured with algae to enhance remediation but unlike activated sludge for secondary effluents treatment, algae can eliminate nitrogen and phosphorus without organic carbon requirement (Aslan and Kapdan, 2006). Microalgae culture or sludge can be used as a diet supplement for live feed culture (Li et al., 2008a; Li et al., 2008b) or digested for bio-ethanol conversion or harvested for biodiesel or biocompounds.

The wastewater treatment utilizing the algal–bacterial system is capable of removing about 80% COD (Weiland, 2010). Utilizing *N. oculata* and *Chlorella* sp., the highest removal of COD (95–98%) and BOD (90–98%) TOC (80–86%) and TN (80%) are achieved after seven days of anaerobic treatment as compared to the treatment without microalgae (Table 8.5) (Ahmad et al., 2014b; Ahmad et al., 2014c). POME treated with anaerobic co-cultivation of *T. suecica* achieves high removal efficiency of COD, BOD, TOC, and TN after three and seven days HRT at 87–95%, 84–95%, 67–90%, 73–80%, respectively (Ahmad et al., 2014a). The lower removal efficiency of COD (53%), BOD (73%), TOC (49%), and TN (48%) are achieved on day 3 of aerobic treatment without microalgae. Filtered POME composition in sea water at different levels (1, 5, 10, 15, and 20%) used as an alternative medium for algal cultivation has attained enhanced cell growth and lipid accumulation (Shah et al., 2014a; Shah et al., 2016). At 10% POME for *N. oculata* and *T. suecica* with high μ_{max} (0.21 and 0.20 d^{-1}) and lipid content (39 and 27%), respectively, were achieved after 16 days of flask cultivation. The POME/seawater media also achieve high removal of COD (93.6–95%), BOD (96–97%), TOC (71–75%), TN (78.8–90.8%), and oil and grease

(92–94.9%). The major fatty acids composition of lipid recovered from *N. oculata* and *T. suecica* cultivated in 10% POME composition with sea water are pentadecanoic acid (C15:0), palmitic acid (C16:0), stearic acid (C18:0) for SFA; and palmitolic acid (C16:1) and oleic acid (C18:1) for MUFA. The total SFA (59.24, 68.74%); MUFA (15.14, 12.26%); and PUFA (9.07, 8.88%) are obtained for *N. oculata* and *T. suecica*, respectively. *N. oculata* contains high palmitic acid (C16:0) at 28.22% and palmitolic (C16:1) at 9.37% while *T. suecica* contains high palmitic acid (C16:0) at 36.48% and pentadecanoic acid (C15:0) at 9.21%. In PUFA profile, the highest percentage of linolenic acid (C18:3) is found in *N. oculata* (4.54%) and *T. suecica* (5.11%). Algal cultivation in 10% filtered POME with sea water therefore is suitable for improved cell growth as well as MUFA and PUFA production. The percentage of fatty acids content of microalgae can be tuned based on the growth phases from which the cultures are harvested. With high saturated and MUFA s, *N. oculata* and *T. suecica* are potential candidates for the production of biodiesel (Shah et al., 2016).

TABLE 8.5   Aerobic and Anaerobic Treatment of POME with and without Microalgae.

Parameters	Removal Efficiency (%)															
	Aerobic Treatment								Anaerobic Treatment							
	Days 3				Days 7				Days 3				Days 7			
	Without algae	N. oculata	Chlorella sp.	T. suecica	Without algae	N. oculata	Chlorella sp.	T. suecica	Without algae	N. oculata	Chlorella sp.	T. suecica	Without algae	N. oculata	Chlorella sp.	T. suecica
pH	7.8	7.9	7.9	7.8	7.7	7.8	7.6	7.7	6	6.3	7.2	7.2	5.7	5.6	6.8	7.1
BOD	73	82	81	83	77	84	88	86	78	90	86	67	87	98	95	90
COD	53	65	73	69	62	90	87	96	73	83	86	87	87	97	98	95
TOC	49	56	67	59	58	65	75	78	62	63	68	67	70	80	86	80
TN	48	60	67	59	61	64	75	64	69	73	59	73	70	80	78	80

Anaerobic cultivation of *C. vulgaris* and *Scenedesmus dimorphus* with POME for eight days HRT removes 50.5 and 86% COD; 61.6 and 86.5% BOD; and 61 and 99.5% TN, respectively (Techobanoglous, 2002). The previously reported removal efficiencies of TOC (76.6%) and TN (84%) are achieved in the treatment of industrial wastewater by *C. vulgaris* (Godos and Bianco, 2009; Hee-Jeong and Seung-Mok, 2012; Zhou, 2012). Increasing *C. vulgaris* content from 1 to 10 gL^{-1} increases the removal rate of BOD to 80–89%, COD to 78–82%, TN to 81–85%, TP to 32–36%, NH_3–N to 99–97%, and PO_4–P to 45–49% (Hee-Jeong and Seung-Mok, 2012). Shorter HRTs of two days have been reported for 89% BOD and 88% COD reduction using *C. vulgaris* grown in seed and animals feed production wastewater at 30 °C (Residua, 2007). The algae-based STP achieves total BOD removal of 82% (Thani et al., 1999) and 76% COD removal from piggery wastewater in high-rate algal ponds (Ranga Rao et al., 2007). A study with *C. protothecoides* similarly achieves the removal efficiency of 78.3% COD, when algae is grown in concentrated soybean wastewater (Zhou, 2012). However, lower COD removal (59–79%) has been reported by combining the high-rate algal pond, using filamentous green algae and an artificial wetland (Beccari et al., 1996).

8.5.3.2 BIOMETHANE PRODUCTION

The most extensive study using anaerobic digestion technology has been on the oil palm wastes, which can be utilized to meet the energy requirement of the palm oil mills (Lorestani, 2006). Through the implementation of CDM, the mills could earn carbon credits as revenue by utilizing the methane gas (Poh and Chong, 2009). The performance of an anaerobic digester for biogas production can be improved by optimizing factors such as the quality of POME, inocula composition, co-substrate addition, pH, temperature, OLR, reactor type and design, retention time, and pre-treatment process. With *Chlorella* sp. after three days HRT, the highest biomethane yield (5276 mLL^{-1} POME d^{-1}) and the highest specific biogas production rate (0.129 m^{3}kg^{-1} COD day^{-1}) are achieved at 2 mLmL^{-1} POME and EFB of 0.12 gmL^{-1} POME. With *N. oculata*, the biomethane is slightly lower (4812 mLL^{-1} POME day^{-1}), but the specific biogas production rate is comparable (0.126 m^{3}kg^{-1} COD d^{-1}) (Ahmad et al., 2014a; Ahmad et al., 2014b; Ahmad et al., 2014c). With reduced amount of EFB (0.06 gmL^{-1} POME), but high mono-algal *N. oculata* and *Chlorella* sp. cultured separately (2 mLmL^{-1} POME), comparable biomethane yield (4443–4524 mLL^{-1} POME d^{-1}) and the specific biogas production rate (0.120–0.122 m^{3}kg^{-1} COD d^{-1}) are obtained. At lower amount of *Chlorella* sp. (1 mLmL^{-1} POME) but high EFB (0.12 gmL^{-1}), lower biomethane yield (3816 mLL^{-1} POME d^{-1}) and specific biogas production rate (0.105 m^{3}kg^{-1} COD d^{-1}) although comparable to *N. oculata*, and both at all times register higher production than *T. suecica*. With anaerobic co-cultivation of *T. suecica*, EFB, and POME, the biomethane production is 3965 mLL^{-1}POME d^{-1} and the specific biogas production is

0.116 m^3kg^{-1} COD d^{-1}. Without microalgae, the specific biogas production is high (0.127 m^3kg^{-1} COD d^{-1}) but with lower biomethane yield (3642 mLL^{-1} POME day^{-1}) (Ahmad et al., 2014a). These, however, are still much higher than the reported methane production of 573–1170 $mLL^{-1}d^{-1}$ from co-digestion of *Scenedesmus* sp. and *Chlorella* sp. separately, with 50% waste paper (Yen and Brune, 2007). Without microalgae, the highest biomethane is 3650.3 mLL^{-1} POME d^{-1} but with equivalent specific biogas production rate of 0.121 m^3kg^{-1} COD d^{-1} at high EFB (0.12 gmL^{-1}) POME. Without both algae and sludge inocula, no biomethane is produced and CO_2 (190 mL CO_2 L^{-1} POME d^{-1}) is much lower with some hydrogen detected (78 mL H_2 L^{-1} POME d^{-1}) although the specific biogas production rate (0.130 m^3kg^{-1} COD d^{-1}) remains the same. High microalgae and EFB co-digestion with POME, at the correct ratio of POME and sludge inocula lead to 1.1–1.4-fold higher biomethane production than without microalgae co-digestion. The specific biogas production rate, however, remains consistent between 0.094 and 0.129 m^3kg^{-1} COD day^{-1} (Ahmad et al., 2014a; Ahmad et al., 2014b; Ahmad et al., 2014c).

Microalgae display major characteristic of methanogens by generating biogas containing methane as an integral part of their energy metabolism, removing over 80% BOD by anaerobic treatment (Satyawali and Balakrishnan, 2008). The lipid content of 23.7–30.4% under slightly above mesophilic conditions at 48 °C has resulted in higher biogas production (0.128 m^3kg^{-1} COD d^{-1}) after three days HRT (Ahmad et al., 2014c). The higher the lipid content of the cell, the higher will be the potential for biomethane production, as these can serve as nutrients for bacteria, and for microalgae to work in tandem with bacteria to breakdown the EFB and POME. Algal biomass containing between 2 and 22% lipid produces methane yield at 0.47–0.80 m^3 CH_4 kg^{-1} VS in anaerobic digestion (Cirne et al., 2007; Li et al., 2011). The productivity can be increased by mixing the proteinaceous algal biomass with carbon rich waste such as primary sewage sludge (Chua and Oh, 2010) and EFB and POME to increase the C/N ratio of digester feeding to balance the high nitrogen concentration of algal sludge for methane production (Yen and Burne, 2007). The C/N ratio of 29:1 is close to the range known to have a positive effect on the biomethane yield which is at the range of 20:1–30:1. The optimized C/N ratio of 20:1–25:1 has been reported for the co-digestion of algae and waste paper (Yen and Burne, 2007). The recommended optimal ratio for anaerobic bacterial growth is 25:1 (Yen and Burne, 2007; Tabatabaei et al., 2011; Saleh et al., 2012; Thong et al., 2012). Co-digestion of EFB with POME enhance microbial biodegradability and has resulted in 25–32% higher methane production at mixing ratios of 0.4:1, 0.8:1, and

2.3:1 on the VS basis than with EFB alone (Thong et al., 2012). An active microbial activity supported by the readily biodegradable organic components in POME enhances hydrolytic capacity of the digestion contributing to higher release of hemicellulose from the lignocellulosic component of EFB (Kaparaju and Felby, 2010).

The addition of microalgae and EFB also enhances the buffering capacity of the digester. Co-digestion is beneficial because potential toxic NH_4 is diluted which allows improved OLR and enhanced biogas yield (Satyawali and Balakrishnan, 2008). With excess VFAs, the acidogens grow rapidly and produce more volatile acids to reduce the pH. Methanogenesis may not occur as it requires pH around 6.5–7.5 and the methanogens may not be able to keep up with the changes and degrade acids as fast as they are generated. These may lead to low methane production (Poh and Chong, 2009). Several studies have also looked at microalgal co-digestion with sludge under mesophilic and thermophilic conditions where methane production at 0.42 Lg^{-1} VS is much higher under mesophilic than thermophilic conditions (0.17–0.32 L g^{-1} VS) (Yadvika et al., 2004). The variations in solids retention times (SRTs) between 11 and 30 days do not affect gas production (Sialve et al., 2009). The major drawback is the energy to maintain thermophilic condition. Economical integrated processes that combine algae cultivation and wastewater treatment system for methane production should therefore be the most suitable approach.

8.6 CONCLUSION AND FUTURE OUTLOOK

Integrated algal wastewater treatment provides GHG abatement with a combination of low energy wastewater treatment, renewable fuel production, high-value biocompounds, and fertilizer recovery. The production of biodiesel, biomethane, biohydrogen, and bio-ethanol as co-products or high-value biocompounds has immense potential of decreasing the total cost of algal biofuels production thereby improving economic feasibility. Mixed bacterial–algal co-immobilized systems in water treatment plants is ideal for methane or hydrogen production. For these, selection of algal strain/consortium, climatic conditions, existing infrastructure, logistic considerations as well as overall availability of waste resources in sufficient quantity are important considerations. A transparent hybrid photobioreactor utilizing both incidents light directly from the sun, as well as collected and distributed light can achieve higher biomass production with heterotrophic/mixotrophic growth systems and waste water media. Hybrid bioreactor that utilizes the design can make up deficiencies for algal growth in systems such as greenhouse reactors, raceway cultivators, and ponds. Novel consideration is the fiber optic photon units strategically placed in the growth section within the bioreactor for possible growth throughout the entire daylight time period, and not just when the sun is directly overhead. Since a wastewater treatment system is viable for efficient tertiary-level treatment, a refined full-scale algal production, harvesting and biofuel conversion technologies should be the major consideration for future green energy generation.

KEYWORDS

- Aerobic treatment
- Algae
- Anaerobic treatment
- Biodiesel
- Bioenergy
- Bio-ethanol
- Biohydrogen
- Biomass utilization
- Biomethane
- Immobilization
- Industrial wastes
- Palm oil mill effluent
- Product purification
- Reactor engineering
- Waste remediation

REFERENCES

Abdel-Raouf, N.; Al-Homaidan, A. A.; Ibraheem, I. B. M. Microalgae and wastewater treatment. *Saudi J Bio Sci* 2012, 19, 3, 257–275.

Abdullah, M. A.; Afzaal, M.; Ismail, Z.; Ahmad, A.; Nazir, M. S.; Bhat. A. H. Comparative study on structural modification of *Ceiba pentandra* for oil sorption and palm oil mill effluent treatment. *Desalin Water Treat* 2015a, 54(11), 3044–3053.

Abdullah, M. A.; Shah, S. M. U.; Ahmad, A.; El-Sayed, H. Algal biotechnology for bioenergy, environmental remediation and high value biochemicals. In *Biotechnology and bioinformatics: Advances and applications for bioenergy, bioremediation, and biopharmaceutical research.* Thangadurai, D., Sangeetha, J., Eds.; Apple Academic Press: New Jersey, USA, 2015b, pp 301–344.

Acién, F. G.; Fernández, J. M.; Magán, J. J.; Molina, E. Production cost of a real microalgae production plant and strategies to reduce it. *Biotechnol Adv* 2012, 30, 1344–1353.

Afzaal, M. Structural modification of *Ceiba pentandra* (kapok) and mode of operation for residual oil removal and palm oil mill effluent treatment. PhD Thesis, Universiti Teknologi PETRONAS, Malaysia, 2014.

Afzaal, M.; Balasubramanian, P.; Abdullah, M. A. Continuous heavy metal removal from palm oil mill effluent using natural *Ceiba pentandra* (L.) Gaertn packed-bed column. *Appl Mech Mat* 2014, 625, 822–825.

Ahlborg, U. G.; Thunberg, T. M. Chlorinated phenols: Occurrence, toxicity, metabolism and environmental impact. *Crit Rev Toxicol* 1980, 7, 1–35.

Ahmad, A. L.; Ismail, S.; Bhatia, S. Water recycling from palm oil mill effluent (POME) using membrane technology. *J Desalin* 2003, 157, 87–95.

Ahmad, A. L.; Mat, Y. N. H.; Derek, C. J.; Lim, J. K. Microalgae as a sustainable energy source for biodiesel production: A review. *Renew Sustain Energ Rev* 2011, 15, 584–593.

Ahmad, A.; Shah, S. M. U.; Buang, A.; Othman, M. F.; Abdullah, M. A. Evaluation of aerobic and anaerobic co-digestion of *Tetraselmis suecica* and oil palm empty fruit bunches by response surface methodology. *Adv Mat Res* 2014a, 925, 243–247.

Ahmad, A.; Shah, S. M. U.; Othman, M. F.; Abdullah, M. A. Aerobic and anaerobic co-cultivation of *Nannochloropsis oculata* with oil palm empty fruit bunch for enhanced biomethane production and palm oil mill effluent treatment. *Desalin Water Treat* 2014b, doi:1 0.1080/19443994.2014.960458.

Ahmad, A.; Shah, S. M. U.; Othman, M. F.; Abdullah, M. A. Enhanced palm oil mill effluent treatment and biomethane production by co-digestion of oil palm empty fruit bunches with *Chlorella* sp. *Can J Chem Eng* 2014c, 92, 1636–1642.

Alejandro, R. M.; Leopoldo, G.; Mendoza-Espinosa, T. S. Growth and nutrient removal in free and immobilized green algae in batch and semi-continuous cultures treating real wastewater. *Bioresour Technol* 2010, 101, 58–64.

Almeida, P. D.; Silva, P. D. The peak of oil production – timings and market recognition. *Energ Policy* 2009, 37, 1267–1276.

Alvarez, P. J. J.; Vogel, T. M. Substrate interactions of benzene, toluene, and para-xylene during microbial degradation by pure cultures and mixed culture aquifer slurries. *Appl Environ Microbiol* 1991, 57, 10, 2981–2985.

Alzate, M. E.; Muñoz, R.; Rogalla, F.; Fdz-Polanco, F.; Perez-Elvira, S. I. Biochemical methane potential of microalgae: Influence of substrate to inoculum ratio, biomass concentration and pretreatment. *Bioresour Technol* 2012, 123, 488–494.

Amélie, T.; Philistin, M.; Ferey, F.; Walenta, G.; Irisson, J. O.; Bernard, O.; Sciandra, A. Effect of gaseous cement industry effluents on four species of microalgae. *Bioresour Technol* 2015, 143, 353–359.

Amin, S. Review on biofuel oil and gas production processes from microalgae. *Energ Convers Manag* 2009, 50, 1834–1840.

Andreasen, T. The AMINODAN system for treatment of palm oil mill effluent. In *Proceeding of regional workshop on palm oil mill technology and effluent treatment*, PORIM: Malaysia, 1982, pp 213–215.

Ankita, K.; Anup, K.; Akhilendra, K. P.; Chandan, G. Carbon dioxide assisted *Spirulina platensis* cultivation using NPK-10:26:26 complex fertilizer in sintered disk chromatographic glass bubble column. *J CO$_2$ Utilization* 2014, 8, 49–59.

April, T. M.; Foght, J. M.; Currah, R. S. Hydrocarbon degrading filamentous fungi isolated from flare pit soils in northern and western Canada. *Can J Microbiol* 2000, 46, 1, 38–49.

Aravindhan, R.; Rao, J. R.; Nair, B. U. Application of a chemically modified green macro alga as a biosorbent for phenol removal. *J Environ Manag* 2009, 90(5), 1877–1883.

Aslan, S.; Kapdan, I. K. Batch kinetics of nitrogen and phosphorus removal from synthetic wastewater by algae. *Ecol Eng* 2006, 28, 64–70.

Ayati, B.; Ganjidoust, H. Comparing the efficiency of UAFF and UASB with hybrid reactor in treating wood fiber wastewater. *Iranian J Environ Health Sci Eng* 2006, 3, 39–44.

Babaee, A.; Shayegan, J. Effect of organic loading rates (OLR) on production of methane from anaerobic digestion of vegetables waste. Proceedings of World Renewable Energy Congress-2011, Sweden: Linköping, 2011.

Bajhaiya, A. K.; Mandotra, S. K.; Suseela, M. R.; Kiran, T.; Ranade, S. Algal biodiesel: The next generation biofuel for India. *As J Exp Biol Sci* 2010, 1(4), 728–739.

Barbosa, B.; Albrecht, M.; Wijffels, R. Hydrodynamic stress and lethal events in sparged microalgae cultures. *Biotechnol Bioeng* 2003, 83, 112–120.

Barbosa, M. J.; Zijffers, J. W.; Nisworo, A.; Vaes, W.; Van, S. J.; Wijffels, R. H. Optimization of biomass, vitamins, and carotenoid yield on light energy in a flat panel reactor using the A-stat technique. *Biotechnol Bioeng* 2005, 89, 233–242.

Bashan, Y.; Inoculants of plant growth-promoting bacteria for use in agriculture. *Biotechnol Adv* 1998, 16(4), 729–770.

Basri, M. F.; Yacob, S.; Hassan, M. A.; Shirai, Y.; Wakisaka, M.; Zakaria, M. R.; Phang, L. Y. Improved biogas production from palm oil mill effluent. *World J Microbiol Biotechnol* 2009, 26(3), 505–514.

Beccari, M.; Bonemazzi, F.; Majone, M.; Riccardi, C. Interaction between acidogenesis and methanogenesis in the anaerobic treatment of olive oil mill effluents. *Water Res* 1996, 30, 183–189.

Bentley, C. D.; Carroll, P. M.; Watanabe, W.O.; Riedel, A. M. Intensive rotifer production in a pilot-scale continuous culture recirculating system using nonviable microalgae and an ammonia neutralizer. *J World Aquacult Soc* 2008, 39, 625–635.

Bertrand, B.; Jean-Paul, B.; Arnaud, M. F.; François, R.; César, N.; Jean-Yves, C.; Alain, G. Separation efficiency of a vacuum gas lift for microalgae harvesting. *Bioresour Technol* 2013, 128, 235–240.

Bilad, M. R.; Discart, V.; Vandamme, D.; Foubert, I.; Muylaert, K.; Vankelecom, I. F. Coupled cultivation and pre-harvesting of microalgae in a membrane photobioreactor (MPBR). *Bioresour Technol* 2014a, 155, 410–417.

Bilad, M. R.; Hassan, A. A.; Vankelecom, L. F. J. Membrane technology in microalgae cultivation and harvesting: A review. *Biotechnol Adv* 2014b, 32, 1283–1300.

Bitog, J. P.; Lee, I. B.; Lee, C. G.; Kim, K. S.; Hwang, H. S.; Hong, S. W.; Mostafa, E. Application of computational fluid dynamics for modeling and designing photobioreactors for microalgae production: A review. *Comp Electro Agri* 2011, 76(2), 131–147.

Borja, R.; Banks, C. J. Kinetics of methane production from palm oil mill effluent in an immobilised cell bioreactor using saponite as support medium. *Bioresour Technol* 1994, 48, 209–214.

Borkenstein, C. G.; Knoblechner, J.; Fruehwirth, H.; Schagerl, M. Cultivation of *Chlorella emersonii* with flue gas derived from a cement plant. *J Appl Phycol* 2011, 23, 131–135.

Borowitzka, M. A. Carotenoid production using microorganisms. In *Single cell oils: Microbial and algal oils,* Cohen, Z., Ratledge, C., Eds.; AOCS Press: Urbana, 2010, pp 225–240.

Borowitzka, M. A. Commercial production of microalgae: Ponds, tanks, tubes and fermenters. *J Biotechnol* 1999, 70(1–3), 313–321.

Borowitzka, M. A. *Phycology, Encyclopedia of Life Sciences (eLS).* Wiley and Sons: London, 2012, doi:10.1002/9780470015902.90000. pub3.

Branyikova, I.; Marsalkova, B.; Doucha, J.; Branyik, T.; Bisova, K.; Zachleder, V.; Vitova, M. Microalgae-novel highly efficient starch producers. *Biotechnol Bioeng* 2011, 108, 766–776.

Brennan, L.; Owende, P. Biofuel from microalgae: A review of technologies for production, processing and extraction of biofuel and co-products. *Renew Sust Energ Rev* 2010, 14, 557–577.

Brown, L. M. Uptake of carbon dioxide from flue gas by microalgae. *Energ Convers Manag* 1996, 37, 1363–1367.

Brown, M. L.; Zeiler, K. G. Aquatic biomass and carbon dioxide trapping. *Energ Convers Manag* 1993, 34, 1005–1013.

Burton, T.; Lyons, H.; Lerat, Y.; Stanley, M.; Rasmussen, M. B. *A review of the potential of marine algae as a source of biofuel in Ireland.* Sustainable Energy Ireland-SEI: Dublin, 2009.

Cabanelas, I. T. D.; Arbib, Z.; Chinalia, F. A.; Souza, C. O.; Perales, J. A.; Almeida, P. F.; Nascimento, I. A. From waste to energy: Microalgae production in wastewater and glycerol. *Appl Energ* 2013, 109, 283–290.

Cao, Y. R.; Liu, Z.; Cheng, G. L.; Jing, X. B.; Xu, H. Exploring single and multi-metal biosorption by immobilized spent *Tricholoma lobayense* using multi-step response surface methodology. *Chem Eng J* 2010, 164(1), 183–195.

Capasso, R.; Evidente, A.; Schivo, L.; Orru, G.; Marcialis, M. A.; Cristinzio, R. Antibacterial polyphenols from olive oil mill waste waters. *J Appl Bacteriol* 1995, 79, 393–398.

Carrère, H.; Dumas, C.; Battimelli, A.; Batstone, D. J.; Delgenes, J. P.; Steyer, J. P.; Ferrer, I. Pretreament methods to improve sludge anaerobic degradability: A review. *J Hazard Mater* 2010, 183, 1–15.

Carvalho, A. P.; Meireles, L. A.; Malcata, F. X. Microalgal reactors: A review of enclosed system designs and performances. *Biotech Prog* 2006, 22, 1490–1506.

Cerniglia, C. E.; Gibson, D. T.; Van, B. C. Oxidation of naphthalene by cyanobacteria and microalgae. *J Gen Microbiol* 1980, 116(2), 495–500.

Chavalparit, O.; Rulkens, W. H.; Mol, A. P. J.; Khaodhair, S. Options for environmental sustainability of the crude palm oil industry in Thailand through enhancement of industrial ecosystems. *Environ Dev Sustain* 2006, 8, 271–287.

Chen, C. Y.; Durbin, E. G. Effects of pH on the growth and carbon uptake of marine phytoplankton. *Mar Ecol Prog Ser* 1994, 109, 83–94.

Chen, C. Y.; Yeh, K. L.; Aisyah, R.; Lee, D. J.; Chang, J. S. Cultivation, photobioreactor design and harvesting of microalgae for biodiesel production: A critical review. *Bioresour Technol* 2011, 102, 71–81.

Chen, P. H. Factors influencing methane fermentation of micro-algae. PhD Thesis, University of California, Berkeley, CA, USA, 1987.

Chen, P. H.; Oswald, W. J. Thermochemical treatment for algal fermentation. *Environ Int* 1998, 24, 889–897.

Chen, P.; Min, M.; Chen, Y.; Wang, L.; Li, Y.; Chen, Q.; Wang, C.; Wan, Y.; Wang, X.; Cheng, Y.; Deng, S.; Hennessy, K.; Lin, X.; Liu, Y.; Wang, Y.; Martinez, B.; Ruan, R. Review of the biological and engineering aspects of algae to fuels approach. *Int J Agric Biol Eng* 2009, 2, 1–30.

Chen, P.; Sanyal, J.; Dudukovic, M. P. Numerical simulation of bubble columns flows: Effect of different breakup and coalescence closures. *Chem Eng Sci* 2005, 60, 1085–1101.

Cheng, J.; Huang, R.; Li, T.; Zhou, J.; Cen, K. Biodiesel from wet microalgae: Extraction with hexane after the microwave-assisted transesterification of lipids. *Bioresour Technol* 2014, 170, 69–75.

Cheng, J.; Zhu, X.; Borthwick, A. Palm oil mill effluent treatment using a two-stage microbial fuel cells system integrated with immobilized biological aerated filters. *Bioresour Technol* 2010, 101, 2729–2734.

Chesapeake Bay Foundation. *Manure's impact on rivers, streams and the Chesapeake Bay.* Chesapeake Bay Foundation: Annapolis, 2004, pp 1–26.

Chisti, Y. Biodiesel from microalgae beats bioethanol. *Trends Biotechnol* 2008, 26(3), 126–131.

Chisti, Y. Biodiesel from microalgae. *Biotechnol Adv* 2007, 25, 294–306.

Chisti, Y. Fuels from microalgae. *Biofuels* 2010, 1(2), 233–235.

Cho, S.; Park, S.; Seon, J.; Yu, J.; Lee, T. Evaluation of thermal, ultrasonic and alkali pretreatments on mixed-microalgal biomass to enhance anaerobic methane production. *Bioresour Technol* 2013, 143, 330–336.

Choi, S. P.; Nguyen, M. T.; Sim, S. J. Enzymatic pretreatment of *Chlamydomonas reinhardtii* biomass for ethanol production. *Bioresour Technol* 2010, 101, 5330–5336.

Choorit, W.; Wisarnwan, P. Effect of temperature on the anaerobic digestion of palm oil mill effluent. *Electron J Biotechnol* 2007, 10, 376–385.

Christenson, L.; Sims, R. Production and harvesting of microalgae for wastewater treatment, biofuels, and bioproducts. *Biotechnol Adv* 2011, 29 (6), 686–702.

Chua, S. C.; Oh, T. H. Review on Malaysia's national energy developments: Key policies, agencies, programs and international involvements. *Renew Sustain Energ Rev* 2010, 14, 2916–2925.

Chynoweth, D. P. Renewable biomethane from land and ocean energy crops and organic wastes. *Hort Sci J* 2005, 40, 283–286.

Cirne, D. G.; Paloumet, X.; Bjornsson, L.; Alves, M. M.; Mattiasson, B. Anaerobic digestion of lipid-rich waste - effects of lipid concentration. *Renew Energ* 2007, 32, 965–975.

Clarens, A. F.; Resurreccion, E. P.; White, M. A.; Colosi, L. M. Environmental lifecycle comparison of algae to other bioenergy feedstocks. *Environ Sci Technol* 2010, 44, 1813–1819.

Cohen, Y. Biofiltration - the treatment of fluids by microorganisms immobilized into the filter bedding material: A review. *Bioresour Technol* 2001, 77(3), 257–274.

Colla, L. M.; Bertolin, T. E.; Costa, J. A. V. Fatty acids profile of *Spirulina platensis* grown under different temperatures and nitrogen concentrations. *Z Naturforsch* 2004, 59, 55–59.

Converti, A.; Casazza, A. A.; Ortiz, E.; Perego, P.; Del, M. B. Effect of temperature and nitrogen concentration on the growth and lipid content of *Nannochloropsis oculata* and *Chlorella vulgaris* for biodiesel production. *Chem Eng Proc* 2009, 48, 1146–1151.

Costa, J. A. V.; De, M. M. G. The role of biochemical engineering in the production of biofuels from microalgae. *Bioresour Technol* 2011, 102(1), 2–9.

Daneshfozoun, S.; Nazir, M. S.; Abdullah, B.; Abdullah, M. A. Surface modification of celluloses extracted from oil palm empty fruit bunches for heavy metal sorption. *Chem Eng Trans* 2014, 37, 679–684.

Danquah, M. K.; Ang, L.; Uduman, N.; Moheimani, N.; Forde, G. M. Dewatering of microalgal culture for biodiesel production: Exploring polymer flocculation and tangential flow filtration. *J Chem Technol Biotechnol* 2009, 84, 1078–1083.

De La Rubia, M. A.; Fernández-Cegrí, V.; Raposo, F.; Borja, R. Anaerobic digestion of sunflower oil cake: A current overview. *Water Sci Technol* 2013, 67, 410–417.

De Morais, M. G.; Costa, J. A. V. Carbon dioxide fixation by *Chlorella kessleri, C. vulgaris, Scenedesmus obliquus* and *Spirulina* sp. cultivated in flasks and vertical tubular photobioreactors. *Biotechnol Lett* 2007, 29(9), 1349–1352.

De-Bashan, L. E.; Bashan, Y. Immobilized microalgae for removing pollutants: Review of practical aspects. *Bioresour Technol* 2010, 101, 1611–1627.

De-Bashan, L. E.; Hernandez, J. P.; Morey, T.; Bashan, Y. Microalgae growth-promoting bacteria as "helpers" for microalgae: A novel approach for removing ammonium and phosphorus from municipal wastewater. *Water Res* 2004, 38, 466–474.

Demirbas, M. F. Biofuels from algae for sustainable development. *Appl Energ* 2011, 88(10), 3473–3480.

Dismukes, G. C.; Carrieri, D.; Bennette, N.; Ananyev, G. M.; Posewitz, M. C. Aquatic phototrophs: Efficient alternatives to land based crops for biofuels. *Curr Opin Biotechnol* 2008, 19, 235–240.

Efremenko, E. N.; Nikolskaya, A. B.; Lyagin, I. V.; Senko, O. V.; Makhlis, T. A.; Stepanov, N. A.; Maslova, O. V.; Mamedova, F.; Varfolomeev, S. D. Production of biofuels from pretreated microalgae biomass by anaerobic fermentation with immobilized *Clostridium acetobutylicum* cells. *Bioresour Technol* 2012, 114, 342–348.

Ekama, G. A.; Wentzel, M. C. Organic material removal. In *Biological wastewater treatment: Principles, modeling, and design*, Henze. M., van Loosdrecht., Ekama, G., Brdjanovic, D., Eds.; IWA Publishing, 2008, pp 1–511.

Ellis, B. E. Degradation of phenolic compounds by microalgae. *Plant Sci Lett* 1977, 8, 213–216.

Environmental Management Guideline for the Palm Oil Industry. Environmental Advisory Assistance for Industry, 1997.

EPA. Inventory of U.S. Greenhouse Gas Emission and Sinks 1990–2009, USA, 2010.

EQA. *Environmental Quality Act 1974*. Government of Malaysia, Kuala Lumpur, 2001.

Eriksen, N. The technology of microalgal culturing. *Biotechnol Lett* 2008, 30(9), 1525–1536.

Fahr, K.; Wetzstein, H. G.; Grey. R.; Schlosser, D. Degradation of 2,4-dichlorophenol and pentachlorophenol by two brown rot fungi. *FEMS Microbiol Lett* 1999, 175, 127–132.

Faisal, M.; Unno, H. Kinetic analysis of palm oil mill wastewater treatment by a modified anaerobic baffled reactor. *Biochem Eng J* 2001, 9, 25–31.

Fan, L.; Zhang, Y.; Cheng, L.; Zhang, L.; Tang, D.; Chen, H. Optimization of carbon dioxide fixation by *Chlorella vulgaris* cultivated in membrane-photobioreactor. *Chem Eng Technol* 2007, 30(8), 1094–1099.

Fang, C.; O-Thong, S.; Boe, K.; Angelidaki, I. Comparison of UASB and EGSB reactors performance, for treatment of raw and deoiled palm oil mill effluent (POME). *J Hazard Mater* 2011, 189(1–2), 229–234.

Feng, P.; Deng, Z.; Hu, Z.; Fan, L. Lipid accumulation and growth of *Chlorella zofingiensis* in flat plate photobioreactors outdoors. *Bioresour Technol* 2011, 102, 10577.

Fernandes, B. D.; Dragoner, G. M.; Teixiera, J. A.; Vicente, A. A. Light regime characterization in an airlift: Photobioreactor for production of microalgae with high starch content. *Appl Biochem Biotechnol* 2010, 161, 218–226.

Fernández-Rodríguez, M. J.; Rincón, B.; Fermoso, F. G.; Jiménez, A. M.; Borja, R. Assessment of two-phase olive mill solid waste and microalgae co-digestion to improve methane production and process kinetics. *Bioresour Technol* 2014, 157, 263–269.

Foo, K. Y.; Hameed, B. H. Insight into the applications of palm oil mill effluent: A renewable utilization of the industrial agricultural waste. *Renew Sustain Energ Rev* 2010, 14, 1445–1452.

Gaffron, H.; Rubin, J. Fermentative and photochemical production of hydrogen in algae. *J Gen Physiol* 1942, 26, 219–240.

Gavrilescu, M.; Chisti, Y. Biotechnology - a sustainable alternative for chemical industry. *Biotechnol Adv* 2005, 23, 471–499.

Gerpen, J. V. Biodiesel processing and production. *Fuel Proc Technol* 2005, 86, 1097–1107.

Glenda, C. B.; Germán, B.; Gloria, M.; Gopalakrishnan, K. A cost-effective strategy for the bio-prospecting of mixed microalgae with high carbohydrate content: Diversity fluctuations in different growth media. *Bioresour Technol* 2014, 163, 370–373.

Godos, I. D.; Bianco, S.; Garcia-Encina, P. A.; Becares, E.; Munoz, R. Long-term operation of high rate algal ponds for the bioremediation of piggery wastewaters at high loading rates. *Bioresour Technol* 2009, 100, 4332–4339.

Golueke, G. C.; Oswald, J. W.; Gotaas, B. H. Anaerobic digestion of algae. *Appl Environ Microbiol* 1957, 5, 47–55.

Gonzalez, L. E.; Bashan, Y. Increased growth of the microalga *Chlorella vulgaris* when co-immobilized and cocultured in alginate beads with the plant-growth-promoting bacterium *Azospirillum brasilense*. *Appl Environ Microbiol* 2000, 66, 1527–1531.

González-Fernández, C.; Molinuevo-Salces, B.; García-González, M. C. Evaluation of anaerobic codigestion of microalgal biomass and swine manure via response surface methodology. *Appl Energ* 2011, 88(10), 3448–3453.

González-Fernández, C.; Sialve, B.; Bernet, N.; Steyer, J. P. Thermal pretreatment to improve methane production of *Scenedesmus* biomass. *Biomass Bioenerg* 2012, 40, 105–111.

Greenwell, H. C.; Laurens, L. M. L.; Shields, R. J.; Lovitt, R. W.; Flynn, K. J. Placing microalgae on the biofuels priority list: A review of the technological challenges. *J Royal Soc Interface* 2010, 7, 703–726.

Grima, E. M.; Camacho, F. G.; Perez, J. A. S.; Sevilla, J. M. F.; Fernandez, F. G. A.; Gomez, A. C. A mathematical model of microalgal growth in light-limited chemostat culture. *J Chem Technol Biotechnol* 1994, 61, 167–173.

Gudin, C.; Chaumont, D. Cell fragility - key problem of microalgae mass production on closed photobioreactors. *Bioresour Technol* 1991, 38, 145–151.

Guiry, M. D.; Guiry, G. M. AlgaeBase. National University of Ireland, Galway, http://www.algaebase.org

Habib, M. A. B.; Yusoff, F. M.; Phang, S. M.; Ang, K. J.; Mohamed, S. Nutritional values of chironomid larvae grown in palm oil mill effluent and algal culture. *Aquacult* 1979, 158, 95–105.

Halleux, H.; Lassaux, S.; Renzoni, R.; Germain, A. Comparative life cycle assessment of two biofuels ethanol from sugar beet and rapessed methyl ester. *Int J LCA* 2008, 13(3), 184–190.

Hankamer, B.; Lehrs, F.; Rupprecht, J.; Mssgnug, J. H.; Posten, C.; Kruse, O. Photosynthetic biomass and H_2 production by green algae: From bioengineering to bioreactor scale-up. *J Plant Physiol* 2007, 131, 10–21.

Harun, R.; Danquah, M. K.; Forde, G. M. Microalgal biomass as a fermentation feedstock for bioethanol production. *J Chem Tech Biotechnol* 2010a, 85(2), 199–203.

Harun, R.; Jason, W. S. Y.; Cherrington, T.; Danquah, M. K. Exploring alkaline pre-treatment of microalgal biomass for bioethanol production. *Appl Energy* 2010b, 88, 3464–3467.

Harun, R.; Singh, M.; Forde, G. M.; Danquah, M. K. Bioprocess engineering of microalgae to produce a variety of consumer products. *Renew Sustain Energ Rev* 2010c, 14, 1037–1047.

Harwood, J. L.; Guschina, I. A. The versatility of algae and their lipid metabolism. *Biochim* 2009, 91, 679–684.

Hassan, M. A.; Yacob, S.; Shirai, Y. Treatment of palm oil wastewaters. In *Handbook of Industrial and Hazardous Wastes Treatment*, Wang, L.K., Hung, Y., Lo, H.H., Yapijakis, C., Eds.; Marcel Dekker, Inc.: New York, 2004, pp 719–736.

Hee-Jeong, C.; Seung-Mok, L. Effects of microalgae on the removal of nutrients from wastewater: various concentrations of *Chlorella vulgaris*. *Environ Eng Res* 2012, 17(S1), S3–S8.

Henrard, A. A.; de Morais, M. G.; Costa, J. A. V. Vertical tubular photobioreactor for semi-continuous culture of *Cyanobium* sp. *Bioresour Technol* 2011, 102, 4897–900.

Hernández, D.; Riaño, B.; Coca, M.; García-González, M. C. Treatment of agro-industrial wastewater using microalgae–bacteria consortium combined with anaerobic digestion of the produced biomass. *Bioresour Technol* 2013, 135, 598–603.

Ho, S. H.; Chen, W. M.; Chang, J. S. *Scenedesmus obliquus* CNW-N as a potential candidate for CO_2 mitigation and biodiesel production. *Bioresour Technol* 2010, 101, 8725–8730.

Ho, S. H.; Huang, S. W.; Chen, C. Y.; Hasunuma, T.; Kondo, A.; Chang, J. S. Bioethanol production using carbohydrate-rich microalgae biomass as feedstock. *Bioresour Technol* 2013, 135, 191–198.

Holliger, C.; Gaspard, S.; Glod, G. Contaminated environments in the subsurface and bioremediation: Organic contaminants. *FEMS Microbiol Rev* 1997, 20(3–4), 517–523.

Hong, Y. W.; Yuan, D. X.; Lin, Q. M.; Yang, T. L. Accumulation and biodegradation of phenanthrene and fluoranthene by the algae enriched from a mangrove aquatic ecosystem. *Mar Poll Bull* 2008, 56, 1400–1405.

Hu, C.; Li, M.; Li, J.; Zhu, Q.; Liu, Z. Variation of lipid and fatty acid compositions of the marine microalga *Pavlova viridis* (Prymnesiophyceae) under laboratory and outdoor culture conditions. *World J Microbiol Biotechnol* 2008, 24, 1209–1214.

Hu, Q.; Kurano, N.; Kawachi, M.; Iwasaki, I.; Miyachi, A. Ultra high-cell-density culture of a marine alga *Chlorococcum littorale* in a flat-plate photobioreactor. *Appl Microbiol Biotechnol* 1998, 46, 655–662.

Huang, J.; Li, Y.; Wan, M.; Yan, Y.; Feng, F.; Qu, X.; Wang, J.; Shen, G.; Li, W.; Fan, J.; Wang, W. Novel flat-plate photobioreactors for microalgae cultivation with special mixers to promote mixing along the light gradient. *Bioresour Technol* 2014, 159, 8–16.

Hulatt, C. J.; Thomas, D. N. Energy efficiency of an outdoor microalgal photobioreactor sited at mid-temperate latitude. *Bioresour Technol* 2011a, 102, 6687.

Hulatt, C. J.; Thomas, D. N. Productivity, carbon dioxide uptake and net energy return of microalgal bubble column photobioreactors. *Bioresour Technol* 2011b, 102, 5775.

Igamberdiev, A. U.; Kleczkowski, L. A. Optimization of CO_2 fixation in photosynthetic cells via thermodynamic buffering. *Biosystems* 2011, 103(2), 224–229.

Ignacio, M. G. Microalgae immobilization: Current techniques and uses. *Bioresour Technol* 2008, 99, 3949–3964.

Ismail, Z. I. *New machine set to boost palm kernel cake industry*. New Straits Times: Malaysia, 2004.

Jacob-Lopes, E.; Scoparo, C. H. G.; Lacerda, L. M. C. F.; Franco, T. T. Effect of light cycles (night/day) on CO_2 fixation and biomass production by microalgae in photobioreactors. *Chem Eng Process* 2009, 48(1), 306–310.

James, C. M.; Al-Khars, A. M. An intensive continuous culture system using tubular photobioreactors for producing microalgae. *Aquacult* 1990, 87(3–4), 381–393.

Janssen, M.; Tramper, J.; Mur, L. R.; Wijffels, R. H. Enclosed outdoor photobioreactors: Light regime, photosynthetic efficiency, scale-up, and future prospects. *Biotechnol Bioeng* 2003, 81(2), 193–210.

Jiménez, C.; Cossio, B. R.; Labella, D.; Xavier.; Niell, F. The feasibility of industrial production of *Spirulina* (Arthrospira) in southern Spain. *Aquacult* 2003, 217(1–4), 179–190.

Jin, Y. M.; Veiga, M. C.; Kennes, C. Bioprocesses for the removal of nitrogen oxides from polluted air. *J Chem Technol Biotechnol* 2005, 80, 483–494.

Johnson, M. B.; Wen, Z. Production of biodiesel fuel from the microalga *Schizochytrium limacinum* by direct transesterification of algal biomass. *Energ Fuel* 2009, 23, 5179–5183.

Kaewpintong, K.; Shotipruk, A.; Powtongsoo, K. S.; Pavasant, P. Photoautotrophic high-density cultivation of vegetative cells of *Haematococcus pluvialis* in airlift bioreactor. *Bioresour Technol* 2007, 98, 288–295.

Kahru, A.; Reiman, R.; Ratsep, A. The efficiency of different phenol-degrading bacteria, and activated sludges in detoxification of phenolic leachates. *Chemosphere* 1998, 37, 301–318.

Kaparaju, P.; Felby, C. Characterization of lignin during oxidative and hydrothermal pre-treatment processes of wheat straw and corn stover. *Bioresour Technol* 2010, 101, 3175–3181.

Kapdan, I. K.; Kargi, F. Bio-hydrogen production from waste materials. *Enz Microbiol Technol* 2006, 38, 569–582.

Kelly, M.; Dworjanyn, S. The potential of marine biomass for anaerobic biogas production: A feasibility study with recommendations for further research. Scottish Association for Marine Science: Scotland, 2008.

Kesaano, M.; Sims, R. C. Algal biofilm based technology for wastewater treatment. *Algal Res* 2014, 5, 231–240.

Keymar, P.; Ruffell, I.; Pratt, S.; Lant, P. High pressure thermal hydrolysis as pre-treatment to increase the methane yield during anaerobic digestion of microalgae. *Bioresour Technol* 2013, 131, 128–133.

Khan, S. A.; Rashmi, A.; Hussain, M. Z.; Prasad, S.; Banerjee, U. C. Prospects of biodiesel production from microalgae in India. *Renew Sust Energ Rev* 2009, 13, 2361–2372.

Khan, Z.; Bhadouria, P.; Bisen, P. S. Nutritional and therapeutic potential of *Spirulina*. *Curr Pharm Biotechnol* 2005, 5, 373–379.

Kirkwood, A. E.; Nalewajko, C.; Fulthrope, P. R. Physiological characteristics of cyanobacteria in pulp and paper waste-treatment systems. *J Appl Phycol* 2003, 15, 325–335.

Kleckner, V.; Kosaric, N. Degradation of phenols by algae. *Environ Technol* 1992, 13, 493–501.

Knothe, G. Dependence of biodiesel fuel properties on the structure of fatty acid alkyl esters. *Fuel Proc Technol* 2005, 86, 1059–1070.

Knuckey, R. M.; Brown, M. R.; Robert, R.; Frampton, D. M. F. Production of microalgal concentrates by flocculation and their assessment as aquaculture feeds. *Aquacult Eng* 2006, 35(3), 300–313.

Kommareddy, A. R.; Anderson, G. A. *Mechanistic modeling of photobioreactor system*. ASAE: St. Joseph, Michigan, 2005.

Kong, Q.; Li, L.; Martinez, B.; Chen, P.; Ruan, R. Culture of microalgae *Chlamydomonas reinhardtii* in wastewater for biomass feedstock production. *Appl Biochem Biotechnol* 2010, 160, 9–18.

Krichnavaruk, S.; Powtongsook, S.; Pavasant, P. Enhanced productivity of *Chaetoceros calcitrans* in airlift photobioreactors. *Bioresource Technology* 2007, 98(11), 2123–2130.

Kvenvolden, K. A.; Cooper, C. K. Natural seepage of crude oil into the marine environment. *Geo-Marine Lett* 2003, 23(3–4), 140–146.

Lakaniemi, A. M.; Hulatt, C. J.; Thomas, D. N.; Tuovinen, O. H.; Puhakka, J. A. Biogenic hydrogen and methane production from *Chlorella vulgaris* and *Duanliella tertiolecta*. *Biotechnol Biofuels* 2011, 4, 34.

Lam, G. P.; Vermuë, M. H.; Olivieri, G.; van den Broek, L. A. M.; Barbosa, M. J.; Eppink, M. H. M.; Wijffels, R. H.; Kleinegris, D. M. M. Cationic polymers for successful flocculation of marine microalgae. *Bioresour Technol* 2014, 169, 804–807.

Lam, M. K.; Lee, K. T. Immobilization as a feasible method to simplify the separation of microalgae from water for biodiesel production. *Chem Eng J* 2012, 191, 263–268.

Lam, M. K.; Lee, K. T. Renewable and sustainable bioenergies production from palm oil mill effluent (POME): Win–win strategies toward better environmental protection. *Biotechnol Adv* 2011, 29(1), 124–141.

Larsdotter, K. Wastewater treatment with microalgae a literature review. *Vatten* 2006, 62(1), 31.

Lau, L. C.; Tan, K. T.; Lee, K. T.; Mohamed, A. R. A comparative study on the energy policies in Japan and Malaysia in fulfilling their nations' obligations towards the Kyoto Protocol. *Energ Policy* 2009, 37, 4771–4778.

Lau, P. S.; Tam, N. F. Y.; Wong, Y. S. Wastewater nutrients (N and P) removal by carrageenan and alginate immobilized *Chlorella vulgaris*. *Environ Technol* 1997, 18, 945–951.

Laurens, L. M.; Dempster, T. A.; Jones, H. D.; Wolfrum, E. J.; Van, W. S.; McAllister, J. S.; Rencenberger, M.; Parchert, K. J.; Gloe, L. M. Algal biomass constituent analysis: Method uncertainties and investigation of the underlying measuring chemistries. *Anal Chem* 2012, 84, 1879–1887.

Leadbeater, B. S. C. The 'Droop equation' - Michael Droop and the legacy of the 'cell-quota model' of phytoplankton growth. *Protist* 2006, 157(3), 345–358.

Leahy, J. G.; Colwell, R. R. Microbial degradation of hydrocarbons in the environment. *Microbiol Rev* 1990, 54(3), 305–315.

Lebeau, T.; Robert, J. M. Biotechnology of immobilized micro-algae: A culture multi-step response surface methodology. *Chem Eng J* 2010, 164, 183–195.

Lee, A. K.; Lewis, D. M.; Ashman, P. J. Microbial flocculation, a potentially low-cost harvesting technique for marine microalgae for the production of biodiesel. *J Appl Phycol* 2008, 21, 559–567.

Lee, J. H.; Lee, J. S.; Shin, C. S.; Park, S. C.; Kim, S. W. Effects of NO and SO_2 on growth of highly-CO_2-tolerant microalgae. *J Microbiol Biotechnol* 2000, 10, 338–343.

Lee, J. S.; Kim, D. K.; Lee, J. P.; Park, S. C.; Koh, J. H.; Cho, H. S.; Kim, S. W. Effects of SO_2 and NO on growth of *Chlorella* sp. KR-1. *Bioresour Technol* 2002, 82, 1–4.

Lehninger, A. L.; Nelson, D. L.; Cox, M. M. *Principles of biochemistry*. W.H. Freeman and Co.: USA, 2004.

Li, F. F.; Yang, Z. H.; Zeng, R.; Yang, G.; Chang, X.; Yan, J. B. Microalgae capture of CO_2 from actual flue gas discharged from a combustion chamber. *Ind Eng Chem Res* 2011a, 50, 6496–6502.

Li, S.; Yue, X.; Jing, Y.; Bai, S.; Dai, Z. Fabrication of zonal thiol-functionalized silica nanofibers for removal of heavy metal ions from wastewater. *Colloids Surf A* 2011, 380, 229–233.

Li, Y.; Chen, Y.; Chen, P.; Min, M.; Zhou, W.; Martinez, B.; Zhu, J.; Ruan, R. Characterization of a microalgae *Chlorella* sp. well adapted to highly concentrated municipal wastewater for nutrient removal and biodiesel production. *Bioresour Technol* 2011, 102(8), 5138–5144.

Li, Y.; Horsman, M.; Wang, B.; Wu, N.; Lan, C. Q. Effects of nitrogen sources on cell growth and lipid accumulation of green alga *Neochloris oleoabundans. Appl Microbiol Biotechnol* 2008a, 81, 629–636.

Li, Y.; Horsman, M.; Wu, N.; Lan, C. Q.; Dubois-Calero, N. Biofuels from microalgae. *Biotechnol Prog* 2008b, 24(4), 815–820.

Li, Y.; Wang, B.; Wu, N.; Lan, C. Q. Effects of nitrogen sources on cell growth and lipid production of *Neochloris oleoabundans. Appl Microbiol Biotechnol* 2008c, 81(4), 629–636.

Lim, S.; Teong, L. K. Recent trends, opportunities and challenges of biodiesel in Malaysia: An overview. *Renew Sustain Energ Rev* 2010, 14, 938–954.

Lima, S. A.; Raposo, M. F. J.; Castro, P. M.; Morais, R. M. Biodegradation of chlorophenol by a microalgae consortium. *Water Res* 2004, 38(1), 97–102.

Liu, C. H.; Chang, C. Y.; Cheng, C. L.; Lee, D. J.; Chang, J. S. Fermentative hydrogen production by *Clostridium butyricum* CGS5 using carbohydrate-rich microalgal biomass as feedstock. *Int J Hydrogen Energ* 2012, 37, 15458–15464.

Liu, C. H.; Chang, C. Y.; Liao, Q.; Zhu, X.; Liao, C. F.; Chang, J. S. Biohydrogen production by a novel integration of dark fermentation and mixotrophic microalgae cultivation. *Int J Hydrogen Energ* 2013, 38(35), 15807–15814.

Liu, J.; Yuan, C.; Hu, G.; Li, F. Effects of light intensity on the growth and lipid accumulation of microalgae *Scenedesmus* spp. under nitrogen limitation. *Appl Biochem Biotechnol* 2012, 166(8), 2127–2137.

Lorestani, A. A. Z. Biological treatment of palm oil mill effluent (POME) using an up-flow anaerobic sludge fixed film (UASFF) bioreactor. PhD Thesis, Universiti Sains Malaysia, 2006.

Lou, H. P.; Al-Dahhan, M. H. Analyzing and modeling of photobioreactors by combining first principles of physiology and hydrodynamics. *Biotechnol Bioeng* 2004, 85, 4.

Ludin, N. Palm oil biomass for electricity generation in Malaysia. Pusat Tenaga Malaysia, 2009, http://www.ptm.org.my/biogen

Ma, A. N. Current status of palm oil processing wastes management. In *Waste management in Malaysia: Current status and prospects for bioremediation*, Yeoh, B.G., Ed.; 1993, pp 111–136.

Ma, A. N. Innovations in management of Palm Oil Mill Effluent. *Planter* 1999, 75, 381–389.

Maa, F.; Hanna, M. A. Biodiesel production: A review. *Bioresour Technol* 1999, 70, 1–15.

Markou, G.; Georgakakis, D. Cultivation of filamentous cyanobacteria (blue–green) in agro-industrial wastes and wastewater: A review. *Appl Energ* 2011, 88(10), 3389–3401.

Mata, M. T.; Martins, A. A.; Caetano, N. S. Micaroalgae for biodiesel production and other applications: A review. *Renew Sust Energ Rev* 2010, 14, 217–232.

Matsumoto, H.; Hamasaki, A.; Sioji, N.; Ikuta, Y. Influence of CO_2, SO_2 and NO in flue gas on microalgae productivity. *J Chem Eng Japan* 1997, 30, 620–624.

McHugh, D. J. A guide to the seaweed industry. FAO Fisheries Technical Papers, 2003, 441, 1–105.

McManamon, C.; Burke, A.; Holmes, J. D.; Morris, M. A. Amine-functionalised SBA-15 of tailored pore size for heavy metal adsorption. *J Colloid Interface Sci* 2012, 369, 330–337.

Medina-Bellver, J. I.; Marín, P.; Delgado, A. Evidence for *in situ* crude oil biodegradation after the Prestige oil spill. *Environ Microbiol* 2005, 7(6), 773–779.

Mendes, A.; Reis, A.; Vasconcelos, R.; Guerra, P.; Lopes, D. S. T. *Crypthecodinium cohnii* with emphasis on DHA production: A review. *J Appl Phycol* 2009, 21, 199–214.

Mendez, L.; Mahdy, A.; Demuez, M.; Ballesteros, M.; González-Fernández, C. Effect of high pressure thermal pretreatment on *Chlorella vulgaris* biomass: Organic matter solubilisation and biochemical methane potential. *Fuel* 2014, 117, 674–679.

Miao, X.; Wu, Q.; Yang, C. Fast pyrolysis of microalgae to produce renewable fuels. *J Anal Appl Pyrol* 2004, 71, 855–863.

Mirón, S. A.; Cerón, G. M. C.; García, C. F.; Molina, G. E.; Chisti, Y. Mixing in bubble column and airlift reactors. *Chem Eng Res Des* 2004, 82(10), 1367–1374.

Miyamoto, K.; Wable, O.; Benemann, J. R. Vertical tubular reactor for microalgae cultivation. *Biotech Lett* 1988, 10, 703–708.

Molina, E. G.; Camacho, F. G.; Fernandez, F. G. A. Production of EPA from *Phaeodactylum tricornutum*. In *Chemicals from microalgae*, Cohen, Z., Ed.; CRC Press: Florida, 1999, pp 57–92.

Molina, G. E.; Belarbi, E. H.; Acien Fernandez, F. G.; Robles Medina, A.; Chisti, Y. Recovery of microalgal biomass and metabolites: Process options and economics. *Biotechnol Adv* 2003, 20(7–8), 491–515.

Molinuevo-Salces, B.; García-González, M. C.; González-Fernández, C. Performance comparison of two photobioreactors configurations (open and closed to the atmosphere) treating anaerobically degraded swine slurry. *Bioresour Technol* 2010, 101, 5144–5149.

Monlau, F.; Barakat, A.; Trably, E.; Dumas, C.; Steyer, J. P.; Carrere, H. Lignocellulosic materials into biohydrogen and biomethane: Impact of structural features and pretreatment. *Crit Rev Environ Sci Technol* 2013, 43(3), 260–322.

Moreira, S. M.; Moreira-Santos, M.; Guilhermino, L.; Ribeiro, R. Immobilization of the marine microalga *Phaeodactylum tricornutum* in alginate for *in situ* experiments: Bead stability and suitability. *Enz Microbiol Technol* 2006, 38, 135–141.

Moreno-Garrido, I. Microalgae immobilization: Current techniques and uses. *Bioresour Technol* 2008, 99, 3949–3964.

Munoz, R.; Guieysse, B. Algal-bacterial processes for the treatment of hazardous contaminants: A review. *Water Res* 2006, 40, 2799–2815.

Naik, S. N.; Goud, V. V.; Rout, P. H.; Dalai, A. H. Prodcution of first and second generation biofuels: A comprehensive review. *Renew Sust Energ Rev* 2010, 14, 578–597.

Naim, R.; Saif, U. R.; Jong-In, H. Rapid harvesting of freshwater microalgae using chitosan. *Proc Biochem* 2013, 48, 1107–1110.

Najafpour, G. D.; Zinatizadeh, A. A. L.; Mohamed, A. R.; Hasnain, M. I.; Nasrollahzadeh, H. High-rate anaerobic digestion of palm oil mill effluent in an upflow anaerobic sludge-fixed film bioreactor. *Proc Biochem* 2006, 41, 370–379.

Nakajima, A.; Horikoshi, T.; Sakaguchi, T. Recovery of uranium by immobilized microorganisms. *Eur J Appl Microbiol Biotechnol* 1982, 1, 88.

Nazir, M. S. Eco-friendly extraction, characterization and modification of microcrystalline cellulose from oil palm empty fruit bunches. PhD Thesis, Universiti Teknologi PETRONAS, Malaysia, 2013.

Nazir, M. S.; Wahjoedi, B. A.; Yussof, A. W.; Abdullah, M. A. Eco-friendly extraction and characterization of cellulose from oil palm empty fruit bunches. *Bioresour* 2013, 8(2), 2161–2172.

Negoro, M.; Hamasaki, A.; Ikuta, Y.; Makita, T.; Hirayama, K.; Suzuki, S. Carbon-dioxide fixation by microalgae photosynthesis using actual flue-gas discharged from a boiler. *Appl Biochem Biotechnol* 1993, 39, 643–653.

Nguyen, T. L.; Lee, D. J.; Chang, J. S.; Liu, J. C. Effects of ozone and peroxone on algal separation via dispersed air flotation. *Colloids Surf B* 2013, 105, 246–250.

Nobre, B. P.; Villalobos, F.; Barragan, B. E.; Oliveira, A. C.; Batista, A. P.; Marques, P. A. S. S.; Gouveia, L. A. Biorefinery from *Nannochloropsis* sp. microalga – extraction of oils and pigments: Production of biohydrogen from the leftover biomass. *Bioresour Technol* 2013, 135, 128–136.

Oh, H. M.; Lee, S. J.; Park, M. H.; Kim, H. S.; Kim, H. C.; Yoon, J. H. Harvesting of *Chlorella vulgaris* using a bioflocculant from *Paenibacillus* sp. AM49. *Biotechnol Lett* 2001, 23, 1229–1234.

Oh, T. H.; Pang, S. Y.; Chua, S. C. Energy policy and alternative energy in Malaysia: Issues and challenges for sustainable growth. *Renew Sust Energ Rev* 2010, 14, 1241–1252.

Olaizola, M. Commercial production of astaxanthin from *Haematococcus pluvialis* using 25,000-liter outdoor photobioreactors. *J Appl Phycol* 2000, 12(3), 499–506.

Oncel, S.; Sukan, V. F. Comparison of two different pneumatically mixed column photobio-reactors for the cultivation of *Artrospira platensis* (*Spirulina platensis*). *Bioresour Technol* 2008, 99, 4755–4760.

Órpez, R.; Martínez, M. E.; Hodaifa, G.; Yousfi, F. E.; Jbari, N.; Sánchez, S. Growth of the mi-croalga *Botryococcus braunii* in secondarily treated sewage. *Desalin* 2009, 246, 625–630.

Oswald, W. J. My sixty years in applied algology. *J Appl Phycol* 2003, 15, 99–106.

Otsuka, K.; Yoshino, A. A fundamental study on anaerobic digestion of sea lettuce, MTS/IEEE Techno-Ocean'04: Bridges across the Oceans – Conference Proceedings, 2004, pp 1770–1773.

Packer, M. Algal culture of carbon dioxide; biomass generation as a tool for greenhouse gas mitigation with reference to New Zealand energy strategy and policy. *Energ Policy* 2009, 37, 3428–3437.

Paepatung, N.; Kullavanijaya, P.; Loapitinan, O.; Songkasiri, W.; Noppharatana, A.; Chaiprasert, P. *Assessment of biomass potential for biogas production in Thailand.* Joint Graduate School of Energy and Environment: Bangkok, Thailand, 2006.

Panda, A. K.; Mishra, S.; Bisaria, V. S.; Bhojwani, S. S. Plant cell reactors - a perspective. *Enz Microbiol Technol* 1989, 11, 386–397.

Park, J. H.; Yoon, J. J.; Park, H. D.; Lim, D. J.; Kim, S. H. Anaerobic digestibility of algal bioethanol residue. *Bioresour Technol* 2012, 113, 78–82.

Park, S.; Li, Y. Evaluation of methane production and macronutrient degradation in the an-aerobic co-digestion of algae biomass residue and lipid waste. *Bioresour Technol* 2012, 111, 42–48.

Passos, F.; Ferrer, I. Microalgae conversion to biogas: Thermal pretreatment contribution on net energy production. *Environ Sci Technol* 2014, 48, 7171–7178.

Perez, R. R.; Benito, G. G.; Miranda, M. P. Chlorophenol degradation by *Phanerochaete chrysosporium*. *Bioresour Technol* 1997, 60, 207–213.

Pienkos, P. T.; Darzins, A. The promise and challenges of micro-algal derived biofuels. *Biofuel Bioprod Bioref* 2009, 3, 431–440.

Pinto, G.; Pollio, A.; Previtera, L.; Temussi, F. Biodegradation of phenols by microalgae. *Biotechnol Lett* 2002, 24, 2047–2051.

Poh, P. E.; Chong, M. P. Development of anaerobic digestion methods for palm oil. *Bioresour Technol* 2009, 100, 1–9.

Powell, E. E.; Hill, G. A. Economic assessment of an integrated bioethanol–biodiesel–microbial fuel cell facility utilizing yeast and photosynthetic algae. *Chem Eng Res Des* 2009, 87, 1340–1348.

Prasertsan, P. *Biomass residues from palm oil mills in Thailand: An overview on quantity and potential usage.* Pergamon Press: USA, 1996, 388.

Procedia, E. Integrated biogas-microalgae from waste waters as the potential biorefinery sources in Indonesia. *Energ Procedia* 2014, 47, 143–148.

Pulz, O. Photobioreactors: Production systems for phototrophic microorganisms. *Appl Microbiol Biotechnol* 2001, 57(3), 287–293.

Pulz, O.; Gross, W. Valuable products from biotechnology of microalgae. *Appl Microbiol Biotechnol* 2004, 65, 635–648.

Pulz, O.; Scheinbenbogan, K. Photobioreactors design and performance with respect to light energy input. *Adv Biochem Eng Biotech* 1998, 59, 123–152.

Qiang, L.; Lin, L.; Rong, C.; Xun, Z. A novel photobioreactor generating the light/dark cycle to improve microalgae cultivation. *Bioresour Technol* 2014, 161, 186–191.

Rahim, B. A.; Raj, R. Pilot plant study of a biological treatment system for palm oil mill effluent. In *Proceedings of regional workshop on palm oil mill technology and effluent treatment,* PORIM: *Malaysia,* 1982, pp 163–170.

Ramasamy, E. V.; Abbasi, S. A. Energy recovery from dairy waste-waters: Impacts of biofilm support systems on anaerobic CST reactors. *Appl Energ* 2000, 65, 91–98.

Ran, C. Q.; Chen, Z. A.; Zhang, W.; Yu, X. J.; Jin, M. F. Characterization of photobiological hydrogen production by several marine green algae. *Wuhan Ligong Daxue Xuebao* 2006, 28(2), 258–263.

Ranga Rao, A.; Dayananda, C.; Sarada, R.; Shamala, T. R.; Ravishankar, G. A. Effect of salinity on growth of green alga *Botryococcus braunii* and its constituents. *Bioresour Technol* 2007, 98, 560–564.

Ranjbar, R.; Inoue, R.; Katsuda, T.; Yamaji, H.; Katoh, S. High efficiency production of astaxanthin in an airlift photobioreactor. *J Biosci Bioeng* 2008, 106(2), 204–207.

Rasoul-Amini, S.; Montazeri-Najafabady, N.; Mobasher, M. A.; Hoseini-Alhashemi, S.; Ghasemi, Y. *Chlorella* sp., a new strain with highly saturated fatty acids for biodiesel production in bubble-column photobioreactor. *Appl Energ* 2011, 88, 3354.

Rawat, I.; Ranjith, K. R.; Mutanda, T.; Bux, F. Dual role of microalgae phycoremediation of domestic wastewater and biomass production for sustainable biofuels production. *Appl Energy* 2011, 88(10), 3411–3424.

Razzak, S. A.; Hossain, M. M.; Lucky, R. A.; Bassi, A.S.; de Lasa, H. Integrated CO_2 capture, wastewater treatment and biofuel production by microalgae culturing - a review. *Renew Sustain Energ Rev* 2013, 27, 622–653.

Reiko, S.; Yoshiaki, M.; Tomoko, Y.; Tsuyoshi, T.; Mitsufumi, M. Seasonal variation of biomass and oil production of the oleaginous diatom *Fistulifera* sp. in outdoor vertical bubble column and raceway-type bioreactors. *J Biosci Bioeng* 2014, 117, 720–724.

Residua. *Anaerobic digestion.* Warmer Bulletin: North Yorkshire, United Kingdom, 2007, pp 1–4.

Richmond, A. *Handbook of microalgal culture: Biotechnology and applied phycology.* Blackwell Science Ltd.: UK, 2004.

Richmond, A. Microalgal biotechnology at the turn of the millennium: A personal view. *J Appl Phycol* 2000, 12(3–5), 441–451.

Richmond, A. Principles for attaining maximal microalgal productivity in photobioreactors: An overview. *Hydrobiol* 2004, 512, 33–37.

Richmond, A.; Cheng-Wu, Z. Optimization of a flat plate glass reactor for mass production of *Nannochloropsis* sp. outdoors. *J Biotechnol* 2001, 85, 259–269.

Richmond, A.; Cheng-Wu, Z.; Zarmi, Y. Efficient use of strong light for high photosynthetic productivity: Interrelationships between the optical path, the optimal population density and cell-growth inhibition. *Biomol Eng* 2003, 20(4–6), 229–236.

Rittmann, B. E. Opportunities for renewable bioenergy using microorganisms. *Biotechnol Bioeng* 2008, 100, 203–212.

Rodolfi, L.; Zittelli, G. C.; Bassi, N.; Padovani, G.; Biondi, N.; Bonini, G. Microalgae for oil: Strain selection, induction of lipid synthesis and outdoor mass cultivation in a low-cost photobioreactor. *Biotechnol Bioeng* 2008, 102(1), 100–112.

Rodriguez, I.; Turnes, M. L.; Mejuto, M. C.; Cela, R. Determination of chlorophenols at the sub-ppb level in tap water using derivatization, solid phase extraction and gas chromatography with plasma atomic emission detection. *J Chromatogr* 1996, 721, 297–304.

Roncarati, A.; Meluzzi, A.; Acciarri, S.; Tallarico, N.; Melotti, P. Fatty acid composition of different microalgae strains (*Nannochloropsis* sp., *Nannochloropsis oculata* (Droop) Hibberd, *Nannochloris atomus* Butcher and *Isochrysis* sp.) according to the culture phase and the carbon dioxide concentration. *J World Aquacult Soc* 2004, 35, 401–411.

Rude, M. A.; Schirmer, A. New microbial fuels: A biotech perspective. *Curr Opin Microbiol* 2009, 12, 274–281.

Ruiz-Marin, A.; Mendoza-Espinosa, L. G. Ammonia removal and biomass characteristics of alginate-immobilized *Scenedesmus obliquus* cultures treating real wastewater. *Fresenius Environ Bull* 2008, 17, 1236–1241.

Rupprecht, J. From systems biology to fuel - *Chlamydomonas reinhardtii* as a model for a systems biology approach to improve biohydrogen production. *J Biotechnol* 2009, 142, 10–20.

Rupprecht, J.; Hankamer, B.; Mussgnug, J. H. Perspectives and advances of biological H$_2$ production in microorganisms. *Appl Microbiol Biotechnol* 2006, 72, 442–449.

Sag, Y.; Nourbakhsh, M.; Aksu, Z.; Kutsal, T. Comparison of Ca-alginate and immobilized Z. *ramigera* as sorbents for copper (II) removal. *Proc Biochem* 1995, 30(2), 175–181.

Saleh, A. F.; Kamarudin, E.; Yaacob, A. B.; Yussof, A. W.; Abdullah, M. A. Optimization of biomethane production by anaerobic digestion of palm oil mill effluent using response surface methodology. *Asia-Pac J Chem Eng* 2012, 7(3), 353–360.

Salim, S.; Vermuë, M. H.; Wijffels, R. H. Ratio between autoflocculating and target microalgae affects the energy-efficient harvesting by bio-flocculation. *Bioresour Technol* 2012 118, 49–55.

Samori, G.; Samori, C.; Guerrini, F.; Pistocchi, R. Growth and nitrogen removal capacity of *Desmodesmus communis* and of a natural microalgae consortium in a batch culture system in view of urban wastewater treatment, part I. *Water Res* 2013, 47, 791–801.

Samson, R.; Leduy, A. Multistage continuous cultivation of blue-green alga *Spirulina maxima* in flat tank photobioreactors. *Can J Chem Eng* 1985, 63, 105–112.

Sander, K.; Murthy, G. S. Life cycle analysis of algae biodiesel. *Int J Life Cycle Assess* 2010, 15, 704–714.

Sarah, M. J.; Jones Susan, S. M.; Harrison, S. T. L. Aeration energy requirements for lipid production by *Scenedesmus* sp. in airlift bioreactors. *Algal Res* 2014, 5, 249–257.

Satyawali, Y.; Balakrishnan, M. Wastewater treatment in molasses-based alcohol distilleries for COD and color removal: A review. *J Environ Manag* 2008, 86, 481–497.

Sawayama, S.; Inoue, S.; Dote, Y.; Yokoyama, S. Y. CO_2 fixation and oil production through microalga. *Energ Convers Manag* 1995, 36, 729–731.

Schafer, T. E.; Lapp, C. A.; Hanes, C. M.; Lewis, J. B.; Wataha, J. C.; Schuster, G. S. Estrogenicity of bisphenol A and bisphenol A dimethacrylate *in vitro*. *J Biomed Mater Res* 1999, 45, 192–197.

Schwede, S.; Rehman, Z. U.; Gerber, M.; Theiss, C.; Span, R. Effects of thermal pretreatment on anaerobic digestion of *Nannocloropsis salina* biomass. *Bioresour Technol* 2013, 143, 505–511.

Semple, K. T.; Cain, R. B. Degradation of phenol and its methylated homologues by *Ochromonas danica*. *FEMS Microbiol Lett* 1997, 152, 133–139.

Semple, K. T.; Cain, R. B.; Schmidt, S. Biodegradation of aromatic compounds by microalgae. *FEMS Microbiol Lett* 1999, 170, 291–300.

Sgroi, M.; Bollito, G.; Saracco, G.; Specchia, S. BIOFEAT: Biodiesel fuel processor for a vehicle fuel cell auxiliary power unit study of the feed system. *J Power Sources* 2005, 149, 8–14.

Shah, S. M. U. Cell culture optimization and reactor studies of green and brown microalgae for enhanced lipid production. PhD Thesis, Universiti Teknologi PETRONAS, Malaysia, 2014.

Shah, S. M. U.; Ahmad, A.; Othman, M. F.; Abdullah, M. A. Effects of palm oil mill effluent media on cell growth and lipid content of *Nannochloropsis oculata* and *Tetraselmis suecica*. *Int J Green Energ* 2016, 13(2), 200–207.

Shah, S. M. U.; Ahmad, A.; Othman, M. F.; Abdullah, M. A. Enhancement of lipid content in *Isochrysis galbana* and *Pavlova lutheri* using palm oil mill effluent as an alternative medium. *Chem Eng Trans* 2014a, 37, 733–738.

Shah, S. M. U.; Che Radziah, C. M.; Ibrahim, Z. S.; Latif, F.; Othman, M. F.; Abdullah, M. A. Effects of photoperiod, salinity and pH on cell growth and lipid content of *Pavlova lutheri*. *Annals Microbiol* 2014b, 64(1), 157–164.

Shen, Y.; Yuan, W.; Pei, Z. J.; Wu, Q.; Mao, E. Microalgae mass production methods. *Trans ASABE* 2009, 52, 1275–1287.

Shigeoka, T.; Sato, Y.; Takeda, Y. Acute toxicity of chlorophenols to green algae *Selenastrum capricornutum* and *Chlorella vulgaris* and quantitative structure–activity relationships. *Environ Toxicol Chem* 1988, 7, 847–854.

Shilton, A.; Mara, D.; Craggs, R.; Powell, N. Solar-powered aeration and disinfection, anaerobic co-digestion, biological CO_2 scrubbing and biofuel production: The energy and carbon management opportunities of waste stabilisation ponds. *Water Sci Technol* 2008, 58(1), 253–258.

Sialve, B.; Bernet, N.; Bernard, O. Anaerobic digestion of microalgae as a necessary step to make microalgal biodiesel sustainable. *Biotechnol Adv* 2009, 27, 409–416.

Sierra, E.; Acien, F. G.; Fernandez, J. M.; Garcia, J. L.; Gonzales, C.; Molina, E. Characterization of a flat plate photobioreactor for the production of microalgae. *Chem Eng J* 2008, 138, 136–147.

Siew, O. B. Simultaneous removal of phenol, ammonium and thiocyanate from coke wastewater by aerobic degradation. *J Hazard Mater* 2006, 137, 1773–1780.

Singh, J.; Gu, S. Commercialization potential of microalgae for biofuels production. *Renew Sust Energ Rev* 2010, 14(9), 2596–2610.

Singh, N.; Dhar, D. Microalgae as second generation biofuel, a review. *Agro Sust Dev* 2011, 31(4), 605–629.

Skjanes, K.; Lindblad, P.; Muller, J. BioCO$_2$ – a multidisciplinary, biological approach using solar energy to capture CO$_2$ while producing H$_2$ and high value products. *Biomol Eng* 2007, 24, 405–413.

Somerville, C. Biofuels. *Curr Biol* 2007, 17, 115–119.

Spolaore, P.; Joannis-Cassan, C.; Duran, E.; Isambert, A. Commercial applications of microalgae. *J Biosci Bioeng* 2006, 101(2), 87–96.

Stasinakis, A. S.; Elia, I.; Petalas, A. V.; Halvadakis, C. P. Removal of total phenols from olive-mill wastewater using an agricultural by-product olive pomace. *J Hazard Mater* 2008, doi:10.1016/j.jhazmat.2008.03.012.

Stephenson, A. L.; Kazamia, E.; Dennis, J. S.; Howe, C. J.; Scott, S. A.; Smith, A. G. Life-cycle assessment of potential algal biodiesel production in the United Kingdom: A comparison of raceways and air-lift tubular bioreactors. *Energ Fuel* 2010, 24, 4062–4077.

Stewart, C.; Hessami, M. A. A study of methods of carbon dioxide capture and sequestration-the sustainability of a photosynthetic bioreactor approach. *Energ Conv Manag* 2005, 46, 403–420.

Sturm, B. S.; Lamer, S. L. An energy evaluation of coupling nutrient removal from wastewater with algal biomass production. *Appl Energy* 2011, 88(10), 3499–3506.

Suali, E.; Sarbatly, R. Potential of CO$_2$ utilization by microalgae in Malaysia. *Int J Global Environ Issues* 2010, 12(2–4), 150–160.

Subramaniam, V.; Ma, A. N.; Choo, Y. M.; Sulaiman, N. M. N. Environmental performance of the milling process of Malaysian palm oil using the life cycle assessment approach. *Am J Environ Sci* 2008, 4, 310–315.

Suh, I. S.; Lee, S. B. A light distribution model for an internally radiating photobioreactor. *Biotech Bioeng* 2003, 82, 180–189.

Sulaiman, A.; Busu, Z.; Tabatabaei, M.; Yacob, S.; Abd-Aziz, S.; Hassan, M. A.; Shirai, Y. The effect of higher sludge recycling rate on anaerobic treatment of palm oil mill effluent in a semi-commercial closed digester for renewable energy. *Am J Biochem Biotechnol* 2009, 5, 1–6.

Tabatabaei, M.; Sulaiman, A.; Nikbakht, A. M.; Yousef, N.; Najafpour, G. Influential parameters on biomethane generation in anaerobic wastewater treatment plants. In *Alternative Fuel*, Manzanera, M., Ed.; InTech: Croatia, 2011, pp 227–262.

Tamagnini, P.; Leitao, E.; Oliveira, P.; Ferriera, D.; Pinto, F.; Harris, D. J.; Heidorn, T.; Lindblad, P. Cyanobacterial hydrogenases: Diversity, regulation and applications. *FEMS Microbiol Rev* 2007, 31, 692–720.

Techobanoglous, G. *Handbook of Solid Waste Management*. The McGraw-Hill Companies, Inc.: New York, 2002.

Thomas, W. H.; Gibson, C. H. Effect of small scale turbulence on microalgae. *J Appl Phycol* 1990, 2, 71–77.

Thong, S. O.; Boe, K.; Angelidaki, I. Thermophilic anaerobic co-digestion of oil palm empty fruit bunches with palmoil mill effluent for efficient biogas production. *Appl Energ* 2012, 93, 648–654.

Tong, S. L.; Bakar, J. Waste to energy: Methane recovery from anaerobic digestion of palm oil mill effluent. *Energ Smart* 2004, 4, 1–8.

Trepanier, C.; Parent, S.; Comeau, Y.; Bouvrette, J. Phosphorus budget as a water quality management tool for closed aquatic mesocosms. *Water Res* 2002, 36, 1007–1017.

Uduman, N.; Qi, Y.; Danquah, M. K.; Forde, G. M.; Hoadley, A. Dewatering of microalgal cultures: A major bottleneck to algae-based fuels. *J Renew Sustain Energ* 2010, 2, 012701.

Ueno, Y.; Kurano, N.; Miyachi, S. Ethanol production by dark fermentation in the marine green alga, *Chlorococcum littorale*. *J Ferment Bioeng* 1998, 86, 38–43.

Ugoji, E. O. Anaerobic digestion of palm oil mill effluent and its utilization as fertilizer for environmental protection. *Renew Energ* 1997, 10, 291–294.

Ugwu, C. U.; Aoyagi, H.; Uchiyama, H. Photobioreactors for mass cultivation of algae. *Bioresour Technol* 2008, 99(10), 4021–4028.

Ulrici, W. Contaminant soil areas, different countries and contaminant monitoring of contaminants. In *Biotechnology - Environmental Process II - Soil Decontamination, Vol. 11b*, Klein, J., Ed., Wiley-VCH: Weinheim, Germany, 2000, pp 5–42.

UNEP. Land degradation in South Asia: Its severity, causes and effects upon the people, World Soil Resources Report, INDP/UNEP/FAO, Rome: FAO, 1994, 78.

US DOE National algal biofuels technology roadmap. Report No.: DOE/EE-0332. U.S. Department of Energy, Office of Energy Efficiency and Renewable Energy, Biomass Program, 2010.

Van den Hoeck, C.; Mann, D. G.; Jahns, H. M. *Algae: An Introduction to Phycology*. Cambridge University Press: Cambridge, 1995.

Vandamme, D.; Foubert, I.; Muylaert, K. Flocculation as a low-cost method for harvesting microalgae for bulk biomass production. *Trends Biotechnol* 2013, 31, 233–239.

Vazquez, I.; Rodriguez, J.; Maranon, E.; Castrillon, L.; Fernandez, Y. Simultaneous removal of phenol, ammonium and thiocyanate from coke wastewater by aerobic degradation. *J Hazard Mater* 2006, 137, 1773–1780.

Velasquez-Ort, S. B.; Garcia-Estrada, R.; Monje-Ramirez, I.; Harvey, A.; Orta Ledesma, M. T. Microalgae harvesting using ozoflotation: Effect on lipid and FAME recoveries. *Biomass Bioenerg* 2014, 70, 356–363.

Vergara-Fernandez, A.; Vargas, G.; Alarcon, N.; Velasco, A. Evaluation of marine algae as a source of biogas in a two-stage anaerobic reactor system. *Biomass Bioenerg* 2008, 32, 338–344.

Vijayaraghavan, K.; Ahmad, D.; Ezani, M. Aerobic treatment of palm oil mill effluent. *J Environ Manag* 2007, 82, 24–31.

Vijayaraghavan, K.; Mahadevan, A.; Sathishkumar, M.; Pavagadhi, S.; Balasubramanian, R. Biosynthesis of Au(0) from Au(III) via biosorption and bioreduction using brown marine alga *Turbinaria conoides*. *Chem Eng J* 2011, 167, 223–227.

Walker, J. D.; Colwell, R. R.; Vaituzis, Z.; Meyer, S. A. Petroleum degrading achlorophyllous alga *Prototheca zopfii*. *Nature* 1957, 254(5499), 423–424.

Wang, L.; Min, M.; Li, Y.; Chen, P.; Chen, Y.; Liu, Y. Cultivation of green algae *Chlorella* sp. in different wastewaters from municipal wastewater treatment plant. *Appl Biochem Biotechnol* 2009, 162(4), 1174–1186.

Wang, S. J.; Loh, K. C. Facilitation of cometabolic degradation of 4-chlorophenol using glucose as an added growth substrate. *Biodegrad* 1999, 10, 261–269.

Wang, S. K.; Hu, Y. R.; Wang, F.; Stiles, A. R.; Liu, C. Z. Scale-up cultivation of *Chlorella ellipsoidea* from indoor to outdoor in bubble column bioreactors. *Bioresour Technol* 2014, 156, 117–122.

Weiland, P. Biogas production current state and perspectives. *Appl Microb Biotech* 2010, 85, 849–860.

Wen-Can, H.; Jong-Duk, K. Cationic surfactant-based method for simultaneous harvesting and cell disruption of a microalgal biomass. *Bioresour Technol* 2013, 149, 579–581.

Wikfors, G. H.; Patterson, G. W. Differences in strains of *Isochrysis* of importance to mariculture. *Aquacult* 1994, 123, 127–135.

Willke, T.; Vorlop, K. D. Industrial bioconversion of renewable resources as an alternative to conventional chemistry. *Appl Microbiol Biotechnol* 2004, 66, 131–142.

Woertz, I.; Feffer, A.; Lundquist, T.; Nelson, Y. Algae grown on dairy and municipal wastewater for simultaneous nutrient removal and lipid production for biofuel feedstock. *J Environ Eng* 2011, 135(11), 1115–1123.

Wong, F. M. A review on the progress of compliance with the palm oil control regulations. *Seminar on advances in palm oil effluent control technology*, Kuala Lumpur, 1980, pp 142–149.

Wong, P. K.; Lam, K. C.; So, C. M. Removal and recovery of Cu(II) from industrial effluent by immobilized cells of *Pseudomonas putida* II-11. *Appl Microbiol Biotechnol* 1993, 39(1), 127–131.

Wong, P. W.; Nik, M. S.; Kshisundaram, N. M.; Balaraman, V. Pre-treatment and membrane ultrafiltration using treated Palm Oil Mill Efflent (POME). *Membr Sci Technol* 2002, 24, 890–898.

Wu, T. Y.; Mohammad, A. W.; Jahim, J. M.; Anuar, N. A holistic approach to managing palm oil mill effluent (POME): Biotechnological advances in the sustainable reuse of POME. *Biotechnol Adv* 2009, 27, 40–52.

Wu, T. Y.; Mohammad, A. W.; Jahim, J. M.; Anuar, N. Palm oil mill effluent (POME) treatment and bioresources recovery using ultrafiltration membrane: Effect of pressure on membrane fouling. *Biochem Eng J* 2007, 35, 309–317.

Wu, T. Y.; Mohammad, A. W.; Jahim, J. M.; Anuar, N. Pollution control technologies for the treatment of palm oil mill effluent (POME) through end-of-pipe processes. *J Environ Manag* 2010, 91, 1467–1490.

Xie, T. M.; Abrahamsson, K.; Fogelqvist, E.; Josefsson, B. Distribution of chlorophenolics in a marine environment. *Environ Sci Technol* 1986, 20, 457–463.

Xuan, J.; Leung, M. K. H.; Leung, D. Y. C.; Ni, M. A. Review of biomass-derived fuel processors for fuel cell systems. *Renew Sust Energ Rev* 2009, 13, 1301–1313.

Yacob, S.; Ali, H. M.; Shirai, Y.; Wakisaka, M.; Subash, S. Baseline study of methane emission from anaerobic ponds of palm oil mill effluent treatment. *Sci Total Environ* 2006, 366, 187–196.

Yacob, S.; Hassan, M. A.; Shirai, Y.; Wakisaka, M.; Subash, S. Baseline study of methane emission from open digesting tanks of palm oil mill effluent treatment. *Chemosphere* 2005, 59, 1575–1581.

Yadvika, S.; Sreekrishnan, T. R.; Kohli, S.; Rana, V. Enhancement of biogas production from solid substrates using different techniques - a review. *Bioresour Technol* 2004, 95, 1–10.

Ya-Ling, C.; Yu-Chuan, J.; Guan-Yu, L.; Shih-Hsin, H.; Kuei-Ling, Y.; Chun-Yen, C. Dispersed ozone flotation of *Chlorella vulgaris*. *Bioresour Technol* 2013, 101, 9092–9096.

Yang, L. R.; Guo, C.; Chen, S.; Wang, F.; Wang, J.; An, Z. T.; Liu, C. Z.; Liu, H. Z. pH sensitive magnetic ion exchanger for protein separation. *Ind Eng Chem Res* 2009, 48, 944–950.

Yen, H. W.; Brune, D. E. Anaerobic co-digestion of algal sludge and waste paper to produce methane. *Bioresour Technol* 2007, 98, 130–134.

Yi, J. C.; Mei, F. C.; Chung, L. L. An integrated anaerobic–aerobic bioreactor (IAAB) for the treatment of palm oil mill effluent (POME): Start-up and steady state performance. *Proc Biochem* 2012, 47, 485–495.

Yi-Ru, H.; Chen, G.; Feng, W.; Shi-Kai, W.; Feng, P.; Chun-Zhao, L. Improvement of microalgae harvesting by magnetic nanocomposites coated with polyethylenimine. *Chem Eng J* 2014, 242, 341–347.

Yoon, J. H.; Choi, S. S.; Park, T. H. The cultivation of *Anabaena variabilis* in a bubble column operating under bubbly and slug flows. *Bioresour Technol* 2011, 110, 430.

Yuan, X.; Wang, M.; Park, C.; Sahu, A. K.; Ergas, S. J. Microalgae growth using high-strength wastewater followed by anaerobic co-digestion. *Water Environ Res* 2012, 84, 396–404.

Zhang, J.; Hu, B. A. novel method to harvest microalgae via co-culture of filamentous fungi to form cell pellets. *Bioresour Technol* 2012, 114, 529–535.

Zhang, K.; Kurano, N.; Miyachi, S. Optimized aeration by carbon dioxide gas for microalgal production and mass transfer characterization in a vertical flat-plate photobioreactor. *Bioproc Biosyst Bioeng* 2002, 25, 97–101.

Zhang, T.; Gong, H.; Wen, X.; Lu, C. Salt stress induces a decrease in excitation energy transfer from phycobilisomes to photosystem II but an increase to photosystem I in the cyanobacterium *Spirulina platensis. J Plant Physiol* 2010, 167, 951–958.

Zhang, Y.; Su, H.; Zhong, Y.; Zhang, C.; Shen, Z.; Sang, W.; Yan, G.; Ji, M. K.; Kim, H. C.; Sapireddy, V. R.; Yun, H. S.; Abou-Shanab, R. A. I.; Choi, J.; Lee, W.; Timmes, T. C.; Inamuddin, B. H. Nutrient removal and biodiesel production by integration of freshwater algae cultivation with piggery wastewater treatment. *Water Res* 2013, 47, 4294–4302.

Zhang, Y.; White, M. A.; Colosi, L. M. Environmental and economic assessment of integrated systems for dairy manure treatment coupled with algae bioenergy production. *Bioresour Technol* 2013, 130, 486–494.

Zhou, X. The effect of bacterial contamination on the heterotrophic cultivation of *Chlorella pyrenoidosa* in wastewater from the production of soybean products. *Water Res* 2012, 46, 5509–5516.

CHAPTER 9

BIOAMBIANT PRESERVATION OF RAW HIDES USING PLANT-BASED MATERIALS – A GREEN TECHNOLOGY TO REDUCE TANNERY WASTE WATER POLLUTION

PRAFULLA NAMDEO SHEDE[1], ASHISH VASANTRAO POLKADE[2], PRADNYA PRALHAD KANEKAR[3], PRASHANT KAMALAKAR DHAKEPHALKAR[4], and SEEMA SHREEPAD SARNAIK[5]

[1]Department of Microbiology, MES' Abasaheb Garware College, Karve Road, Pune 411004, India
[2]Microbial Culture Collection, National Centre for Cell Science, First floor, Central Tower, Sai Trinity Building Garware Circle, Sutarwadi, Pashan, Pune 411021, India
[3]Department of Biotechnology, Modern College of Arts Science and Commerce, Shivajinagar, Pune 411005, India
[4]Bioenergy, MACS Agharkar Research Institute, G.G. Agarkar Road, Pune 411004, India
[5]Microbial Sciences Division, MACS Agharkar Research Institute, G.G. Agarkar Road, Pune 411004, India

CONTENTS

9.1   INTRODUCTION

In looking for a covering material for himself, his hut and food, primitive man turned either to leaves from plants or to the skins of animals he killed. The latter were usually chosen for clothing, as they were bigger, stronger, and warmer. As per British Standard Definitions, hide is the raw skin of a mature or fully grown animal of the larger kinds, such as cattle and horses; also camels, rhinoceroses and whales and skin is the raw skin of a mature, fully grown animal of the smaller kinds, like goats, sheep, pig, and others. The flayed skins or hides of dead animals are used as raw materials for leather making. Leather is nothing but the skins or hides whose protein has been stabilized through the process of tanning.

9.1.1   MAIN SOURCES OF RAW MATERIAL FOR LEATHER MANUFACTURING

The outer flayed covering, obtained from literally any animal and containing considerable amount of specific proteins can possibly serve as raw material for leather manufacturing. The skins could be acquired from animals like sheep, goat, pig, reptile mostly snakes and lizards, crocodile, birds and fishes, or the immature animals of the larger species, such as calves and colts. The hides could be gained from buffalo, bull, cow, horse, camel, elephant, and whale. The biological subfamily Bovine, comprising ten different genera of medium to large-sized animals including buffalo, is the most prominent contributing toward the hides produced worldwide. According to the Food and Agriculture Organization Statistics, world bovine (cattle and buffalo) population is constantly increasing. In 2013, the cumulative heads of cattle and buffalos available as livestock alone were 1.69 billion. Its worldwide distribution is detailed in Figure 9.1.

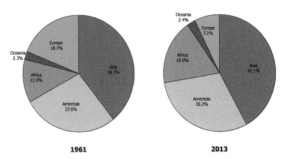

FIGURE 9.1   Worldwide distribution of bovine population (Source: Food and Agriculture Organization of The United Nations Statistics Division).

9.1.2 STRUCTURE AND COMPOSITION OF BOVINE HIDE

The bovine hides processed worldwide are mainly acquired from the African and Asian water buffalos scientifically known as *Bovidae bubalis*. The bovine hide can be mainly divided in to two principle layers from leather making point of view as 'epidermis' and 'corium'. Epidermis or the outer layer also called the striated epithelium and the corium, inner layer also called as dermis, cutis vera or true skin. These two layers are quite different in their structure and functions. The epidermis, which is a very thinner layer covering the underneath corium is removed during the leather processing. Corium, the main layer of hide constitutes about 98% of its thickness and comprises mainly of the fibrous protein bundles. About 90–95% of the corium layer of most of the hides is converted into the leather. The main components of the skins/hides are shown in Figure 9.2.

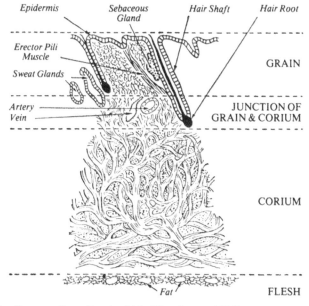

FIGURE 9.2 Cross-section of bovine hide (Sharphouse, 1983).

A fresh hide consists of water, protein, fatty materials, and some mineral salts. Of these the most important for leather making is the protein. This protein may consist of many types of structural (fibrous) and non-structural (globular) forms. The important one is collagen, the fibrous protein that on tanning gives leather. Corium consists of a network of collagen fibers very intimately woven and joined together. The approximate composition of a freshly flayed hide is given in Figure 9.3. Overall, the freshly flayed hide could be a feast for bacteria if it is not processed instantly in the beam house and/or tannery to prepare leather from it.

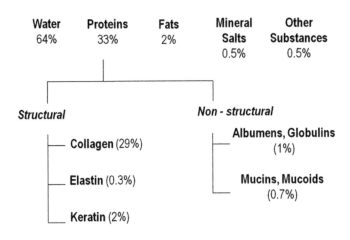

FIGURE 9.3   Approximate composition of bovine hide.

9.1.3   CHIEF PROCESSES USED IN LEATHER MANUFACTURING

Leather manufacturing in particular is an age-old skill acquired by our ancestors. This whole process is roughly divided into three parts as before tannage, tannage, and after tannage processes. Although modern leather processing is very much dependent on automated machineries, the process remains the same. These chief processes are described in Table 9.1. There are many variations on this simple line of processing. Their choice determines the character of the leather produced. Their in-depth study forms a firm base for 'Leather Technology'.

TABLE 9.1   Chief Processes Used in Leather Manufacture* (Shede, 2007).

Chief Process	Sub Processes	Function
Before tannage	Flaying	Removes the skin from animal
	Curing	Preserves skins during transport or storage
	Washing	Restores cured skins to natural raw condition
	Liming	Loosens hair, fat and flesh
	Unhairing	Removes hair
	Fleshing	Cut away unwanted fat and flesh
	Deliming	Neutralizes alkali from liming process
	Bating	Make the skin softer and cleaner
	Pickling	Brings the skin to right acidity for tannage
Tannage	Tanning	Hides are tanned by whichever method appropriate such as using vegetable tans or chrome or alum or oil
After tannage	Washing	Removes surplus tan
	Setting Out	Removes wrinkles and flatten the leather
	Oiling	Makes the grain flexible and of good colour
	Stuffing	Waterproofing
	Drying	Removes surplus water
	Rolling	Makes leather firm and flat
	Splitting	Achieves uniform thickness
	Washing	Makes free from surplus tan
	Neutralizing	Adjusts acidity
	Dyeing	Obtains required color
	Fatliquoring	Achieves softness
	Setting Out	Removes creases and surplus moisture
	Staking	Softens the leather
	Seasoning	Improves appearance
	Glazing	Pressurized polishing
	Plating	Gives smooth, flat surface
	Embossing	Produces fancy design

*Not all these processes may be given to a particular type of hide or skin; the order of after tannage processes varies for different leathers.

9.1.4   SOCIOECONOMIC INFLUENCE OF 'LEATHER INDUSTRY'

The word 'industry' in its own has a meaning. It not only conveys vision, creativity, business, trade, capital, raw materials, hard work, and employment all the same also wastes, pollution, and waste management irrespective of the prefix added to it and the 'leather industry' is no exception for this. The figures stated for raw material of leather and finished leather products import-exports with respect to Indian and Global perspectives are striking to both economists and sociologists (Fig. 9.4). The revenue generated in the import export of leather raw material and finished products is considered to be influencing developing country's economy.

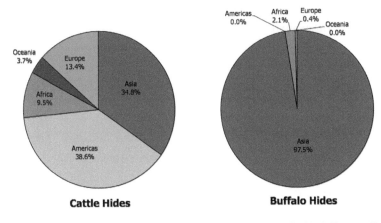

FIGURE 9.4   Worldwide distribution of bovine hide production in 2013 (Source: Food and Agriculture Organization of The United Nations Statistics Division).

The total number of small scale and large-scale tanneries worldwide is immense. The report commissioned by ETSU on behalf of Department of Trade and Industry, Government of India on the topic 'Evaluation of energy from waste investment opportunities in India' and prepared by FEC Consultants Ltd., and Delphi International Ltd., in the year 1997 estimated 1935 tanneries in India. It also states that in India, tanneries have developed mostly in a concentrated fashion around specific centers for historical and social reasons. The state-wise distribution of tanneries in India in the year 2006 is given in Table 9.2.

The Indian leather sector receives direct investment from foreign countries and is a source for 10% global leather requirement with a production value over US $4 billion (annual export value over US $2 billion). As per the Council for Leather Export (CLE), India, the tanneries in India have more than 2.5 million employees, 30% of which are women. The industrial setups and/or tanneries distributed throughout the globe on one hand offer employment while in many cases create environmental problems by means of waste generated.

TABLE 9.2   State-wise Distributions of Large-Scale Tanneries in India.

State	Area	Number of Tanneries
Tamil Nadu	Chennai, Ranipet, Ambur, Pernampet, Vaniyambadi, Dindigul, Trichy	520
West Bengal	Tangra, Topsia, Tiljala	400
Uttar Pradesh	Kanpur, Unnaon, Agra	200
Maharashtra	Mumbai, Dharavi, Aurangabad, Pune	200
Karnataka	Banglore, Dharwar, Belgaum, Humnabad, Bijapur	180
Rajasthan	Beawar, Bhinwal	150
Punjab	Jalandhar, Nakodar	50
Gujarat	Idar, Bhavnagar, Ahmdabad, Rajkot, Wadhwan	50
Andhra Pradesh	Hyderabad, Vijaynagaram, Karimnagar, Warangal	30
Haryana	Ambala, Jind	20
Bihar	Muzaffarpur, Mekemaghat	15
Jammu and Kashmir		10

9.2   POLLUTION PROBLEMS FACED BY LEATHER INDUSTRY

Leather industry makes use of hides and skins, which are believed to be byproducts of meat industry. Considering the global production of these hides and skins, leather industry becomes the world's largest industry based on byproducts. As stated earlier, leather processing in general involves a series of unit operations that can be classified into three groups: (i) beam house operations or pre-tanning which clean the hides or skins, (ii) tanning which permanently stabilizes the skin or hide matrix, and (iii) post-tanning and finishing operations where aesthetic value is added to leather. At each stage, various chemicals are used and a variety of materials are expelled, which ultimately lead to the environmental stresses (Suresh et al., 2001; Thanikaivelan et al., 2003; Sivakumar et al., 2005; Song et al., 2011). This could be the reason for placing leather industry in the list of most polluting industries worldwide.

In terms of biological oxygen demand (BOD), chemical oxygen demand (COD), and total dissolved solids (TDS), almost 70% of the pollution originates from pre-tanning operations. Almost all sulfide emissions originate from dehairing operations. In addition, chrome-tanning operations result in large emissions of chromium and sulfate ions. With the total annual global processing capacity of 9×10^9 kg hides and skins, it is estimated that $30-40 \times 10^{10}$ L of liquid effluent is generated (Thanikaivelan et al., 2004). Salt used for preserving the hides and skins generates huge amounts of pollution in terms of TDS and chlorides (Boopathy and Sekaran, 2014). According to Kanagaraj et al. (2005a), the effluent generated in the soaking operation alone accounts for 40% of chlorides and 55% of TDS in the entire leather processing operations.

The discharge limit for TDS is quite stringent in India and the restrictions from the global leather buyers and countries are also severe. The socio-economic status of leather industry and the pressure from various pollution monitoring agencies has forced the leather sector to look for various cleaner pre-tanning and tanning methodologies. These methodologies are based on the partial or complete replacement of tanning agents, dehairing chemicals, and the curing agents, especially the alternative preservative methods that will eliminate the use of salt as a preservative.

9.3   ALTERNATIVE METHODS TO SALT CURING – CURRENT SCENARIO

In olden days, every man would tan or dress each skin/hide as he acquired it and before it putrefied or the expert tanner would do it for him. As the demand for leather and the special qualities in it became greater, the processes became more complicated and required special skills, tools, processing plant and machinery. This led to the growth of tanneries collecting skins/hides over large areas, and producing and selling their special products all over the country and eventually the world.

This raised a special problem that if transport of the raw skins/hides took days, weeks or months before they were tanned, putrefaction would start and the skins/hides would be useless when they arrive at the tannery. The need arose for simple methods of stopping this putrefaction, methods which would not so alter the skins from the raw state that the tanner could not tan or color them, as he required. These are referred to as curing or preserving methods and are well designed considering the type of material to be preserved and the factors leading to its putrefaction. These curing methods are almost same for all types of skins/hides with certain exceptions. The most favored preservative throughout the world was 'common salt', chemically termed as sodium chloride or NaCl. The traditional usage of this common salt in the process of preservation has got certain limitations in present situation as discussed earlier. Salt is one of the most notorious components of any effluent to treat and it is said that the only cost effective way to remove it from an effluent is to avoid putting it in the first place. Secondly, the rate of salt penetration in the hides also decides the curing efficiency of the salting process (Cooper et al., 1972). Moreover, halophiles are one more factor that demands an alternative to salt. The halophilic bacteria capable of growing at high salt concentrations were evident on salt cured hides and skins (Stuart and Frey, 1938; Woods et al., 1971; Kallenberger, 1984) and this is generally called as 'red heat' amongst the tanners. Considering this, various attempts were made by different workers (Kanagaraj and Chandra Babu, 2002) to use compounds other than salt as preservatives of the raw hides and skins or to engage different physical methods to achieve preservation of raw material.

9.3.1 PHYSICAL METHODS OF CURING

These are the methods wherein the skins and hides were preserved by physical treatments. One more conventionally used method of preservation other than salting is drying of skins and hides. Drying, in particular is such method where the putrefaction is stopped by removal of water so that only 10–14% moisture is retained in hide, which eventually ceases the bacterial activity. Roughly the drying could be achieved by two ways as sun drying – the hides are hung in the sun and shed drying – the hides are dried in open sided, covered shed, designed to keep off the direct heat of sun however allows good ventilation. According to Sharphouse (1983), drying has got certain limitations if the drying is too slow which is possible in cold and wet climate. The putrefaction may occur before the moisture content is low enough to stop the bacterial action and if the drying is too fast and the temperature is too high, part of the wet skins will start to gelatinize to glue like material making the skins hard and brittle. Another disadvantage of drying is dried hides require careful handling and packing, as they are hard and mere bending may cause them to crack. Apart from this, the dried hides and skins are susceptible to insect attack. Hence drying could not be the method of choice for the committed tanner.

Storing the raw skins and hides at low temperatures termed as chilling or freezing is also considered as the alternative to salt curing. Haines (1981) has studied the effect of chilling and freezing on preservation of cattle hides. He found that the hides can be safely held at 0°C for three weeks without any additional treatment, whereas at −10°C the hides could be preserved for three months. Chilling and freezing, although a common practice in developed countries for preservation of meat and byproducts of meat. It is highly impossible to use this method in any developing country, not even India for its demand of high cost and energy inputs in the total process of preservation. Considering the share of underdeveloped and developing countries in the production of raw hides and skins which is more than half of the total global production capacity, one should not solely depend on such an expensive method. Even if the developed countries favor such an energy intensive and pricey method to cope up with the stringent pollution control norms of their own countries or province, these methods do have some lacunae in those countries as well. The failure of machinery or the interruption in power supply to the machinery maintaining such low temperature could lead to the putrefaction of chilled and frozen hides and skins. Secondly, it becomes very difficult to transport huge amounts of raw hides and skins in chilled or frozen conditions. Hence chilling and freezing could not be the methods of choice for the tanner from the undeveloped and/or developing countries with certain exceptions from the developed countries.

9.3.2 CHEMICAL METHODS OF CURING

These are the methods where both proprietary products and standard chemicals were used in place of salt as biocides or biostatic compounds to preserve the raw skins and hides. Reports for use of such chemicals in place of common salt dates back to 80's however in actual sense, the intellect to minimize the pollution caused by salt is unintentionally owned by Kritzinger and Vanzyl (1954). They have reported the preservation of hides with used salt. Their main concern was to make available the salt to curers when there was less salt supply due to excessive rains or drought. Hence they proposed reuse of salt as preservative thereby decreasing the concentration of salts in soak liquors and eventually in the tannery effluent. Although, they failed to mention their achievement in their research article, no doubt it was the first indirect attempt to minimize the pollution caused by salt in the tannery effluents. Further, various attempts were made directly to minimize the pollution problems caused by the salt as a raw skin and hide preservative.

The approach used by most of the researchers in this case was, use of the chemicals with known antibacterial activities. In most of the cases, these chemicals possessed bactericidal activities with few exceptions where the chemicals used were bacteriostatic in nature. Several of them were structurally similar to one another and were from the same class or family of chemicals however poses different properties. Mostly the inorganic and organic salts of different metals such as zinc, sodium, potassium, nickel, and mercury were engaged in these studies. These biocides were effective in terms of preservation of raw skins and hides. They could preserve the skins and hides for the period of two days to six months, depending on the chemical used. Table 9.3 portrays the attempts made by various researchers for replacement of common salt with other chemicals as short-term preservatives of raw hides and skins. The above-mentioned chemicals are yet again problematic with the environment and health perspective. Majority of these substances comes under the hazardous chemical categories as toxic or very toxic, harmful or irritant, corrosive and dangerous for the environment (Table 9.4).

TABLE 9.3 Evaluation of Chemical Preservatives for Skin and Hide Preservation.

Chemicals	Raw Material	Period (days)	References
Sodium bisulfate DextrasetUN*	Hides	8–14	Hopkins et al. (1973)
Benzalkonium chloride Sodium trichlorophenate Sodium pentachlorophenate Sodium silicofluoride Sodium chlorite Zinc dimethyldithiocarbamate Zinc chloride Zinc borate Nickel chloride Mercuric chloride p-Chloro-m-cresol	Skins	2–14	Sivaparvathi and Nandy (1974)
Acetic acid Sodium sulfite Sodium bisulfite	Hides	6–30	Hopkins and Bailey (1975)
Sodium chlorite Gloquat C* (a quaternary compound) Glokill 77* (linear and cyclic hydroxylamine) Vantocil IB* (polymeric biguianide) Vantoc CL* (lauryl dimethyl benzyl ammonium chloride)	Hides	6–8	Haffner and Haines (1975)
Organic biocide[#] Inorganic biocide[#]	Skins/Hides	90	Venkatachalam et al. (1982)
Sodium carbonate	Hides	8	Rao and Henrickson (1983)
Phenol, Borax	Skins/Hides	180	Selvarangan and Shanmugasundaram (1984)

TABLE 9.3 *(Continued)*

Chemicals	Raw Material	Period (days)	References
p-Chloro-m-cresol Zinc dimethyl dithiocarbamate B-hydroxyquinoline Cetrimide Benzalkonium chloride Boric acid Sodium pentachlorophenate	Skins	12–14	Sivaparvathi et al. (1993)
Potassium chloride	Hides	180	Bailey (1995)
Silica gel, p-Chloro-m-cresol	Skins	14	Kanagaraj et al. (2001)
Boric acid, Sodium metabisulphate	Skins	14–30	Kanagaraj et al. (2005a), Kanagaraj et al. (2005b)
Cetyl trimethyl ammonium bromide	Skins	12	Ganesh Babu et al. (2009)

*Brand name of the product; #Chemical nature not revealed.

TABLE 9.4 Safety Hazards Involved in the Use of the Alternative Chemical Preservatives.

Toxic/Very Toxic	Harmful/Irritant	Corrosive	Dangerous for the Environment
Sodium chlorite, nickel chloride, mercuric chloride, p-chloro-m-cresol, phenol	Zinc dimethyldithiocarbamate, sodium sulfite, sodium bisulfite, hydroxylamine, biguianide, sodium carbonate, borax, cetrimide, 8-hydroxyquinoline	Sodium bisulfate, benzalkonium chloride, zinc chloride, acetic acid, hydroxylamine	Zinc chloride, nickel chloride, mercuric chloride, hydroxylamine, cetrimide

Secondly the expenses involved in the use of such chemical preservative is very high as the chemicals to be used are required in bulk that too in the maximum pure form to achieve utmost possible efficacy in preservation. In nearly all of the above cases, increased cost and decreased preservative efficiency was observed as compared to common salt. Another disadvantage of these alternative chemicals is the difficulty in the handling of the chemicals by the curers. The ease experienced in the handling of common salt is not possible for these chemicals, as they may require some special skills and care by handlers or the curers, being hazardous in nature. Most importantly the main goal of salt less preservation could not be achieved, as these chemical substances themselves are the salts of certain metals. Their use as an alternative preservative in the tannery definitely reduced the TDS problem from soak liquors although could not eliminate it completely.

9.3.3 BIOLOGICAL METHODS OF CURING

These are the methods wherein compounds obtained from biological medium is used as preservatives for raw skins and hide and are termed as 'biopreservatives'. One such compound is 'bacteriocin' or 'antibiotic', a small peptide with antibacterial activity. Several bacteria are reported to produce various bacteriocins. Hanlin et al. (1995) has used bacteriocins, nisin, and pediocin AcH obtained from lactic acid bacteria as antibacterial for the preservation of cattle hides. While concluding, they have mentioned that if a shift in the bacterial population occurs on rawhide, the bacteriocins will need to be supplemented with other components. In these studies, they have used sodium dodecyl sulfate (SDS) at 0.25% w/v of the mixture. Addition of this detergent in the preservative mixture makes it unfavorable from the viewpoint of environment conscious tanners. An approach to inhibit the collagenolytic test bacterium *Vibrio alginolyticus* that could degrade the fibrous matrix from hide was used by Berwick et al. (1990). In their studies they made use of β-lactam, aminoglycosides, and tetracycline type antibiotics. Tetracycline type antibiotics were most effective against the test organisms and they also reduced the hide biodeterioration rate significantly to store the raw hide till ten days.

Antibiotics as preservative agents offer certain advantages over the methods discussed previously. These advantages include their biodegradability, efficacy at low concentrations and their relative ease of production by bacterial fermentation. On the contrary, if the equipments and facilities required for fermentation are not owned by the tanners, then the use of antibiotics as preservatives will increase the cost of leather manufacturing. The possibility, that the small-scale tanners could not necessarily afford the setup and running cost of fermentation facilities, has to be considered. In such cases the tanners will have to depend entirely on the commercial antiseptic formulations from the pharmaceutical company. The bulk purchase of antibiotics may reduce the cost of hide preservation however the nature of antibiotics in many cases will force the tanners to make necessary arrangements for their storage at low temperatures that again involves capital investment. Most importantly the tannery effluent containing the unused and washed antibiotics if introduced in natural habitats and/or water bodies may create health problems. The microorganisms, susceptible to these antibiotics will be constantly exposed to their high dosage, making them resistant to such antibiotics. All these factors make it a non-user-friendly method from the point of view of attentive tanners.

In 1981, the very first attempt was made to preserve the raw skins and hides using plant-based material. Venkatachalam et al. (1981) have used the herb *Decalepis hamiltonii* for this purpose. The boiled root extract when applied on flesh side could preserve the hides and skins for 3–5 days, whereas the soaked hides and skins in the decoction were found to remain in good condition for at least two weeks. Further they found that to make it more effective, this aryl alcohol has to be mixed with inorganic salts and bases. This mixture could preserve the hides and skins for a month. The preservation period of two weeks is not sufficient to count as a short-term storage period with respect to tanner's logic of leather processing, whereas addition of inorganic salts and bases to the extract deflects the main aim of salt less preservative. Thus, by making it practically unsuitable as biopreservative, the research on these lines was brought to a standstill. Although the search of biopreservative for raw skins and hides to entirely replace the salt was kicked off in last two decades, for one reason or the other, it has been surrendered to the baselines where it started way back in the early 90s.

9.4   CONTEMPORARY BIOTECHNOLOGY FOR LEATHER

The biological methods engaging the biopreservatives were thought to have advantage over salt, alternative chemicals, and the physical methods. They do not just reduce the TDS, salinity, and chlorides from soak liquors however they completely eliminate such problems from the soak liquors. The efforts taken by the researchers to discover and develop the biopreservatives of raw skins and hides are of great significance. Though their hard work could not give any realistic method of biopreservation, it was good enough to suggest the right approach to invent the effective biopreservative.

The approach talked about could be termed as 'biotechnological approach'. In the tanning industry, biotechnology is not a new concept. For several years, enzymes have been extensively used at all stages in the leather making, with the exception of curing and actual tanning process. At present, the biotechnological methods are being used with great accuracy in soaking, dehairing, bating and degreasing procedures. With this situation, salt less biopreservation of raw skins and hides is not less than a challenge. Thanikaivelan et al. (2004) in their review on recent trends in leather processing, has discussed a few biotechnological products and processes along with probability of their success. Bioproducts for ambient preservation of raw skins and hides is amongst them.

In the task of inventing any preservation method, one needs to know the nature of material to be preserved, adverse changes occurring in the material when stored without any preservative and the components responsible for such changes. The information available on raw buffalo hide as a material to be preserved has to be gathered through various ways, such as the changes occurring on the raw buffalo hide during ambient storage. The components responsible for these changes could be studied by considering the protein-aceous nature of raw buffalo hides. In this quest the naturally occurring anti-microbial compounds could be used. Olasupo et al. (2003) has reported few such compounds active against the *Escherichia coli* and *Salmonella enterica* serovar *Typhimurium*. Curing agent, which is a plant-based substance and advantageous in terms of biodegradability and eco friendly nature over alter-nate chemicals other than sodium chloride and expensive biocides/antibiot-ics, could be identified from the available pool of knowledge. Its efficacy as a biopreservative for raw buffalo hide could be checked before concluding it as a worthy biopreservative.

9.5 COMPONENTS RESPONSIBLE FOR PUTREFACTION OF RAW BUFFALO HIDE

Occurrence of putrefaction of the raw skins and hides, if not cured by physical or chemical method, is a traditional understanding. One of the components responsible for the putrefaction is a bacterium, was the vital information made available after the inception of bacteriology. Leather making is one such trade where the solution to problem was available far before the prob-lem was identified. Application of common salt was the solution to avoid the problem of growth of putrefying bacteria on the raw hides and skins.

As mentioned earlier, buffalo hide is considered to be a by-product of meat industry. The process of flaying involves handling of hides and contact of hide with flaying knifes. The flayed hides are washed to remove dirt and blood. Even in the process of flaying, plenty of water is being continuously poured on the half flayed carcasses of buffaloes. All these factors could be the potent sources of bacterial contamination. Apart from this, the state of animal to be slaughtered, its post- and pre-mortem history such as feeding habits and diseases harbored also decides the load of microorganisms on the flayed hides.

However the work done in this context and emphasizing the above-mentioned factors are not related to the raw buffalo hides. For instance, Venkatesan et al. (1973) studied microbial flora associated with goat skins and found facultative anaerobes on fresh goat skin. Botteldoorn et al. (2003) showed high degree of carcass cross-contamination with *Salmonella* sp. from the slaughterhouse environment in case of slaughtered pigs. How the environment of slaughterhouse influence microbial contamination could be understood by the studies done by Adeleye and Adebiyi (2003) on microbiological assessment of a Nigerian abattoir. They observed that almost throughout the year the runoff from slaughtering slabs consisted 10^8cfu/mL bacteria, which may lead to contamination of hides and animal carcasses. *Acinetobacter* sp. was reported from the fresh beef samples by Eribo and Jay (1985). One more factor stated above is pre-mortem history of the animal to be slaughtered may produce contamination of hides which was proved by Byrne et al. (2000) where they washed the cattle to be slaughtered with power hose for 3 min to remove all visible faecal matter on the live animal. They observed that this kind of pre-slaughter washing reduced incidences of *Escherichia coli* O157:H7 transfers from hide to the carcass during slaughter.

Apart from these bacterial components, the enzymes produced by these bacteria and the autolytic enzymes were considered to be important factors that lead to putrefaction of skins and hides. The protein rich raw buffalo hides are prone to attack by various proteolytic enzymes. Various bacteria are reported to produce proteases that could damage the raw buffalo hides. Few of them are known to posses the ability to produce specific group of enzymes that selectively degrade a particular element from the raw hide. With this perspective the enzymes collagenase, elastase, keratinase, and others could be of grave importance in putrefaction of raw buffalo hide. The collagenase produced by *Clostridium histolyticum* is one such enzyme that degrades the triple helical region of native collagen under physiological conditions. Yoshida and Noda (1965), Kono (1969), Markel et al. (1975), Dreisbach and Markel (1978) are some researchers who have worked on bacterial collagenase. However, Vanwart and Steinbrink (1981) and Barrett et al. (1989) have discussed continuous spectrophotometric and fluorimetric assay for bacterial collagenase. Koshy et al. (1999) has suggested an alternative assay method measuring collagenolytic activity using H–acetylated collagen. Huang and Abramson (1975) and Yoshioka et al. (1987) are the ones amongst many who have demonstrated presence of mammalian collagenase. Keratinase is one such enzyme which degrades keratin and the organisms producing keratin have been reported by Onifade et al. (1998), Wang and Shih (1999), Gradisor et al. (2000), Yamamura et al. (2002), Gessesse et al. (2003), Singh (2003) and Macedo et al. (2005). Few of these researchers have discussed the use of keratinase produced by these organisms for the process of dehairing however hardly any of them has considered their role in putrefaction of raw buffalo hide. Elastase is another enzyme, which could degrade the elastin from raw buffalo hide. He et al. (2004) have described the bacteria producing this enzyme. Along with these enzymes, there are many more which can act on one or more than one element of raw buffalo hide leading to unwanted changes. These are esterase (Deasy et al., 1968), pepsin (Steven, 1966), lipases, phosphatase, oxidase, dehydrogenase, and glycolytic enzymes. Polkade (2007) has extensively studied the microflora of raw buffalo hides by both culture dependent and culture independent approach. Polkade et al. (2013) developed a suitable nutrient medium containing skin extract to study microflora of buffalo hide. The studies identified 65 distinct morphotypes of bacteria. Many of these organisms were found to be proteolytic, amylolytic, and lipolytic becoming cause of hide putrefaction.

The changes occurring in the hide due to these components will lead to release of certain degradation products. Protein being the major constituent, main degradation products of raw buffalo hides or skins could be the amino acids. They have been detected from the preserved skins and hides to evaluate the effective preservation (Sivaparvathi et al., 1993). Another degradation product from the raw buffalo hide could be free fatty acids released as a result of fat hydrolysis. Shede et al. (2008) have studied changes occurring in number and types of bacteria in raw buffalo hide on standing. Their activity contributes to the putrefaction of the hide. Shede et al. (2009) showed that 'keeping qualities' of hides are always dependent on the total microbial flora associated with the freshly flayed and/or stored hides and the biochemical changes brought about by these microorganisms during short-term storage at ambient temperature.

9.6 PLANT-BASED MATERIALS AS IDEAL BIOAMBIANT PRESERVATIVES

It is estimated that there are 250,000–500,000 species of plants on earth (Borris, 1996), a very small share of which is used by humans and animals as their food. Customarily many of them are also used as a source of medicine. Finding healing powers in plants is an ancient idea. One such possession shown by plentiful plant species is antimicrobial activity.

Plants have an almost limitless ability to synthesize secondary metabolites. In many cases, these substances serve as plant defense mechanisms against predation by microorganisms, insects, and herbivores. Functional antimicrobial phytochemicals can be divided into several categories as phenolics (Proestos et al., 2005) and polyphenols, quinones, flavones, flavonoids and flavonols, tannins, coumarins, alkaloids, lactins, and polypeptides. Other than these, there are few complex mixtures containing chemicals that prove to be antimicrobial in nature. Latex of papaya is one such complex mixture containing enzyme, alkaloid, and terpenoids. Plant derived oils also possess antimicrobial activities (Varel and Miller, 2001). Different parts of the plant, viz. root, leaves, fruits, stem, bark, flowers, seeds, and fruits may contain such kind of activities.

Several researchers have tested numerous plant species for their antimicrobial activities. In most cases, they used organic solvents for extraction of different plant parts with certain exceptions wherein aqueous extracts have been used. *Aloe vera* (Pawar et al., 2005), *Thespesia populnea* (Shastry et al., 2005), *Eclipta alba* (Kothari and Shrivastava, 2005), *Euphorbia fusiformis* (Natarajan et al., 2005), *Zingiber officinale, Bosenbergia pandurata, Curcouma longa* (Thongson et al., 2004), *Puerariae radix* (Kim and Fung, 2004), *Cymbopogon citratus, Ocimum basilicum, Ocimum gratissimum, Thymus vulgaris* (Nguefack et al., 2004), *Larrea divaricata* (Quiroga et al., 2004), *Eucalyptus maculata* (Takahashi et al., 2004), *Aquilegia vulgaris* (Bylka et al., 2004), *Erythrina poeppigiana* (Sato et al., 2003) and *Withania somnifera* (Poonkothai et al., 2005) are few plants screened for their antibacterial and antifungal activities. The test organisms for their studies were from the genera *Staphylococcus, Streptococcus, Pseudomonas, Proteus, Listeria, Salmonella, Bacillus, Klebsiella, Escherichia, Micrococcus, Enterococcus, Bordetella, Aspergillus, Trichoderma, Curvularia, Pichia, Saccharomyces,* and *Candida*. Most of them were pathogenic in nature. The plant parts used in these studies were different for each screened plant. In many of the above cases it was observed that the single plant material shows antimicrobial activity against a wide array of organisms. This could strengthen the idea of using plant substances as antimicrobial agents to preserve the raw buffalo hide.

Typically the trend pursued to explore plants with antibacterial activities is based on the traditional knowledge involving ethno-botanical information. Indigenous medicinal plants and their antibacterial activities were of great interest to scientists from Iran (Bonjar et al., 2003, 2004a, 2004b, 2004c, 2004d), Ivory Coast (Okepekon et al., 2004), East Africa (Fabry et al., 1998), South Africa (Elgorashi et al., 2003), Asia (Almas, 2001), Taiwan (Wang and Huang, 2005) and Yemen (Ali, 2001). They found almost each plant tested was having the antibacterial and/or antifungal activity for screened microbial cultures. In the Indian continent, there are plenty of herbs, shrubs, and trees with strong antibacterial activities. However keeping in mind the proposed global use of selected plant materials in leather industries *Plumbago* sp., *Lawsonia* sp., and *Azadirachta* sp. should be the plants of preference.

9.6.1 AZADIRACHTA SP.

The genus *Azadirachta* consists of two species of Indomalayan origin. The species *Azadirachta indica* A. Juss. (syn. *Melia azadirachta* Linn.) is common in India. Vernacularly it is known as Neem tree (English – Margosa tree). It is a large, evergreen tree of about 12–18 m high. Leaves of these plants are imparipinnate, alternate, 20–38 cm long with 8–19 leaflets. The plant is distributed throughout India (Almeida and Almeida, 2005), in deciduous forests and also widely cultivated. It is a hardy tree, grows well in saline soils and drought conditions and is propagated from seeds. It can grow in all types of soil. Within one year, the seedling grows up to a height of 120 cm. Rapid multiplication through leaf culture has been found successful. The parts used of this plant are bark, leaf, flower, seed, and oil. About 100 compounds, mostly triterpenoids of protolimonoids, limonoids, tetranortriterpenoid-γ-hydroxy butenolides, pentanortriterpenoids, a hexanortriterpenoid apart from a few nontriterpenoid constituents have been reported from various parts (Sharma et al., 2000a). The plant shows different pharmacological activities as anticancer, antiviral, spasmogenic, antibacterial, antineoplastic, antifungal, antihelminthic, mosquito larvicidal, hypoglycemic, pesticidal, insecticidal, nematicide, vermicide, antitubercular, antimicrobial, diuretic, antiprotozoal, and antimalarial. The plant substances are nontoxic in nature. Neem is a hardy tree, grows well in saline soils and drought conditions, propagated from seeds, root suckers and stem cuttings. It can be grown in all types of soil. Rapid multiplication through leaf culture has been found successful.

Different researchers have reported the antibacterial and antifungal activities of neem oil. Das et al. (1999), Baswa et al. (2001), Coventry and Allan (2001), and Fandohan et al. (2004) are few of them. SaiRam et al. (2000) have demonstrated antiviral activity of neem against Poliovirus along with antibacterial and antifungal activities. Parida et al. (2002) showed inhibitory effect of neem leaves on dengue virus. Biswas et al. (2002) and Subapriya and Nagini (2005) have commented on medicinal properties and different activities of neem leaves and antiseptic nature is one of them. Antimicrobial effect of neem on certain foliar pathogens was studied by Bipte and Musaddiq (2005). The above-mentioned studies were carried out mainly on the pathogenic cultures and *Staphylococcus aureus*, *Escherichia coli*, *Pseudomonas aeruginosa*, *Xanthomonas* sp., and *Candida albicans* were few of them (Okemo et al., 2001).

The phytochemical investigations of *Azadirachta* sp. have revealed presence of various chemical constituents. Tetracyclic triterpenoids from the leaves of *Azadirachta indica* have been reported by Siddiqui et al. (2004); azadirachtin-A, nimbin and salannin by Babu et al. (2006). The leaf extract of neem has been reported to be nontoxic, non-mutagenic and possesses immuno-stimulant, hepato-protective, anti-oxidant, anti-genotoxic, and anti-cancer activities (Gangar et al., 2006). The leaves of neem mainly yield quercetin (flavonoid) and nimbosterol (β-sitosterol) as well as number of liminoids (nimbin and its derivatives). Fresh matured leaves yield an odorous viscous essential oil, which exhibits antifungal activity. The principal constituents of neem leaf include protein (7.1%), carbohydrates (22.9%), minerals, calcium, phosphorus, vitamin C, carotene, and others. However they also contain glutamic acid, tyrosine, aspartic acid, alanine, proline, glutamine, and cystine like amino acids, and several fatty acids (dodecanoic, tetradecanoic, elcosanic, and others). Shede (2007) has demonstrated use of leaf extract of *Azadirachta* for preservation of buffalo hide.

9.6.2 PLUMBAGO SP.

The genus *Plumbago* consists of ten species distributed in tropical and warm region. Out of these, three are reported in India as *Plumbago zeylanica* Linn., *P. indica* and *P. auriculata*. Vernacularly it is known as Chitrak (English – Ceylon leadwort, White flowered leadwort). These are perennial, sub-scandent shrubs with 60–120 cm height. *P. zeylanica* has white flowers and *P. indica* (syn. *P. rosea*) has red flowers. Cultivated as ornamental plant throughout India, wild in Peninsular India and West Bengal. The parts used of this plant are root and root bark. Plant contains number of naphthoquinone derivatives, viz. plumbagin, 3-chloroplumbagin, 3,3'-biplumbagin, elliptinone, chitranone, zeylinone, isozeylinone, droserone, plumbagic acid, plumbazeylanone, naphthalenone, and isoshinanolone (Sharma et al., 2000b). The plant shows different pharmacological activities as antipyretic, antibacterial, antifungal, anticancer, anticoagulant, antitumor, and hepato-protective. The plant substances are nontoxic in nature. The plant is propagated by seeds or by cuttings of side shoots. Well-drained sunny situation and mild climate are preferable. It can also be propagated through tissue culture technique.

Quite a lot of scientists have described chitrak as an antibacterial plant. Beg and Ahmad (2000) has tested *Plumbago zeylanica* against multidrug resistant bacteria *Staphylococcus aureus*, *Salmonella paratyphi*, *Escherichia coli* and *Shigella dysenteriae* of clinical origin. Chatterjee and Pakrashi (1995), Paiva et al. (2003) and Singh et al. (2004) also gave such information about the antimicrobial activity of chitrak. The plant *Plumbago* sp. has been studied extensively for its chemical composition and their specific activities. Presence of binaphthoquinone in *Plumbago zeylanica* was explained by Okamoto et al. (2001). Lupenone, lupeol acetate, plumbagin and trilinolein are few constituents isolated from *Plumbago zeylanica* by Nguyen et al. (2004). Lin et al. (2003) described the cytotoxic effect of constituents like naphthoquinones and plumbagic acid glucosides from chitrak. DNA cleavage by plumbagin was illustrated by Fujii et al. (1992). Further Knecht et al. (2000) has reported inhibition of enzyme dihydroorotate dehydrogenase (EC 1.3.99.11) by plumbagin. Production of plumbagin from cell cultures of *Plumbago rosea* L. was also reported by Komaraiah et al. (2001, 2002, 2003) and Panichayupakaranant and Tewtrakul (2002). Shede (2007) has extensively studied effect of root powder extracts of *Plumbago* sp. on microflora as well as their hydrolytic enzymes suggesting its potential in bioambient preservation of hides.

9.6.3 LAWSONIA SP.

The genus *Lawsonia* is monotypic, that means it possess only one species, *Lawsonia inermis* Linn. (syn. *Lawsonia alba* Lamk.). Vernacularly it is known as Mehndi (English – Henna). The plant is a glabrous, much branched shrub or small tree. Geographically distributed in North Africa, tropical and old world (Mabberley, 2000). Cultivated as ornamental plant and naturalized all over India. The parts used of this plant are root, leaf, flower, and seeds. The plant contains lawsone, esculetin, fraxetin, isoplumbagin, scopoletin, betulin, betulinic acid, hennadiol, lupeol and its related compounds, lacoumarin, laxanthone I, II and III, flavone glycosides and two pentacytic triterpenes (Sharma et al., 2000c). The plant shows different pharmacological activities as antibacterial, antifungal, anti-inflammatory, antituberular, and antitumor. The plant substances and the compounds present in it are nontoxic in nature (Marzin and Kirkland, 2004). Mehndi grows on any type of soil. It tolerates little alkalinity in the soil. Propagation is done using seeds and cuttings.

The plant is reported to inhibit certain bacteria (Musa et al., 2011a), viz. *Micrococcus luteus*, *Serratia marcescens*, *Klebsiella pneumonia* and *Bordetella bronchiseptica* (Bonjar, 2004d), *Escherichia coli* (Bonjar, 2004b), *Staphylococcus aureus* (Bonjar, 2004a), *Xanthomonas* sp. (Satish et al., 1999), and few water pathogens (Kumar and Gopal, 1999). Malekzadeh (1968) and Habbal et al. (2005) has also shown the antibacterial activity of *Lawsonia* sp. Muhammad and Muhammad (2005) have used *Lawsonia inermis* in the management of burn wound infections.

The literature stated various constituents from *Lawsonia* sp. and the activities of these constituents have been also studied in great details (Musa et al., 2011b, 2011c, 2012). Takeda (1988) reported phenolic glucosides and Bakkali et al. (1997) showed presence of lawsone in *Lawsonia inermis* cell cultures. Leung and Foster (1996) have mentioned presence of lawsone (2-hydroxy-1,4-naphthoquinone) 1,4-naphthoquinone, gallic acid and tannin with antimicrobial activities from mehndi plant. The genotoxic effect of this lawsone from mehndi was assessed by Kirkland and Marzin (2003). Also there are reports of enzyme inhibition by lawsone. Trypsin (Yogisha et al., 2002) and dihydroorotate dehydrogenase (Knecht et al., 2000) were found to be inhibited by lawsone. Extensive studies have been done by Shede (2007) on antimicrobial activity of *Lawsonia* sp. leaf extract. The microflora of raw buffalo hide and their hydrolytic enzymes were inhibited by *Lawsonia*. The bioambiant preservation of raw buffalo hide using *Lawsonia* sp. leaf extract was demonstrated.

9.7   CONCLUSION AND FUTURE PERSPECTIVES

Leather industry, holding a concrete place in Indian and Global economy is under severe threat because of the waste generated and the upcoming stringent laws and regulations of waste disposal posed by concerned governing bodies. To survive in such conditions, leather sector was provoked to move away from traditional processing techniques. Use of bioprocessing methods instead of chemical processes was one such move, which was exercised in the process of soaking, dehairing, bating, and degreasing and reduction in the amount of hazardous waste was achieved. Nevertheless salting which contributes to almost half the TDS from composite tannery waste continues to be a traditional processing routine. Attempts made to minimize these pollution problems by replacing salt with other chemicals are appreciable although found inadequate to cope up with two major fronts as cost of leather production and quality of leather produced.

Although the plant materials are studied in great details for their chemical constituents, recently new additions to list of constituents has been reported (Siddiqui and Kardar, 2001; Nguyen et al., 2004). This supports our assumption that few or many more new constituents with varied biological activities could be discovered from these plant materials.

Previous studies (Shede, 2007) specify that the organic extracts of chitrak root and mehndi leave possess antibacterial activities. The solubility patterns of identified antibacterial principle from these plants also support this data. Plumbagin and lawsone, the major antibacterial constituents from chitrak and mehndi, respectively are known to be soluble more in non-polar solvents than that of polar solvents. Considering this, the antibacterial activity shown by aqueous (polar solvent) extract in the recent studies is of immense importance. Investigation of solubility pattern of antibacterial principle(s) from aqueous extracts in organic solvents revealed that their solubility is more in ethyl acetate than chloroform. This conclusion is based on the percent yield of residues obtained from solvent fractions after extraction. Nevertheless, the extent of antibacterial activity shown by them is nearly comparable. Investigation of antibacterial activities, in plant substances after exhaustive Soxhlet extraction revealed that the antibacterial activity from chitrak root powder and mehndi leaf powder is reduced after ethanol extraction whereas, chloroform extraction could not affect the activity. This is evidence in support of the information that the active antibacterial principle(s) from chitrak root and mehndi leaf powders used in recent studies are more soluble in polar solvent than that of non-polar solvent. These pieces of information strengthen the possibility of antibacterial principle(s) being different than known naphthoquinone derivatives.

The conservative way to preserve the raw buffalo hides is use of common salts leading to high TDS ratio in the tannery waste. The approach to replace this common salt with other chemicals was not appreciated at tannery levels. The literature states the defects and changes occurring in the buffalo hide when preserved with common salt and the alternative chemicals. There are no reports on the changes in hide stored at ambient temperature without any preservatives. Hence it is required to study these kinds of changes, which could be of help in evaluating the efficiency of any alternative preservative. The work carried out on the alternative chemical, biological, and physical methods of preservation states that many of these methods yield inferior quality leather. Hence it becomes of immense importance that the alternative preservation method devised yields good quality leather.

Biopreservation, the process employing bio-based preservatives for preservation of raw buffalo hides has not been explored to its best. The insight in biopreservation gives an enormous prospect to plant substances with the biocidal activities. Such plant substance was reported to preserve the hide but with aid of chemicals. Hence more promising plant substances are needed to be screened. Plenty of plant substances are known to have the bactericidal as well as the bacteriostatic activities. The basis for plant selection should be their abundance, ease of propagation, inexpensive cultivation methods, and broad-spectrum antibacterial activities. Considering this, *Azadirachta* sp., *Plumbago* sp., and *Lawsonia* sp. possibly could be the finest source of long waited biopreservatives.

In the light of socio-economic and environmental aspects of Leather industry, biotechnological intervention appears to play an important role. Preservation of animal skins and hides before they reach tannery for leather making is an important step. Environmental friendly biotechnological method of preservation of skins and hides becomes a desirable choice. The literature available on preservative activity of different plant species opens up a field of invention. Considering this benefit and the socio-economic relevance, Government of India launched a research project in 'Biotechnology for Leather' under Council for Scientific and Industrial Research – New Millennium Indian Leadership Initiative program during 2002–2011. Discovering new plant species and developing plant material based products for preservation of skins and hides could be the future areas of research in leather industry.

KEYWORDS

- **Ambient storage**
- **Antimicrobial activity**
- ***Azadirachta* sp.**
- **Bioambient preservation**
- **Hide degradation**
- **Hide microflora**
- **Hides and skins**
- **Lawsone**
- ***Lawsonia* sp.**
- **Leather deterioration**

- **Leather industry**
- **Leather making**
- **Microbial degradation**
- **Microbial succession**
- **Napthoquinone**
- **Natural products**
- **Phenolics**
- **Pickling of hides**
- **Plant-based materials**
- **Plumbagin**
- ***Plumbago* sp.**
- **Salting of hides**
- **Slaughter house**
- **Tannery**
- **Tannery waste**

REFERENCES

Adeleye, I. A.; Adebiyi, A. A. Physicochemical and microbiological assessment of oko-oba - a Nigerian abattoir. *J Environ Biol* 2003, 24(3), 309–313.

Ali, A. N. A.; Julich, W. D.; Kusnick, C.; Lindequist, U. Screening of yemeni medicinal plants for antibacterial and cytotoxic activities. *J Ethnopharmacol* 2001, 74, 173–179.

Almas, K. The antimicrobial effects of seven different types of Asian chewing sticks. *Odontostomatol Trop* 2001, 96, 17–20.

Almeida, S. M.; Almeida, M. R. *Dictionary of generic names of flowering plants and ferns in Maharashtra*, Thomas Paul Almeida, Shreejee Enterprises: Mumbai, 2005.

Babu, V. S.; Narsimhan, S.; Nair, G. M. Bioproduction of azadirachtin-A, nimbin and salannin in callus and cell suspension cultures of neem (*Azadirachta indica* Juss.). *Curr Sci* 2006, 91(1), 22–24.

Bailey, D. G. Preservation of cattle hides with potassium chloride. *J Am Leather Chem Assoc* 1995, 90, 13–21.

Bakkali, A. T.; Jaziri, M.; Foriers, A.; Heyden, Y. V.; Vanhaelen, M.; Homes, J. Lawsone accumulation in normal and transformed culture of henna, *Lawsonia inermis*. *Plant Cell Tissue Organ Cult* 1997, 51, 83–87.

Barrett, A. J.; Knight, C. G.; Brown, M. A.; Tisljar, U. A continuous fluorimetric assay for clostridial collagenase and Pz-peptidase activity. *Biochem J* 1989, 260, 259–263.

Baswa, M.; Rath, C. C.; Dash, S. K.; Mishra, R. K. Antibacterial activity of karanj (*Pongamia pinnata*) and neem (*Azadirachta indica*) seed oil: A preliminary report. *Microbios* 2001, 105(412), 183–189.

Beg, A. Z.; Ahmad, I. Effect of *Plumbago zeylanica* extract and certain curing agents on multidrug resistant bacteria of clinical origin. *World J Microbiol Biotechnol* 2000, 16, 841–844.

Berwick, P. G.; Gerbi, S. A.; Russsell, A. E. Antibiotics to control green hide biodeterioration. *J Soc Leather Technol Chem* 1990, 74, 142–151.

Bipte, S.; Musaddiq, M. Studies on antimicrobial activity of *Azadirachta indica* L. on certain foliar pathogens. *J Microbial World* 2005, 7(1), 28–31.

Biswas, K.; Chattopadhyay, I.; Banerjee, R. K.; Bandyopadhyay, U. Biological activities and medicinal properties of neem (*Azadirachta indica*). *Curr Sci* 2002, 82(11), 1336–1345.

Bonjar, G. H. S. Evaluation of antibacterial properties of Iranian medicinal-plants against *Micrococcus luteus*, *Serratia marcescens*, *Klebsiella pneumonia* and *Bordetella bronchoseptica*. *Asian J Plant Sci* 2004d, 3(1), 82–86.

Bonjar, G. H. S. Inhibition of three isolates of *Staphylococcus aureus* mediated by plants used by Iranian native people. *J Med Sci* 2004a, 4(2), 136–141.

Bonjar, G. H. S. Screening for antibacterial properties of some Iranian plants against two strains of *Escherichia coli*. *Asian J Plant Sci* 2004b, 3(3), 310–314.

Bonjar, G. H. S.; Aghighi, S.; Nik, A. K. Antibacterial and antifungal survey in plants used in indigenous herbal-medicine of south east regions of Iran. *J Biol Sci* 2004c, 4(3), 405–412.

Bonjar, G. H. S.; Nik, A. K.; Heydari, M. R.; Ghasemzadeh, M. H.; Farrokhi, P. R.; Moein, M. R.; Mansouri, S.; Foroumadi, A. Anti-pseudomona and anti-bacilli activity of some medicinal plants of Iran. *DARU* 2003, 11(4), 157–163.

Boopathy, R.; Sekaran, G. Separation of sodium chloride from the evaporated residue of the reverse osmosis reject generated in the leather industry - optimization by response surface methodology. *Environ Technol* 2014, 35(13–16), 1858–1865.

Borris, R. P. Natural products research: Preservatives from a major pharmaceutical company. *J Ethnopharmacol* 1996, 51, 29–38.

Botteldoorn, N.; Heyndrickx, M.; Rijpens, N.; Grijspeerdt, K.; Herman, L. *Salmonella* on pig carcasses: Positive pigs and cross contamination in the slaughterhouse. *J Appl Microbiol* 2003, 95, 891–903.

Bylka, W.; Szaufer-Hajdrych, M.; Matlawska, I.; Goslinska, O. Antimicrobial activity of isocytisoside and extracts of *Aquilegia vulgaris* L. *Lett Appl Microbiol* 2004, 39, 93–97.

Byrne, C. M.; Bolton, D. J.; Sherldan, J. J.; McDowell, D. A.; Blair, I. S. The effects of pre-slaughter washing on the reduction of *Escherichia coli* O157:H7 transfer from cattle hides to carcasses during slaughter. *Lett Appl Microbiol* 2000, 30, 142–145.

Chatterjee, A.; Pakrashi, S. C. *Plumbaginaceae: The treatise on Indian medicinal plants, Vol. 4*, Publication and Information Directorate, CSIR: New Delhi, 1995, pp 53–56.

Cooper, D. R.; Galloway, A. C.; Woods, D. R. A new look at delayed curing based on the rate of salt penetration and bacterial activity. *J Soc Leather Trade Chem* 1972, 56, 127–138.

Coventry, E.; Allan, E. J. Microbiological and chemical analysis of neem (*Azadirachta indica*) extracts: New data on antimicrobial activity. *Phytoparasitica* 2001, 29 (5), 441–450.

Das, B. K.; Mukherjee, S. C.; Sahu, B. B.; Murjani, G. Neem (*Azadirachta indica*) extract as an antibacterial agent against fish pathogenic bacteria. *Indian J Exp Biol* 1999, 37(11), 1097–1100.

Deasy, C.; Holmstrom, P.; Mushaben, M.; Studer, S.; Alexander, M. A. Effect of esterases on corium collagen. *J Am Leather Chem Assoc* 1968, 63(7), 454–456.

Dreisbach, J. H.; Markel, J. R. Induction of collagenase production in *Vibrio* B-30. *J Bacteriol* 1978, 135(2), 521–527.

Elgorashi, E. E.; Taylor, J. L. S.; Maes, A.; Staden, J.; Kimpe, N.; Verschaeve, L. Screening of medicinal plants used in South African traditional medicine for genotoxic effects. *Toxicol Lett* 2003, 143, 195–207.

Eribo, B. E.; Jay, J. M. Incidence of *Acinetobacter* sp. and other gram-negative, oxidase-negative bacteria in fresh and spoiled ground beef. *Appl Environ Microbiol* 1985, 49(1), 256–257.

Fabry, W.; Okemo, P. O.; Ansorg, R. Antibacterial activity of East African medicinal plants. *J Ethnopharmacol* 1998, 60, 79–84.

Fandohan, P.; Gbenou, J. D.; Gnonlonfin, B.; Hell, K.; Marasas, W. F.; Wingfield, M. J. Effect of essential oils on the growth of *Fusarium verticillioides* and fumonisin contamination in corn. *J Agric Food Chem* 2004, 52(22), 6824–6829.

Fujii, N.; Yamashita, Y.; Arima, Y.; Nagashima, M.; Nakano, H. Induction of topoisomer-ase II-mediated DNA cleavage by the plant naphthoquinones plumbagin and shikonin. *Antimicrob Agents Chemother* 1992, 36(12), 2589–2594.

Ganesh Babu, T.; Nithyanand, P.; Chandra Babu, N. K.; Karutha Pandian, S. Evaluation of cetyltrimethylammonium bromide as a potential short-term preservative agent for stripped goat skin. *World J Microb Biotechnol* 2009, 25(5), 901–907.

Gangar, S. C.; Sandhir, R.; Rai, D. V.; Koul, A. Modulatory effects of *Azadirachta indica* on benzo(a)pyrene-induced for stomach tumorigenesis in mice. *World J Gastroenterol* 2006, 12(17), 2749–2755.

Gessesse, A.; Hatti-Kaul, R.; Gashe, B. A.; Mattiasson, B. Novel alkaline proteases from alkaliphilic bacteria grown on chicken feather. *Enzyme Microb Technol* 2003, 32, 519–524.

Gradisor, H.; Kern, S.; Friedrich, J. Keratinase of *Doratomyces ricrosporus*. *Appl Microbiol Biotechnol* 2000, 53, 196–200.

Habbal, O. A.; Al-Jabri, A. A.; El-Hag, A. H.; Al-Mahrooqi, Z. H.; Al-Hashmi, N. A. *In vitro* antimicrobial activity of *Lawsonia inermis* Linn (henna) - a pilot study on the omani henna. *Saudi Med J* 2005, 26(1), 69–72.

Haffner, M. A.; Haines, B. M. Short-term preservation. *J Soc Leather Technol Chem* 1975, 59, 114–122.

Haines, B. M. Conservation of cattle hides by freezing. *J Soc Leather Technol Chem* 1981, 65, 41–54.

Hanlin, M. B.; Field, R. A.; Ray, B. Characterization of predominant bacteria in cattle hides and their control by a bacteriocin based preservative. *J Am Leather Chem Assoc* 1995, 90, 308–320.

He, G.; Chen, Q.; Ju, X.; Shi, N. Improved elastase production by *Bacillus* sp. EL31410 - further optimization and kinetics studies of culture medium for batch fermentation. *J Zhejiang Univ Sci* 2004, 2, 149–156.

Hopkins, W. J.; Bailey, D. G. Preservation of hides with sulfite. I. concentration and application effects on small-scale experiments with cattle hides. *J Am Leather Chem Assoc* 1975, 70, 248–260.

Hopkins, W. J.; Bailey, D. G.; Weaver, E. A.; Korn, A. H. Potential short-term techniques for the preservation of cattle hides. *J Am Leather Chem Assoc* 1973, 78, 426–436.

Huang, C. C.; Abramson, M. Purification and characterization of collagenase from guinea pig skin. *Biochim Biophys Acta* 1975, 384, 484–492.

Kallenberger, W. E. Halophilic bacteria in brine curing. *J Am Leather Chem Assoc* 1984, 79, 104–114.

Kanagaraj, J.; Chandra Babu, N. K. Alternatives to salt curing techniques - a review. *J Sci Ind Res* 2002, 61, 339–348.

Kanagaraj, J.; Chandra Babu, N. K.; Sadulla, S.; Rajkumar, G. S.; Visalakshi, V.; Kumar, N. C. Cleaner techniques for the preservation of raw goat skins. *J Clean Prodn* 2001, 9, 261–268.

Kanagaraj, J.; John Sundar, V.; Murlidharan, C.; Sadulla, S. Alternatives to sodium chloride in prevention of skin protein degradation - a case study. *J Clean Prodn* 2005a, 13, 825–831.

Kanagaraj, J.; Sastry, T. P.; Rose, C. Effective preservation of raw goat skins for the reduction of total dissolved solids. *J Clean Prodn* 2005b, 13, 959–964.

Kim, S.; Fung, D. Y. C. Antibacterial effect of crude water-soluble arrowroot (*Puerariae radix*) tea extracts on food-borne pathogens in liquid medium. *Lett Appl Microbiol* 2004, 39, 319–325.

Kirkland, D.; Marzin, D. An assessment of genotoxicity of 2-hydroxy-1,4-naphthoquinone, the natural dye ingredient of henna. *Mutat Res* 2003, 537, 183–199.

Knecht, W.; Henseling, J.; Loffler, M. Kinetics of inhibition of human and rat dihydroorotate dehydrogenase by atovaquone, lawsone derivatives, brequinar sodium and polyporic acid. *Chem Biol Interact* 2000, 124, 61–76.

Komaraiah, P.; Naga, A. R.; Kavi, K. P.; Ramakrishna, S. Elicitor enhanced production of plumbagin in suspension cultures of *Plumbago rosea* L. *Enzyme. Microb Technol* 2002, 31, 634–639.

Komaraiah, P.; Kavi Kishore, P. B.; Ramakrishna, S. V. Production of plumbagin from cell cultures of *Plumbago rosea* L. *Biotechnol Lett* 2001, 23, 1269–1272.

Komaraiah, P.; Ramakrishna, S. V.; Reddanna, P.; Kavi Kishor, P. V. Enhanced production of plumbagin in immobilized cells of *Plumbago rosea* by elicitation and *in situ* adsorption. *J Biotechnol* 2003, 101, 181–187.

Kono, T. Roles of collagenase and other proteolytic enzymes in the dispersal of animal tissues. *Biochim Biophys Acta* 1969, 178, 397–400.

Koshy, P. J. T.; Rowan, A. D.; Life, P. F.; Cawston, T. E. 96-well plate assays for measuring collagenase activity using ³H-acetylated collagen. *Anal Biochem* 1999, 275, 202–207.

Kothari, A.; Shrivastava, N. Antimicrobial activity of *Eclipta alba*. *Ind Drugs* 2005, 42(3), 133–135.

Kritzinger, C. C.; Vanzyl, J. H. M. The preservation of hides with used salt. *J Am Leather Chem Assoc* 1954, 49, 207–213.

Kumar, S.; Gopal, K. Screening of plant species for inhibition of bacterial population of raw water. *J Environ Sci Health A Tox Hazard Subst Environ Eng* 1999, A34(4), 975–987.

Leung, A. Y.; Foster, S. Henna. *Encyclopedia of common natural ingredients used in food, drugs and cosmetics*, 2nd Ed. Wiley-Interscience: New York, 1996, 297–299.

Lin, L. C.; Yang, L. L.; Chou, C. J. Cytotoxic naphthoquinones and plumbagic acid glucosides from *Plumbago zeylanica*. *Phytochem* 2003, 62, 619–622.

Mabberley, D. J. *The plant book: A portable dictionary of the vascular plants*, 2nd Ed. Cambridge University Press: Cambridge, 2000.

Macedo, A. J.; Silva, W. O. B.; Gava, R.; Drimeier, D. Novel keratinase from *Bacillus subtilis* S14 exhibiting remarkable dehairing capabilities. *Appl Environ Microbiol* 2005, 71(1), 594–596.

Malekzadeh, F. Antimicrobial activity of *Lawsonia inermis* L. *Appl Microbiol* 1968, 16(4), 663–664.

Markel, J. R.; Dreisbach, J. H.; Ziegler, H. B. Collagenolytic activity of some marine bacteria. *Appl Microbiol* 1975, 29(2), 145–151.

Marzin, D. K.; Kirkland, D. 2-hydroxy-1,4-naphthoquinone, the natural dye of henna, is non-genotoxic in the mouse bone marrow micronucleus test and does not produce oxidative DNA damage in Chinese hamster ovary cells. *Mutat Res* 2004, 560, 41–47.

Muhammad, H. S.; Muhammad, S. The use of *Lawsonia inermis* Linn. (Henna) in the management of burn wound infections. *Afr J Biotechnol* 2005, 4(9), 934–937.

Musa, A. E., Tamil, S.; Aravind, R.; Madhan, B.; Raghava Rao, J.; Chandrasekaran, B. Evaluation of antimicrobial activity of *Lawsonia inermis* (henna) against microbial strain isolated from goat skin/leather. *J Am Leather Chem Assoc* 2011a, 106(5), 170–175.

Musa, A. E.; Aravind, R.; Madhan, B.; Raghava, J.; Chandrasekaran, B. Henna-aluminum combination tannage: A greener alternative tanning system. *J Am Lether Chem Assoc* 2011b, 106(6), 190–199.

Musa, A. E.; Gasmelseed, G. A. Characterization of *Lawsonia inermis* (henna) as vegetable tanning material. *J Forest Prod Ind* 2012, 1(2), 35–40.

Musa, A. E.; Madhan, B.; Aravind, R.; Kanth, S. V.; Raghava Rao, J., Chandrasekaran, B.; Gasmelseed, G. A. Studies on henna - glutaraldehyde combination tanning system. *J Am Leather Chem Assoc* 2011c, 106, 92–101.

Natarajan, D.; Britto, S. J.; Srinivasan, K.; Nagamurugan, N.; Mohanasundari, C.; Perumal, G. Anti-bacterial activity of *Euphorbia fusiformis* - a rare medicinal herb. *J Ethnopharmacol* 2005, 102, 123–126.

Nguefack, J.; Budde, B. B.; Jakobsen, M. Five essential oils from aromatic plants of Cameroon: Their antibacterial activity and ability to permeabilize the cytoplasmic membrane of *Listeria innocua* examined by flow cytometry. *Lett Appl Microbiol* 2004, 39, 395–400.

Nguyen, A. T.; Malonne, H.; Duez, P.; Vanhaelen-Fastre, R.; Vanhaelen, M.; Fontaine, J. Cytotoxic constituents from *Plumbago zeylanica*. *Fitoterapia* 2004, 75, 500–504.

Okamoto, I.; Doi, H.; Kotani, E.; Takeya, T. The aryl-aryl coupling reaction of 1-naphthol with $SnCl_4$ for 2,2'-binaphthol synthesis and its application to the biomimetic synthesis of binaphthoquinone isolated from *Plumbago zeylanica*. *Tetrahedron Lett* 2001, 42, 2987–2989.

Okemo, P. O.; Mwatha, W. E.; Chhabra, S. C.; Fabry, W. The kill kinetics of *Azadirachta indica* a. Juss. (Meliaceae) extracts on *Staphylococcus aureus*, *Escherichia coli*, *Pseudomonas aeruginosa* and *Candida albicans*. *Afr J Sci Technol* 2001, 2(2), 113–118.

Okepekon, T.; Yolou, S.; Gleye, C.; Roblot, F.; Loiseau, P.; Bories, C.; Grellier, P.; Frappier, F.; Laurens, A.; Hocquemiller, R. Antiparasitic activities of medicinal plants used in Ivory coast. *J Ethnopharmacol* 2004, 90, 91–97.

Olasupo, N. A.; Fitzgerald, D. J.; Gasson, M. J.; Narbad, A. Activity of natural antimicrobial compounds against *Escherichia coli* and *Salmonella enterica* serovar *typhimurium*. *Lett Appl Microbiol* 2003, 36, 448–451.

Onifade, A. A.; Al-Sane, N. A.; Al-Mussallam, A. A.; Al-Zarbam, S. Potential for biotechnological application of keratin-degrading microorganisms and their enzymes for nutritional improvement of feather and other keratins as livestock feed resources. *Bioresour Technol* 1998, 66, 1–11.

Paiva, S. R.; Figueiredo, M. R.; Aragao, T. V.; Kaplan, M. A. C. Antimicrobial activity *in vitro* of plumbagin isolated from *Plumbago* species. *Mem Inst Oswaldo Cruz* 2003, 98(7), 959–961.

Panichayupakaranant, P.; Tewtrakul, S. Plumbagin production by root cultures of *Plumbago rosea*. *Electronic J Biotechnol* 2002, 5(3), 228–231.

Parida, M. M.; Upadhyay, C.; Pandya, G.; Jana, A. M. Inhibitory potential of neem (*Azadirachta indica* Juss) leaves on dengue virus type-2 replication. *J Ethnopharmacol* 2002, 79(2), 273–278.

Pawar, V. C.; Bagatharia, S. B.; Thaker, V. S. Antibacterial activity of *Aloe vera* leaf gel extracts against *Staphylococcus aureus*. *Indian J Microbiol* 2005, 45(3), 227–229.

Polkade, A. Molecular and phylogenetic studies on microbial diversity of raw buffalo hide to be used in leather manufacturing. Ph.D. Thesis, University of Pune, India, 2007.

Polkade, A.; Shede, P.; Kanekar, P.; Dhakephalkar, P.; Sarnaik, S. Designing and evaluation of skin extract agar for isolation of microflora from raw buffalo hide. *IIOAB Letters* 2013, 3(1), 8–13.

Poonkothai, M.; Hemaiswarya, S.; Kavitha, D. Antibacterial activity of *Withania somnifera*. *J Microb World* 2005, 7(1), 97–99.

Proestos, C.; Chorianopoulos, N.; Nychas, G. J. E.; Komaitis, M. RP-HPLC analysis of the phenolic compounds of plant extracts - investigation of their antioxidant capacity and antimicrobial activity. *J Agric Food Chem* 2005, 53, 1190–1195.

Quiroga, E. N.; Sampietro, A. R.; Vattuone, M. A. *In vitro* fungitoxic activity of *Larrea divaricata* Cav. extracts. *Lett Appl Microbiol* 2004, 39(7), 12.

Rao, B. R.; Henrickson, R. L. Short-term preservation of cattle hide. *J Am Leather Chem Assoc* 1983, 78, 48–53.

SaiRam, M.; Ilavazhagan, G.; Sharma, S. K.; Dhanraj, S. A.; Suresh, B.; Parida, M. M.; Jana, A. M.; Devendra, K.; Selvamurthy, W. Anti-microbial activity of a new vaginal contraceptive NIM-76 from neem oil (*Azadirachta indica*). *J Ethnopharmacol* 2000, 71(3), 377–382.

Satish, S.; Raveesha, K. A.; Janardhana, G. R. Antibacterial activity of plant extracts on phytopathogenic *Xanthomonas campestris* pathovars. *Lett Appl Microbiol* 1999, 28, 145–147.

Sato, M.; Tanaka, H.; Yamaguchi, R.; Oh-Uchi, T.; Etoh, H. *Erythrina poeppigiana* - derived phytochemical exhibiting antimicrobial activity against *Candida albicans* and methicillin-resistant *Staphylococcus aureus*. *Lett Appl Microbiol* 2003, 37, 81–85.

Selvarangan, R.; Shanmugasundaram, K. A. Saltless curing and preservation of hides and skins. *Leather Sci* 1984, 31(9), 241–244.

Sharma, P. C.; Yelne, M. B.; Dennis, T. J. *Azadirachta indica (nimba), database on medicinal plants used in ayurveda, Vol. 1*, Central Council for Research in Ayurveda and Siddha, Government of India: New Delhi, 2000a, 289–310.

Sharma, P. C.; Yelne, M. B.; Dennis, T. J. *Lawsonia inermis (madayantika), database on medicinal plants used in ayurveda, Vol. 1*, Central Council for Research in Ayurveda and Siddha, Government of India, New Delhi: 2000c, 253–263.

Sharma, P. C.; Yelne, M. B.; Dennis, T. J. *Plumbago zeylanica (chitraka), database on medicinal plants used in ayurveda, Vol. 1*, Central Council for Research in Ayurveda and Siddha, Government of India, New Delhi: 2000b, 102–113.

Sharphouse, J. H. *Leather Technician's Handbook*. 2nd Ed. Leather Producers Association: London, 1983.

Shastry, C. S.; Aravind, M. B.; Joshi, S. D.; Ashok, K.; Joshi, B. Antibacterial and antifungal activity of *Thespesia populnea* L. *Indian Drugs* 2005, 42(2), 81–83.

Shede, P. Short term preservation of buffalo hide using plant based materials as antimicrobial substances - a new approach to leather processing. Ph.D. Thesis, University of Pune, India, 2007.

Shede, P. N.; Kanekar, P. P.; Polkade, A. V.; Sarnaik, S. S.; Dhakephalkar, P. K. Bacterial succession on raw buffalo hide and their degradative activities during ambient storage. *Int Biodet Biodeg* 2008, 62, 65–74.

Shede, P. N.; Kanekar, P. P; Polkade, A. V.; Sarnaik, S. S; Dhakephalkar, P. K.; Chiplonkar, S. A.; Nilegaonkar, S. S. Effect of microbial activities on stored raw buffalo hide. *J Environ Biol* 2009, 30(6), 983–988.

Siddiqui, B. S.; Afshan, F.; Gulzar, T.; Hanif, M. Tetracyclic triterpenoids from the leaves of *Azadirachta indica*. *Phytochem* 2004, 65(16), 2363–2367.

Siddiqui, B. S.; Kardar, M. N. Triterpenoids from *Lawsonia alba*. *Phytochem* 2001, 58, 1195–1198.

Singh, C. J. Optimization of an extracellular protease of *Chrysosporium keratinophilum* and its potential in bioremediation of keratinic wastes. *Mycopathologia* 2003, 156, 151–156.

Singh, J.; Mishra, N. P.; Jain, S. P.; Singh, S. C.; Sharma, A.; Khanuja, S. P. S. Traditional uses of *Plumbago zeylanica* (Chitraka). *J Med Aromatic Plant Sci* 2004, 26, 795–800.

Sivakumar, V.; John Sundar, V.; Rangasamy, T.; Murlidharan, C.; Swaminathan, G. Management of total dissolved solids in tanning process through improved techniques. *J Clean Prodn* 2005, 13, 699–703.

Sivaparvathi, M.; Mahadeswaraswamy.; Ramesh, R.; Nandy, S. C. Short term preservation of goat skins. *J Ind Leather Technol Assoc* 1993, 14(3), 77–84.

Sivaparvathi, M.; Nandy, S. C. Evaluation of preservatives for skin preservation. *J Am Leather Chem Assoc* 1974, 69, 349–362.

Song, J.; Tao, W.; Chen, W. Kinetics of enzymatic unhairing by protease in leather industry. *J Clean Prod* 2011, 19(4), 325–331.

Steven, F. S. The depolymerising action of pepsin on collagen, molecular weight of the component polypeptide chain. *Biochim Biophys Acta* 1966, 130, 190–195.

Stuart, L. S.; Frey, R. W. On the spoilage of stored salted calf skins. *J Am Leather Chem Assoc* 1938, 33, 198–203.

Subapriya, R.; Nagini, S. Medicinal properties of neem leaves: A review. *Curr Med Chem Anticancer Agents* 2005, 5(2), 149–156.

Suresh, V.; Kanthimathi, M.; Thanikaivalen, P.; Rao, J. R.; Nair, B. U. An improved product-process for cleaner chrome tanning in leather processing. *J Clean Prod* 2001, 9, 483–491.

Takahashi, T.; Kokubo, R.; Sakaino, M. Antimicrobial activities of eucalyptus leaf extracts and flavonoids from *Eucalyptus maculata*. *Lett Appl Microbiol* 2004, 39, 60–64.

Takeda, Y. New phenolic glucosides from *Lawsonia inermis*. *J Nat Prod* 1988, 51(4), 725–729.

Thanikaivelan, P.; Rao, J. R.; Nair, B. U.; Ramasami, T. Approach towards zero discharge tanning: Role of concentration on the development of eco-friendly liming-reliming processes. *J Clean Prodn* 2003, 11, 79–90.

Thanikaivelan, P.; Rao, J. R.; Nair, B. U.; Ramasami, T. Progress and recent trends in biotechnological methods for leather processing. *Trends Biotechnol* 2004, 22(4), 181–188.

Thongson, C.; Davidson, P. M.; Mahakaranchanakul, W.; Weiss, J. Antimicrobial activity of ultrasound-assisted solvent-extracted spices. *Lett Appl Microbiol* 2004, 39, 401–406.

Vanwart, H. E.; Steinbrink, D. R. A continuous spectrophotometric assay for *Clostridium histolyticum* collagenase. *Anal Biochem* 1981, 113, 356–365.

Varel, V. H.; Miller, D. N. Plant-derived oils reduce pathogens and gaseous emissions from stored cattle waste. *Appl Environ Microbiol* 2001, 67(3), 1366–1370.

Venkatachalam, P. S.; Sadulla, S.; Duraiswamy, B. Further experiments in saltless curing. *Leather Sci* 1982, 29(6), 217–221.

Venkatachalam, P. S.; Sadulla, S.; Duraiswamy, B. Preliminary curing trials without the use of common salt. *Leather Sci* 1981, 28, 151–154.

Venkatesan, R. A.; Nandy, S. C.; Sen, S. N. Anaerobic microbial flora and their population associated with goatskin during pretanning operations. *J Am Leather Chem Assoc* 1973, 78, 437–446.

Wang, J. J.; Shih, J. C. H. Fermentation production of keratinase from *Bacillus licheniformis* PWD-1 and *B. subtilis* FDB-29. *J Ind Microbiol Biotechnol* 1999, 22, 608–616.

Wang, Y. C.; Huang, T. L. Screening of anti-*Helicobacter pylori* herbs deriving from Taiwanese folk medicinal plants. *FEMS Immunol Med Microbiol* 2005, 43, 295–300.

Woods, D. R.; Welton, R. L.; Cooper, D. R. The microbiology of curing and tanning processes, Part V, anaerobic studies on halophobic and halophilic bacteria from cured hides. *J Am Leather Chem Assoc* 1971, 66, 496–503.

Yamamura, S.; Morita, Y.; Hasan, Q.; Yokoyama, K.; Tamiya, E. Keratin degradation: A cooperative action of two enzymes from *Stenotrophomonas* sp. *Biochem Biophys Res Commun* 2002, 294, 1138–1143.

Yogisha, S.; Samiulla, D. S.; Prashanth, D.; Padmaja, R.; Amit, A. Trypsin inhibitory activity of *Lawsonia inermis*. *Fitoterapia* 2002, 73, 690–691.

Yoshida, E.; Noda, H. Isolation and characterization of collagenases I and II from *Clostridium histolyticum*. *Biochim Biophys Acta* 1965, 105, 562–574.

Yoshioka, H.; Oyamada, I.; Usuku, G. An assay of collagenase activity using enzyme-linked immunosorbent assay for mammalian collagenase. *Anal Biochem* 1987, 166, 172–177.

CHAPTER 10

PHYTOREMEDIATION OF ORGANIC CHEMOPOLLUTANTS

PRADNYA PRALHAD KANEKAR[1], SEEMA SHREEPAD SARNAIK[2], PARAG AVINASH VAISHAMPAYAN[3], and PRAFULLA NAMDEO SHEDE[4]

[1]*Department of Biotechnology, Modern College of Arts, Science and Commerce, Shivajinagar, Pune 411005, India*

[2]*Microbial Sciences Division, MACS Agharkar Research Institute, G.G. Agarkar Road, Pune 411004, India*

[3]*Biotechnology and Planetary Protection Group, Jet Propulsion Laboratory, California Institute of Technology, Pasadena, CA 91109, USA*

[4]*Department of Microbiology, MES' Abasaheb Garware College, Karve Road, Pune 411004, India*

CONTENTS

10.1   INTRODUCTION

Over the past 50 years, rapid growth of population, mining and industrialization significantly contributed to extensive soil, air, and water contamination.
Many industrial wastes contain toxic chemicals, both organic compounds,
such as petrochemicals, polyaromatic hydrocarbons (PAHs), dyes, pesticides,
explosives and others, and inorganic compounds like heavy metals, nitrates,
sulfates and chlorides, and others which are harmful to various forms of life.
Therefore it becomes necessary to remediate these contaminated sites. Some
of the remediation methods include chemical methods, such as use of different chemicals for removal of these contaminants, physico-chemical methods
include techniques like reverse-osmosis, membrane filtration, and others.
However, these treatment methods are costly and difficult to adapt in case of
soil contaminated sites. Therefore biological methods of remediation generally called as bioremediation technologies become suitable in such cases.
Some of organic pollutants are toxic and difficult to degrade compounds.
Because of their insolubility, they accumulate in sediments and thus become
a serious problem.

Bioremediation is a biotechnology that uses microorganisms to degrade and detoxify pollutants causing environmental contamination. This
technology is applied for treatment of wastes. The bioremediation can be
carried out at the site of contamination (*in situ*) or the wastes are transported at some suitable site for treatment (*ex situ*). Exploration of plants
for remediation is an emerging cost-effective approach. The strategies
involving plants are commonly called phytotechnologies which include
phytoremediation. Phytotechnologies are defined as the use of plants to
remediate, treat, stabilize, or control contaminated substrates, and phytoremediation is one of these, specifically dedicated to the removal or
destruction of the contaminant. Phytotechnologies and phytoremediation
exploit the natural plant physiological processes. Environmental pollution
by organic nitro- and chlorinated compounds, hydrocarbons, pesticides,
and others, is caused mainly by human activities (Megharaj et al., 2011).
Microorganisms are employed to develop bioremediation technology for
degradation of organic pollutants. This technology suffers from some
limitations in applications to contaminated sites. Other technologies like
composting, phytoremediation, and related others are suggested. Finding
sources of pollutants is important to know their status in sediments. Perelo
(2010) has suggested adapting new efficient strategies like phytoremediation, bioaugmentation, and others.

10.2 PHYTOREMEDIATION

Phytoremediation describes the treatment of environmental problems (through the use of plants that mitigate the environmental problem) without the need to excavate the contaminant material and dispose of it elsewhere. For treatment of sites contaminated with pesticides, heavy metals, explosives, hydrocarbons, and other related chemopolluants, plant systems that can degrade or stabilize them are used to develop phytoremediation technology. Though phytoremediation is a promising alternative to physical and chemical remediation method it is still a nascent technology. More fundamental research is also urgently required to better understand, control, and fully exploit the metabolic diversity of plants. Certain plant species are endowed with the property to accumulate or degrade the pollutants. Their physiological processes like absorption and translocation can be employed. An integrated process involving microbial activity, plant physiology, environmental engineering, and few others may be fruitful. Phytoremediation is a cost effective, environmental friendly green technology (Schwitzguebel et al., 2002; Schwitzguebel, 2004).

Adler (1996) suggest cleanup of polluted water and soil using plants. McCutcheon and Jørgensen (2008) described phytoremediation as the use of green plants to treat and control wastes in water, soil, and air which is an important part of the new field of ecological engineering. Various factors like characteristics of pollutants and contaminated soil or water, nutrient availability, hydrology, toxicity of the pollutants to the plants, bioavailability, and others, govern the application of phytoremediation process. Long time and area treatment is restricted to rhizosphere and shallow water bodies. At low concentrations of pollutants, plant ecosystems like wetlands, grasslands can be effective.

10.2.1 ADVANTAGES AND LIMITATIONS OF PHYTOREMEDIATION

10.2.1.1 ADVANTAGES

Phytoremediation technology seems to be more advantageous technology as compared to other ones. The phytoremediation technology uses the natural resources like plants, therefore it becomes a cost effective technology as it dose not require high amount of capital for raising the technology. The plants can be easily developed and monitored for their tolerance to paricular contaminant. The training can be given to layman and no highly qualified personnel are required. When this phytoremediation technology is used for remediation of sites contaminated with heavy metals, the plants can be used for recovery of economically important or highly precious metals. These valuable recovered metals can be used anywhere. This technique is also known as 'phytomining'. Since this technology uses the natural resources like plants and naturally occurring organisms in their vicinity, this technology preserves the environment in a more natural state. Many a times this technology can be applied directly at the site of contamination either soil or water and no need of *ex situ* application or excavation of soil. One more advantage of phytoremediation technology is environmental friendliness. Traditional methods disrupt soil structure and reduce soil productivity, whereas phytoremediation can clean up the soil without causing any kind of harm to soil quality.Thus it is a clean, efficient, inexpensive, and environmentally non-disruptive technology as against other chemical or other technologies.

10.2.1.2   LIMITATIONS

Although it is a clean and green technology, this technology also has some limitations. Phytoremediation is limited to the surface area and depth occupied by the roots. Since growth of plants is slow and comparatively low biomass is obtained from them, this technology requires long time for treatment and thus requires a long-term commitment. Leaching of pollutants leading to groundwater pollution can not be fully controlled since the plant systems are used in tretament. Thus the efficiency of phytoremediation of groundwater is affected. Tolerance of the plants is different in case of different contaminants, therefore the survival of the plants is affected by the toxicity of the contaminated land and the general condition of the soil or water site. Since there is bio-accumulation of contaminants, especially metals, into the plants which then pass into the food chain, such plants cannot be used in food chain and require the safe disposal of such affected plant material. This remediation process is dependent on a plant's ability to grow and thrive in an environment that is not ideal for normal plant growth. Phytoremediation may be applied wherever the soil or static water environment has become polluted or is suffering from ongoing chronic pollution.

10.2.2   PROCESSES INVOLVED IN PHYTOREMEDIATION

During phytoremediation, different plant physiological processes are exploited to remove or destroy the contaminants from the contaminated sites (Cunningham et al., 1995, 1996; Singh and Jain, 2003). A range of processes mediated by plants or algae is useful in treating environmental problems. Phytoremediation can occur through the following routes.

10.2.2.1   PHYTOEXTRACTION

Certain plants have ability to take up pollutants from soil, water bodies, and others. The contaminants are absorbed through root system and transported to leaves or stem. This plant physiological process is termed as phytoextraction. Some plants can accumulate pollutants in large quantity till the time they are harvested. Compared to the volume of contaminated site, plant matter that extracts the pollutants is small and therefore easy for disposal. Thus the phytoextraction process is being globally employed. After the process, the cleaned soil can support other vegetation. 'Mining with plants' or phytomining is also being experimented. Willow plants were used by Greger and Landberg (1999) for phytoextraction purpose.

10.2.2.2   PHYTOSTABILIZATION

In this process, the mobility of contaminants in the environment is reduced from the contaminated site. Chemical modification due to plant metabolic activity of the pollutant may also result in immobilization, like phytostabilization of the pollutant. Phytostabilization focuses on long-term stabilization and containment of the pollutant. The pollutants are adsorbed or accumulated in root zone of the plant system, where they are precipitated and stabilized. The pollutants are restricted to roots only and therefore are not available to animals or human beings. In mining regions, the mine tailings are covered by vegetation for stabilization of the pollutants (Mendez and Maier, 2008).

10.2.2.3   PHYTOTRANSFORMATION

This process involves chemical modification of the pollutant as a direct result of plant metabolism, often resulting in their inactivation, or degradation, like phytodegradation. In phytotransformation process, toxic compounds like explosives, pesticides are transformed to nontoxic compounds by some plants, like Cannas. However, there is no complete degradation of the pollutants in phytotransformation. Microorganisms present in rhizosphere may degrade the organic compounds. Thus, plants primarily reduce toxicity of the pollutants. Subramanian et al. (2006) have studied phytotransformation of TNT.

10.2.2.4   PHYTOSTIMULATION

In this process, the microorganisms associated with plant roots enhance the soil microbial activity for the degradation of contaminants. This process is essentially a rhizodegradation of pollutants. Some aquatic plants support growth of microorganisms and stimulate them to degrade the chemical compounds. Rupassara et al. (2002) have studied degradation of atrazine via phytostimulation by the plant hornwort.

10.2.2.5   PHYTOVOLATILIZATION

In this plant physiological process, the pollutants are first transformed to volatile compounds which are released into the air.

10.2.2.6   RHIZOFILTRATION

This process involves filtering water through a mass of roots to remove toxic substances or excess nutrients. The pollutants remain absorbed in or adsorbed by the roots. Plants capable of absorbing pollutants are cultivated and used to minimize concentration of pollutants. However, it is likely that the pollutants are incorporated in plant biomass and returned to environment with death or harvesting of plants.

10.2.3   PLANTS USED IN PHYTOREMEDIATION

A variety of plants are used for the purpose of phytoremediation. These include both aquatic plants and terrestrial plants. Most commonly used aquatic plants include species of *Hydrilla, Pistia, Wolffia, Echornia*, such as water hyacinth, *Ceratophyllum* (hornwort), *Salvinea* (watermoss), *Vallisneria* (eelgrass, tape grass), *Najas*, and *Potamogeton* (pondweed). The terrestrial plants used in phytoremediation technique include species of *Ambrosia, Apocynum, Barley, Brassica, Cannabis, Carex, Festuca, Helianthus, Lupinus, Melilotus, Moringa, Phalaris, Phragmites, Poplar, Polypogon, Salix, Thlaspi, Tradescantia, Tree bog, Vicia* and *Vetiver*, and others.

10.3 CONSTRUCTED WETLANDS

Wetlands are transitional areas between land and water. The boundaries between wetlands and uplands or deep water are therefore not always distinct. The wetlands include a wide variety of ecosystems like marshes, wet meadows, floodplains, swamps, and others. The pollutants like metals are filtered in natural wetlands. This mechanism of biofiltration is simulated in constructed wetland (CW). Wetlands are very productive ecosystems with efficient photosynthesis and transpiration. Phytoremediation in wetlands has been described by Williams (2002).

Different processes like plant–microbe interactions, response of microbes to soil environment and interaction of all these elements in the natural ecosystem like wetlands are yet to be fully understood. Wetlands also serve the purpose of offering shelter to wildlife, land reclamation, restoring habitats disturbed by storm water runoff, mining wastes, and related others. The mechanisms of plant–microbe interaction in wetlands have been studied by Stottmeister (2003).

10.4 PHYTOINDICATORS AND TEST SYSTEMS

Plant species vary in their response to various pollutants occurring in their environment. Effects of various chemopollutants on plant systems have been studied (Dixit and Merle, 1985; Sarbhoy et al., 1991). Before selecting plant species for the purpose of developing phytoremediation technology, it is essential to understand the behavior of plant systems in presence of pollutants. Some plant species, such as *Allium sepa* L. has been evaluated as a test system for industrial pollution (Dixit and Merle, 1985). Examples of plant species studied as indicators of pollution or test systems for evaluating toxicity of pollutants are summarized in Table 10.1.

TABLE 10.1   Plant Species Used as Indicators of Pollution and Test Systems to Evaluate Toxicity of Pollutants.

Organic Pollutant	Plant Species	References
Atrazine (herbicide)	Leguminous plants (*Vigna unguiculata*)	Vaishampayan and Kanekar (2007)
Mancozeb (fungicide)	*Gliricidia sepium* (Jacg.)	Kanekar et al. (1998)
Methyl violet dye	*Wolffia arrhiza* (L.) Wimmer, *Spirodella polyrrhiza* (L.) Schleiden, *Acacia nilotica* (L.) Del, *Casuarina aquisetifolia* Forst	Kanekar et al. (1993a, 1993b)
Thallium	*Lemna minor*	Kwan and Smith (1988)
Pesticides	*Pisum* sp.	Sarbhoy et al. (1991)
Industrial effluents	*Allium cepa* (L.)	Dixit and Merle (1985)

10.5   APPLICATIONS OF PHYTOREMEDIATION

One of the major applications of phytoremediation is to treat polluted soil or water bodies. This process has been demonstrated for treatment of metal mining wastes that contaminate soil and water. Phytoremediation has been shown to be studied globally for pollutants like pesticides, nitroexplosives, metals, crude oil, and related others. Some of the plants, such as mustard, hemp, and alpine pennycress have been found to be successful in accumulation of pollutants occurring in high concentration at contaminated sites. Phytoremediation has been demonstrated as an efficient process in restoration metal mining sites, dumping places of polychlorinated biphenyls (PCBs). This technology is becoming a desirable fruitful option in last twenty years for remediation of sites contaminated with heavy, toxic metals.

10.5.1   PHYTOREMEDIATION OF ORGANIC POLLUTANTS

Environmental pollution with xenobiotics is a global problem and the development of phytoremediation technologies for the plant-based cleanup of contaminated soils is therefore of significant interest. Since it is a new technique, meager data are available on actual use of plants for remediation of wastewaters containing pollutants like pesticides, dyes, nitroaromatics, and other hazardous compounds or remediation of soils contaminated with these toxic and chemicals. However, researchers over the globe studied some plant species for phytoremediation of organic pollutants. The work is summarized in Table 10.2. This research forms basis to develop phytoremediation technologies.

TABLE 10.2   Plant Species Studied for Phytoremediation of Organic Chemopollutants.

Organic Chemopollutant	Plant Species Studied for Phytoremediation	References
Atrazine	Corn and sorghum	Shimabukuro (1968)
Atrazine	Sorghum	Lamoureux et al. (1973)
Atrazine	*Panicum dichotomiflorum*	De Prado et al. (1995)
Atrazine	Poplar	Burken and Schnoor (1997)
Nitroexplosive TNT	Switch grass, brome grass	Peterson et al. (1998)
Nitroexplosive RDX	*Myriophyllum aquaticum, Catharanthus roseus*	Bhadra et al. (2001)
Atrazine	Corn	Cherifi et al. (2001)
Nitroexplosive TNT	*Abutilon avicennae* (Indian mallow)	Chang et al. (2004)
Herbicide atrazine	*Vetiveria zizanioides* L.	Vaishampayan (2004)
Antibiotic tetracycline	*Pistia* sp.	Gujarathi (2005)
Tannery waste water	*Pharagmites australis*	Calheiros et al. (2007)
Phenol	*Brassica napus* (hairy roots)	Coniglio et al. (2008)
RDX	Constructed wetland	Low et al. (2008)
Nitroexplosive TNT	*Vetiveria* sp.	Das et al. (2010)
Textile dye malachite green	*Blumea malcolmii* Hook. f. (cell culture)	Kagalkar et al. (2011)
Naphthalene	*Eichhornia crassipes*	Nesterenko-Malkovskaya et al. (2012)
Textile effluent	*Leucaena leucocephala*	Jayanthi et al. (2014)

10.5.1.1   NITRO COMPOUNDS INCLUDING NITROEXPLOSIVES AND WASTE WATERS CONTAINING NITRATES

Nitroexplosives are difficult to degrade and exert toxicity. They cause large scale contamination of land and groundwater. Under this situation, phytoremediation becomes a method of choice. Phytoremediation of groundwater/sites contaminated with explosives was described by Best et al. (1997, 2006). Peterson et al. (1998) studied the germination of switch grass and brome grass, in soil contaminated with Trinitrotoluene (TNT). Results of their studies indicated that switch grass was more tolerant of TNT than smooth brome grass, but the establishment of both species may be limited to soil containing less than 50 mg/kg of extractable TNT.

Both aquatic and terrestrial plants have been studied for phytoremediation of nitroexplosives like TNT, HMX, RDX, and others. Plants namely pondweed, arrowroot, contrail, poplar have been employed for remediating TNT contamination in a CW (Rodgers and Bunce, 2001). The plant system could degrade 0.019 mg/L TNT per day. TNT removal rates increased with increased plant density and removal kinetics increased with increasing temperature up to 34°C (Medina et al., 2000). *Algae nitella* (stonewort) and *Myriophyllum aquaticum* (parrot feather) exhibited 30 mg/L rate of TNT removal in a hydroponic experiment. Several agricultural and indigenous terrestrial plants were examined for their capacity to accumulate and degrade High Melting Explosive (HMX). Traces of mononitroso-HMX were detected in contaminated soil extracts and leaf extracts. Mechanism for HMX translocation and accumulation in foliar tissue was concluded to be aqueous transpirational flux and evaporation (Groom et al., 2002). These reports brighten the potential of applying phytoremediation for explosives. Phytoremediation is more rugged than microbial bioreactors with respect to physical conditions and changes in contaminant loading.

Vanek et al. (2007) applied phytoremediation techniques for selected explosives. Work on phytoremoval of TNT was carried out by Medina et al. (2000) and Hannink et al. (2001). Groom et al. (2002) studied the accumulation of HMX by indigenous and agricultural plants grown in HMX contaminated water. Kanekar et al. (2003) have mentioned use of some aquatic and terrestrial plant species for phytoremediation of TNT, HMX, and others.

In study by Bhadra et al. (2001), *M. aquaticum* removed RDX from the aqueous medium. RDX levels decreased by approximately 75% in the presence of live plants compared to about 10% in the presence of dead plant matter. RDX disappearance in the presence of dead plant matter typically represents that fraction was sorbed into biomass. This showed that biological processes contributed to the depletion of RDX on exposure to *M. aquaticum*. However, HMX was not metabolized in 'natural' systems by *M. aquaticum*.

Rylott et al. (2009) developed an environment friendly, low-cost phytoremediation technique by applying recent knowledge of the biochemistry underlying endogenous plant detoxification systems and the use of genetic engineering to combine bacterial explosives-detoxifying genes with the phytoremediatory benefits of plants, such as poplar and perennial grasses. In view of slow and incomplete phytoremediation of organic pollutants, such as explosives, Aken (2009) suggested an innovative approach of developing transgenic plants to accomplish the phytoremediation at accelerated rate.

According to his work, bacterial genes encoding enzymes involved in derivatization of explosives has to be introduced in higher plants.

Chang et al. (2004) carried out experiments on phytoremediation of soil contaminated with TNT by growing Indian mallow (*Abutilon avicennae*) in a soil column reactor with 2 kg of TNT contaminated soil (120 mg TNT/ kg) in the top and 18 kg of uncontaminated soil in the bottom. The results showed that planting Indian mallow in TNT contaminated soil enhanced TNT reduction both by stimulating microbial activity that enhances microbial TNT transformation, and by direct uptake and phytotransformation of TNT. Das et al. (2010) in their earlier experiments have shown the high affinity of Vetiver grass for TNT and the catalytic effectiveness of urea in enhancing plant uptake of TNT in hydroponic media. Further the authors have demonstrated complete removal of TNT in urea-treated soil was accomplished by Vetiver at the low initial soil-TNT concentration (40 mg/kg), in soil-pot-experiment masking the effect of urea. When the TNT concentration was doubled (80 mg/kg), there was significant increase in removal of TNT by Vetiver. Thus the authors have concluded that Vetiver grass in the presence of urea effectively removes TNT from soil.

Podlipná et al. (2015) thought that the 2,4-dinitrotoluene (2,4-DNT), which is used mostly as explosive, belongs to the hazardous xenobiotics and soils and waters contaminated with 2,4-DNT may be cleaned by phytoremediation using suitable plant species. The authors studied the ability of crop plants namely hemp, flax, sunflower, and mustard to germinate and grow on soils contaminated with 2,4-DNT. The authors found the lethal concentration of 2,4-DNT for growth of these plant species around 1 mg/g. In hydroponicic experiments, the authors have observed that, the above-mentioned plant species were able to tolerate concentration of 2,4-DNT as high as 200 mg/L, which is close to maximal solubility of 2,4-DNT in water. The authors have recommended the possible use of these crop plants for phytoremediation of nitroaromatic compounds. The uptake and removal of nitrobenzene by *Mirabilis jalapa* L. has been reported by Zhou et al. (2012). The plants could grow in presence of 10 mg nitrobenzene per kg soil without any adverse effect. Thus the plant has potential in phytoremediation of soil contaminated with nitrobenzene.

Nitramines are thought to be metabolized by plants. Transformation of RDX by plant tissue cultures is suggested by Aken et al. (2004a) via a three step process involving a light-independent reduction of RDX to MNX (hexahydro-1-nitroso-3,5-dinitro-1,3,5-triazine) and DNX (-dinitroso-) derivatives by intact plant cells; a plant/light mediated cleavage of heterocyclic ring of RMX, MNX, or DNX into CH_2O and CH_3OH and further mineralization of the chloride metabolites by intact plant cells. In nature, plant–microbe symbiotic association occurs which contributes to transformation of xenobiotics. Symbiotic bacterium *Methylobacterium* sp. isolated from Poplar plant tissues was found to mineralize HMX (Aken et al., 2004b). Thus plant–microbe symbiotic phytoremediation of nitroexplosives can be possible.

Ability of agricultural and terrestrial plant species to degrade HMX was explored by Groom et al. (2002). The growth of wheat and ryegrass was found to be rapid in presence of HMX. However these plants being edible plants, their use in phytoremediation may not be advisable. Terrestrial plant species namely *Poplar* and *Glyricidia* were explored for phytoremediation of HMX wastewater (Dautpure, 2007). These plants removed nitrate from microbially treated HMX wastewater to considerable extent. These studies indicated that an integrative process involving microbial degradation followed by phytoremediation can be exploited for treatment of nitroexplosive wastewaters.

Nitrate removal by plants was studied by a few research workers. Phytoremediation of nitrates was described by Bose and Srivastava (2001). Dautpure (2007) has reported the use of phytoremediation technique for removal of nitrates from high nitrate containing wastewater generated during production of HMX. In the studies, the effect of high nitrate containing wastewater on germination of some legume seeds was seen. The terrestrial plant species studied were Silver Oak (*Grevillea robusta*) and Poplar (*Populus salicaceae*) and the aquatic plant species included Hydrilla (*Hydrilla verticillata* Casp.), Water lettuce (*Pistia stratiotes* Linn.) and Duckweed (*Lemna minor* Linn.). The authors found that the silver oak and Hydrilla plants could tolerate high concentration of nitrate present in the wastewater containing HMX. In some other studies, Dautpure (2007) found the aquatic plants like *Ceratophyllum*, *Vallisneria*, *Salvinia* and *Hydrilla* could be the suitable candidates for developing phytoremediation technology for treatment of high nitrate containing wastewaters generated during production of HMX. Kanekar et al. (2014) have reviewed work on bioremediation of nitroexplosive waste.

10.5.1.2   PESTICIDES

Pesticides are widely used for protection of crops and contribute significantly to environmental pollution. Atrazine is one of the herbicides applied in the agricultural fields to get rid of the weeds. Atrazine is a toxic herbicide and therefore remediation of sites contaminated with atrazine is necessary. Vaishampayan (2004) studied extensively plant–microbe interactive degradation of Atrazine. *Arthrobacter* sp. isolated from rhizosphere of a grass plant *Vetiveria zizanioides* was found to be efficient in degradation of atrazine (Vaishampayan et al., 2007). The grass plant Vetiver (*Vetiveria zizanioides*) with its robust root system was found to be resistant to atrazine. Pot culture studies showed tolerance of Vetiver plant up to 10,000 ppm (Vaishampayan, 2004). Vetiver–*Arthrobacter* sp. culture interactive remediation of soil spiked with atrazine was effective and faster as compared to that by the plant or the bacterial culture alone.

Merini et al. (2009) in their *in vitro* studies on tolerance to atrazine by different plant species and found *Lolium multiflorum* as a novel tolerant species which was able to germinate and grow in the presence of 1 mg/kg of the herbicide. The authors also studied the mechanisms involved in atrazine tolerance such as mutation in *psbA* gene, enzymatic detoxification via P_{450} or chemical hydrolysis through benzoxazinones, and others and demonstrated that atrazine tolerance is conferred by enhanced enzymatic detoxification via P_{450}. Due to its atrazine degradation capacity in soil and its agronomical properties, *L. multiflorum* is a candidate for designing phytoremediation strategies for atrazine contaminated agricultural soils, especially those involving run-off.

Amaya-Chávez et al. (2006) studied the removal efficiency of methyl parathion (MeP) by a common cattail *Typha latifolia* L. from water and artificial sediments. The authors observed high efficiency of cattails for removal of MeP from water and sediments relative to controls. Cattails may thus prove to be a good candidate for developing a phytoremediation system for MeP-contaminated water and artificial sediments. The plant species *Solanum nigrum* L. was found to have potential in remediation of a persistent, leachy fungicide metalaxyl (Teixeira et al., 2011). The plant was found to complete its life cycle in the presence of metalaxyl stress. It accumulates the fungicide in the plant part above ground and thus the plant tissue with high concentration of the fungicide can be disposed off. The authors also suggest that antioxidant response in form of proline accumulation, increase in activities of guaiacol peroxidase and glutathione-s-transferase could be the mechanism of tolerance of the plant to the fungicide. Thus the plant *Solanum nigrum* is a potential candidate for phytoremediation of a fungicide metalaxyl.

Álvarez et al. (2012) studied removal of the organochlorine pesticide, lindane from minimal medium (MM) by two *Streptomyces* native strains, while growing on maize root exudates (REs) as a primary carbon and energy source. Their studies showed that phytostimulation of organochlorine pesticide degrading actinobacteria by maize REs may be a successful strategy for the remediation of lindane-contaminated environments. Mimmo et al.

(2015) developed a plant-based biotest (RHIZOtest) to study the phytoextraction capacity of Italian ryegrass (*Lolium multiflorum* L.) to remove a trichloro azine herbicide, terbuthylazine (TBA) from aqueous solution at three different concentrations of TBA, that is 0.5, 1.0, and 2.0 mg/L. The authors have observed that the ryegrass can remove TBA and concluded that RHIZO test could be adequate in testing removal of herbicides from aqueous/soil solutions.

Accelerated mineralization of organic compounds was shown in rhizospheric soil (Piutti et al., 2002). There was no attempt to establish plant–microbe interaction for remediation of organic pollutants especially atrazine. Vetiver–microbial interactions for remediation of atrazine-contaminated soil were extensively studied by Vaishampayan (2004). Following the successful screening and establishing potential of Vetiver in atrazine removal from soil, a plant microbial interaction effect was studied in remediation of soil contaminated with atrazine.

During the course of the bench scale studies, an atrazine degrading culture *Arthrobacter* sp. strain MCM B-436 was isolated from rhizosphere of Vetiver plant. To establish a plant–microbial interaction model, Vetiver and the microbial culture isolated from its rhizosphere was selected. Soil used in this experiment was dried and passed through 2 mm sieve and 500 g black cotton soil was filled in 1 L capacity glass bottles. Vetiver plants of average 3 months age were procured from a nursery nearby Pune. Then, 100 g biomass of Vetiver plants were planted in 500 g black cotton soil filled in 1 L capacity glass bottles (Fig. 10.1).

FIGURE 10.1   Vetiver-microbial interactions for removal of atrazine from soil: (1) soil + 25 ppm atrazine + vetiver, (2) soil + vetiver (without atrazine), (3) soil + 25 ppm atrazine + vetiver + culture, (4) sterile soil + 25 ppm atrazine, (5) sterile soil + 25 ppm atrazine + culture, (6) soil + 25 ppm atrazine + culture (7) soil + 25 ppm atrazine (Vaishampayan, 2004).

Plants were allowed to grow for 15 days to acclimatize the experimental conditions. All the bottles were kept in the glasshouse conditions. All the plants were watered regularly throughout the experiment. The following experimental sets were run with five replicates for each treatment: (1) Soil + 25 ppm atrazine + Vetiver; (2) Soil + Vetiver (without atrazine); (3) Soil + 25 ppm atrazine + Vetiver + microbial culture; (4) Sterile soil + 25 ppm atrazine; (5) Sterile soil + 25 ppm atrazine + culture; (6) Soil + 25 ppm atrazine + culture; (7) Soil + 25 ppm atrazine. Vetiver plants grown in soil without atrazine exposure served as control to compare with Vetiver plants in atrazine spiked soil for normal growth. Atrazine spiked in sterile and non-sterile soil served as control to determine atrazine losses due to physico-chemical and microbial factors, respectively. For sterile soil controls, soil was sterilized by autoclaving at 121°C for 20 min for three times, at an interval of 24 h. Sterility of the soil was determined by plating soil slurry on Standard plate count agar. Plates were incubated at 37°C for 24 h and after incubation period presence of bacterial colonies was observed. No bacterial growth on the plates suggested proper sterilization of the soil.

After acclimatization of plants for 15 days, soil was exposed to atrazine. Soil was spiked with commercial grade atrazine dissolved in water to attain 25 ppm concentration in the soil by spraying. Immediately after atrazine treatment, 5 g of black cotton soil based bioinoculum of *Arthrobacter* sp. strain MCM B-436 was added to 500 g soil to attain culture density to 10^8 CFU/g. After addition of bioinoculum and spraying of atrazine, soil was watered to ensure uniform distribution of atrazine and the culture. All the glass bottles were kept under glass house condition with average 30 ± 2°C temperature and natural light conditions. Soil samples were collected from all the experimental sets for 4 days with interval of 24 h. From each soil sample (50 g) atrazine was extracted with dichloromethane extraction method and residual atrazine concentration was determined by HPLC.

After 4 days of incubation, the soil planted with Vetiver along with bacterial culture showed complete removal of atrazine. Vetiver plants alone showed 31% atrazine removal from the soil. Bacterial culture was able to remove 94% and 90.8% atrazine from unsterile and sterile soil, respectively.

Plant microbial interaction showed rapid atrazine removal when compared with culture and plants alone. When the results of the 3rd day, were compared, culture and plants alone showed 65% and 24% atrazine removal respectively, while plant microbial interaction showed 80 % atrazine removal. Sterile soil spiked with atrazine showed 6% atrazine removal after 5 days of exposure, which could be attributed to adsorption and other physical factors. Normal flora of soil contributed to 14% atrazine removal (Fig. 10.2).

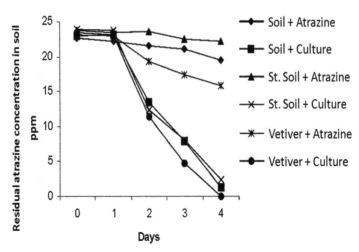

FIGURE 10.2 Vetiver-microbial interactions for removal of atrazine from contaminated soil (St. = sterile) (Vaishampayan, 2004).

Field level experiment was carried out in Sugarcane farm nearby Pune. The field site selected for the experiment had no history of atrazine application. Design of field scale treatment protocols was on the basis of results of the initial laboratory scale experiments. The dimensions of the experimental plots were 1m × 1m × 0.5m (Fig. 10.3). Soil was excavated from the plots and a plastic (polythene) lining was given to cover the entire plot area. The distance between two adjacent plots was 1 m. The objective of the plastic lining was to (1) avoid contamination of the soil adjacent to the experimental plots with atrazine, (2) avoid percolation/seepage of atrazine down the soil column with water, and (3) define soil volume used in the experiment. The entire volume of soil removed was replaced back after the plastic lining was laid. The total soil volume per experimental plot was found to be approximately 500 kg. Ten experimental plots were prepared for Vetiver-microbial interaction experiment. Vetiver plants approximately 7 months old were procured from a nursery nearby Pune. Vetiver plants having biomass of approximately 500 g were planted in each plot. Plants were allowed to grow for 15 days for acclimatization to field conditions.

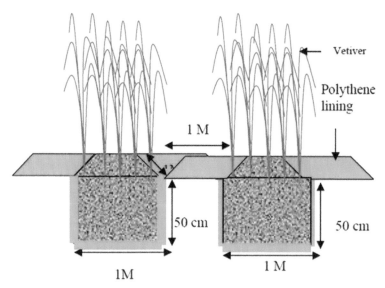

FIGURE 10.3 Diagrammatic representation of field experimental plots (Vaishampayan, 2004).

A field-scale study consisting of ten (0.5 m³) treatment plots was de-signed to test five treatment protocols in duplicate. A random design was used to assign plots for different treatments to avoid bias in any environ-mental factor. Experimental sets were as follows: (1) Soil not spiked with atrazine and planted with Vetiver plant (Vegetation control); (2) Soil spiked with atrazine and planted with Vetiver plants; (3) Soil spiked with atrazine, planted with Vetiver and atrazine degrading culture; (4) Soil spiked with atrazine and atrazine degrading culture; and (5) Soil spiked with atrazine (no Vetiver and no atrazine degrading culture). Soil planted with Vetiver plants without atrazine served as vegetation control. Soil was spiked with aque-ous solution of commercial grade atrazine (25 ppm) by spraying. Bacterial suspension was sprayed to attain *Arthrobacter* sp. strain MCM B-436 load 1×10^8 CFU/g soil. The bacterial culture was added in the soil without any nutrient amendments. Soil spiked with atrazine without culture and Vetiver plantation served as control to determine the atrazine removal by native mi-croflora, chemical degradation and soil bound un-extractable atrazine. All the plots were irrigated regularly throughout the experiment. The experiment was carried out under natural conditions (temperature and light). Soil spiked with atrazine without culture treatment and Vetiver plantation showed 6% atrazine removal from the soil within 8 days. This loss of atrazine may be at-tributed to the role of indigenous microflora, chemical degradation, and soil bound non-extractable atrazine.

Experimental plots with Vetiver plants showed 45.07% loss of atrazine. Soil spiked with *Arthrobacter* sp. strain MCM B-436 culture showed 90.24% removal on the 8th day (Fig. 10.4). Vetiver–culture (*Arthrobacter* sp. strain MCM B-436) interaction contributed to 96% removal of atrazine from the soil. Though the difference in atrazine removal was marginal in plant–microbe interaction and culture treatment, the rate of atrazine degradation in the plant microbial treatment was significant. As on the 4th day, atrazine removal was 27.53% in the plots treated with culture alone and 53.39% in plots with plant-culture together.

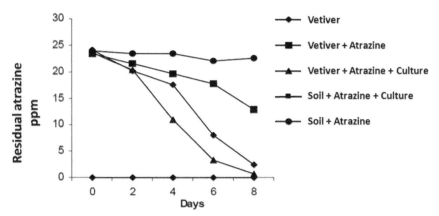

FIGURE 10.4   Vetiver-microbial interactions for the removal of atrazine from field soil (Vaishampayan, 2004).

10.5.1.3   OTHER ORGANIC COMPOUNDS

Dyes and dye industry wastewaters contribute significantly to environmental pollution. Both aquatic and terrestrial plants respond to different dyes. Kanekar et al. (1993a, 1993b) tested aquatic plants and terrestrial plants to evaluate toxicity of methyl violet wastewater. Aquatic plants *Wolffia arrhiza* (L.) Wimmer and *Spirodela polyrhiza* (L.) Schleiden indicated toxicity of untreated methyl violet wastewater. Terrestrial plant species *Acacia nilotica* (L.) Del and *Casuarina equisetifolia* Forts could be grown on microbially treated methyl violet waste-water. Decolorization of textile industry waste containing different dyes was studied by Kagalkar et al. (2011) using cell cultures of *Blumea malcolmii* Hook. 93.4% of decolorization was observed in case of malachite green within 24 h. Thus the plant cell culture shows potential in phytoremediation of dyes. Plant biomass of *Leucaena leucocephala* was found to be useful in treating textile effluent as reported by Jayanthi et al. (2014).

Cunningham et al. (1996) carried out studies on phytoremediation of soils contaminated with organic pollutants. Phenolic compounds present in the drainage from several industries are harmful pollutants and represent a potential danger to human health. The hairy roots of *Brassica napus* possess enzymes such as peroxidases which oxidize phenols. The hairy roots were found to be tolerant to high concentration of phenol and hence could be used for remediation of water containing phenol (Coniglio et al., 2008). Jha et al. (2013) attempted degradation of phenol, a common pollutant found in wastewater of many industries using hairy roots, induced in ornamental variety of *Helianthus annuus* (sunflower). The authors also found complete removal of phenol by these hairy roots and proposed a possible pathway of rhizodegradation of phenol based on the data collected by identification of metabolites by GC-MS. Shi et al. (2012) investigated the toxicity, uptake, accumulation, and removal of 2,4-dichlorophenol (2,4-DCP) in four *Salix matsudana* clones and feasibility of phytoremediation using *S. matsudana* clones in hydroponic system. The authors observed removal of 2,4-DCP from aquatic environment rapidly and efficiently by these clones without any adverse effect on the plants.

Maqbool et al. (2012) reported the effect of bioaugmentation by free and immobilized bacterial culture on the rhizodegradation of petroleum-polluted soil using *Sesbania cannabina* plant. The authors have observed higher and rapid rhizodegradation with indegenous bacteria than bioaugmented bacterial cultures and concluded that natural plant–microbe interaction in the rhizosphere of *S. cannabina* was more efficient in degrading TPH to the petroleum hydrocarbons. Al-Baldawi et al. (2013) developed a phytotoxicity test of bulrush (*Scirpus grossus*) to assess its ability to phytoremediate diesel contamination in simulated wastewater at different concentrations (0, 8700, 17,400, and 26,100 mg/L). The authors observed that *S. grossus* was able to reduce total petroleum hydrocarbon (TPH) by 70.0 and 80.2% at the concentrations of 8700 mg/L and 17,400 mg/L, respectively. The concentration of diesel > 26,100 mg/L was found to be lethal for *S. grossus* after 14 days. The plant *Scirpus grossus* was found to survive in water contaminated with diesel as evidenced by increase in biomass. Al-Baldawi et al. (2014) investigated the optimum conditions for TPH removal from diesel-contaminated water using phytoremediation treatment with *Scirpus grossus* through a Box-Behnken Design. The optimum conditions were found to be a diesel concentration of 0.25% (V_{diesel}/V_{water}), a retention time of 63 days and no aeration with an estimated maximum TPH removal from water and sand of 76.3 and 56.5%, respectively. The authors recommended the use of *S. grossus*, a Malaysian native plant for remediation of wastewater containing hydrocarbons. Polycyclic aromatic hydrocarbons (PAHs) represent one of the major groups of organic contaminants in the aquatic environment. Duckweed (*Lemna minor* L.) is a common aquatic plant widely used in phytotoxicity tests for xenobiotic substances. Zezulka et al. (2013) studied physiological, biochemical and other changes occurring in duckweed plants exposed to fluoranthene for 4–10 days. The authors observed that FLT did not influence the growth and content of photosynthetic pigments in duckweed, however, histological changes occurred only at the level of root cells and tissues. They further observed that FLT treatment had no adverse effect on the chlorophyll fluorescence parameters and activity of some enzymes like ascorbate peroxidase, catalase, superoxide dismutase, and related others, was found to be enhanced in presence of FLT. Kirk et al. (2005) studied rhizospheric microbial community to enhance degradation of organic contaminants. They investigated the effect of ryegrass in combination with alfalfa in soil contaminated with 31,000 ppm of petroleum hydrocarbons.The denaturing gradient gel electrophoresis (DGGE) analysis of PCR-amplified partial 16S rDNA sequences showed increase in the number of rhizosphere bacteria especially petroleum degrading bacteria in the hydrocarbon-contaminated soil with ryegrass and ryegrass/alfalfa mixture. The authors had concluded that plants altered the microbial population which were plant-specific and could contribute to degradation of petroleum hydrocarbons in contaminated soil.

Nesterenko-Malkovskaya et al. (2012) investigated the potential of an aquatic plant, the water hyacinth (*Eichhornia crassipes*) devoid rhizospheric bacteria, to reduce naphthalene (a polyaromatic hydrocarbon) present in wastewater and wetlands. The plants enhanced the removal of pollutants through their consumption as nutrients and also through microbial activity of their rhizospheric bacteria. Naphthalene removal by water hyacinth coupled with natural rhizospheric bacteria was 100% after 9 days; however without rhizospheric bacteria, the removal was only up to 45% during 7 days. In the removal of naphthalene from water, the contribution of aquatic plant *Eichhornia crassipes* is much more than its rhizospheric bacteria.

Among the persistent organic pollutants, dioxins get bioaccumulated in food chain causing a considerable risk to human health. Removal of dioxins using plants was suggested. Uptake of 2,3,7,8-tetra-chlorinated dibenzo-p-dioxin (TCDD) by the plant *Arabidopsis thaliana* was found to be more in roots than shoots. The phytotoxicity of the compound was observed in the plant in form of biological damages (Hanano et al., 2014). The authors thought that better understanding of the plants ability to detoxify dioxins would help to improve their use as a safe bioremediators.

Macek et al. (2000) reviewed various phytoremediation technologies, paying special attention to removal of organic xenobiotics like polychlorinated biphenyls (PCBs) and the application of *in vitro* systems for basic research in the role of plants for the remediation of sites polluted by PCBs and enhancement in their effectiveness.The authors have mentioned various aspects of metabolism of xenobiotic compounds in plant cells, the role of enzymes involved, and the cooperation of rhizospheric microorganisms in accelerating remediation of these toxic organic recalcitrant compounds. They have also discussed the application of this phytoremediation approach as well as the possibility of introduction of foreign genes into plant genome that can enhance the rate of the bioremediation of these toxic organic compounds. Meagher (2000) reported the removal of organic pollutants such as PCBs from soils by transgenic plants containing genes for bacterial enzymes.

Wang et al. (2014) studied the effect of polybrominated diphenyl ether congener (BDE-47) on the growth and antioxidative responses of the seedlings of *Kandelia obovata* and *Avicennia marina* in hydroponic system and observed that *Avicennia* was more tolerant to BDE-47 than *Kandelia*, as its antioxidative enzymes could better counter-balance the oxidative stress caused by the pollutant.

Lao et al. (2003) investigated the metabolic fate of [UL-^{14}C]-3,4-dichloroaniline (DCA) in *Arabidopsis* root cultures and soybean plants over a 48 h period following treatment via the root media. DCA was rapidly taken up by both species and metabolized, predominantly to N-malonyl-DCA in soybean and N-glucosyl-DCA in *Arabidopsis*. The difference in the routes of DCA detoxification in the two plants could be explained partly by the relative activities of the respective conjugating enzymes, DCA-N-malonyltransferase activity in soybean plants. In *Arabidopsis* sp., DCA-N-glucosyl transferase activity was dominant. The studies reveal differential regulation of DCA-N-glucosyltransferase enzyme in the two plant species.

Gujarathi (2005) demonstrated the use of *Pistia* sp. as bioindicators for effective and complete treatment of wastewater generated during production of antibiotics like tetracycline. Dordio et al. (2009) conducted studies to assess ability of *Typha* sp. to withstand and remove clofibric acid (CA), a metabolite of blood lipid regulator drugs from water. Exposure to higher CA concentrations did not affect *Typha*'s photosynthetic pigments but there was overall increase in enzyme activity (ascorbate and guaiacol peroxidases, catalase, and superoxide dismutase). The authors have suggested ability of *Typha* for phytoremediation of CA contaminated waters.

10.6 APPLICATIONS OF CONSTRUCTED WETLANDS FOR DIFFERENT ORGANIC POLLUTANTS

Vymazal (2009) reviewed the use of horizontal flow constructed wetlands (HFCW) for wastewater treatment for more than 30 years. The author has also mentioned that municipal HFCW focus not only on common pollutants but also on special parameters such as pharmaceutical compounds, chemicals causing endocrine disruption or linear alkylbenzene sulfonates (LAS). At present, HFCW are used to treat many other types of industrial wastewaters arising from oil industry, pulp and paper mill, textile industry, tannery, distillery and winery industries, and others. In particular, the use of HFCW is becoming very common for treatment of food-processing wastewaters, and various runoff waters from agriculture, airports, highway, greenhouses, plant nurseries, and others, as well as to treat leachate arising from landfill. In addition to the use of HFCW as a single unit, they are also employed in combination with other types of CWs. The enzymes peroxidases of *Phragmites australis* (POD) were found to degrade Acid Orange 7 (AO7), an azo dye in a vertical flow constructed wetland (VFCW) indicating potential of the plant in tretament of the dye waste (Davies et al., 2005).

Calheiros et al. (2007) studied the survival of different plant species, such as *Typha latifolia, Canna indica, Stenotaphrum secundatum, Phragmites australis*, and *Iris pseudacorus* in subsurface HFCW receiving tannery wastewater. Only *Phragmites australis* and *Typha latifolia* could be established successfully resulting in high removal of organic content from the wastewater. Davies et al. (2009) studied phytoremediation of a textile azo dye (AO7) in a pilot scale CW using most widely used plant in CWs, such as *Phragmites australis*, and found the removal of the dye to be feasible even at concentrations of AO7 as high as 748 ± 42 mg/L. The authors have also carried out studies on the gene-expression for the *Phragmites australis* while integrated in the AO7 wastewater treatment.

Low et al. (2008) investigated the ability of down-flow CWs to remediate hexahydro-1,3,5-trinitro-1,3,5-triazine (RDX) contaminated water at varying loading rates over a period of 2 years. Significant RDX removal occurred (89–96%) for all loading rates ranging from 160 to 1600 mg/m^2/day at a hydraulic retention time of approximately 2 days. RDX degradation occurred in both NO_3^- and SO_4^{2-} dominated electron acceptor zones. These results support the use of CWs for the remediation of low-level RDX-contaminated water. Darryl et al. (2008) had also applied down-flow CW mesocosms for treatment of RDX.

Verlicchi and Zambello (2014) reviewed the occurrence of 137 pharmaceutical compounds in the effluent from various types of CW treating urban wastewater. Investigations of 136 treatment plants including free water systems, horizontal and vertical subsurface flow beds, were reported by the authors. The uptake of selected pharmaceuticals occurring in sediments and gravel by common macrophytes was also reviewed suggesting novel approach for removal of persistent organic compounds.

10.7 CONCLUSION AND FUTURE PERSPECTIVES

Phytoremediation is an emerging technology which uses plant systems for degradation and detoxification of chemopollutants. Terrestrial plant species like *Poplar*, *Vetiver*, *Arabidopsis* are found to be effective in removal of pollutants. Constructed wetlands are natural ecosystems that are commonly used for filtration of pollutants from agricultural runoffs. Among the organic pollutants, nitroexplosives, like TNT, RDX, HMX, and pesticides namely atrazine are extensively studied for their removal from contaminated sites using phytoremediation technologies.

Breeding programs and genetic engineering are powerful methods for enhancing natural phytoremediation capabilities, or for introducing new competence into aquatic and terrestrial plants. Genes for such improved phytoremediation capacity may originate from a microorganism or may be transferred from one plant to another variety better adapted to the environmental conditions at the cleaning site. In past genes specifically encoding a nitroreductase from a particular bacterium were inserted into tobacco and showed faster removal of TNT and enhanced resistance to the toxic effects of TNT. Researchers have also discovered a mechanism in plants that allows them to grow even when the pollution concentration in the soil is lethal for untreated plants. Some naturally occurring biodegradable compounds such as exogenous polyamines allow plants to tolerate concentrations of pollutants 500 times higher than untreated plants and to absorb more pollutants.

Plants are known to accumulate and translocate particular types of contaminants; plants can also be effectively used as biosensors to detect subsurface contamination, thereby allowing investigators to quickly demarcate the trail of contaminants. Phytoscreening may lead to more optimized site investigations and reduce contaminated site cleanup costs. Phytoremediation can become an ecological solution to remediate sites contaminated with organic chemicals.

KEYWORDS

- Atrazine
- Bioaugmentation
- Dichloroaniline
- Dioxins
- Hydrilla
- Hydrocarbons
- Methyl
- Parathion
- Naphthalene
- Nitroexplosives
- Pesticides
- Petroleum hydrocarbons
- Phenolic compounds
- Phytoaccumulation
- Phytodegradation
- Phytoextraction
- Phytoremediation
- Phytoscreening
- Phytostabilization
- Phytostimulation
- Phytotoxicity
- Phytotransformation
- Phytovolatilization
- Plant microbe interaction
- Poplar
- Rhizodegradation
- Rhizofiltration
- Vetiver

REFERENCES

Adler, T. Botanical cleanup crews: Using plants to tackle polluted water and soil. *Sci News* 1996, 150, 42–43.

Aken, B. V.; Yoon, J. M.; Just, C. L.; Schnoor, J. L. Metabolism and mineralization hexa-hydro-1,3,5-trinitro-1,3,5-triazine inside poplar tissues (*Populus deltoides × nigra* DN-34). *Environ Sci Technol* 2004a, 38, 4572–4579.

Aken, B. V.; Yoon, J. M.; Schnoor, J. L. Biodegradation of nitro-substituted explosives 2,4,6-trinitrotoluene, hexahydro-1,3,5-trinitro-1,3,5-triazine, and octahydro-1,3,5,7-tetranitro-1,3,5,7-tetrazocine by a phytosymbiotic *Methylobacterium* sp. associated with poplar tissues (*Populus deltoides × nigra* DN 34). *Appl Environ Microbiol* 2004b, 70(1), 508–517.

Aken, B. V. Transgenic plants for enhanced phytoremediation of toxic explosives. *Curr Opin Biotechnol* 2009, 20(2), 231–236.

Al-Baldawi, I. A. W.; Sheikh Abdullah, S. R.; Hasan, H. A.; Suja, F.; Anuar, N.; Mushrifah, I. Optimized conditions for phytoremediation of diesel by *Scirpus grossus* in horizontal subsurface flow constructed wetlands (HSFCWs) using response surface methodology. *J Environ Manage* 2014, 140, 152–159.

Al-Baldawi, I. A.; Sheikh Abdullah, S. R.; Anuar, N.; Suja, F.; Mushrifah I. A phytotoxicity test of bulrush (*Scirpus grossus*) grown with diesel contamination in a free-flow reed bed system. *J Hazard Mater* 2013, 252–253, 64–69.

Álvarez, A.; Yañez, M. L.; Benimeli, C. S.; Amoroso, M. J. Maize plants (*Zea mays*) root exudates enhance lindane removal by native *Streptomyces* strains. *Int Biodeter Biodegr* 2012, 66(1), 14–18.

Amaya-Chávez, A.; Martínez-Tabche, L.; López-López, E.; Galar-Martínez, M. Methyl para-thion toxicity to and removal efficiency by *Typha latifolia* in water and artificial sediments. *Chemosphere* 2006, 63(7), 1124–1129.

Best, E. P.; Zappi, M. E.; Fredrickson, H. L.; Sprecher, S. L.; Larson, S. L.; Ochoa, M. Screening of aquatic and wetland plants for phytoremediation of explosives - contaminated groundwa-ter from the Iowa Army Ammunition Plant. *Ann N Y Acad Sci* 1997, 829, 179–194.

Best, E. P. H.; Geter, K. N.; Tatem, H. E.; Lane, B. K. Effects, transfer, and fate of RDX from aged soil in plants and worms. *Chemosphere* 2006, 62, 616–626.

Bhadra, R.; Wayment, D. G.; Williams, R. K.; Barman, S. N.; Stone, M. B.; Hughes, J. B.; Shanks, J. V. Studies on plant-mediated fate of the explosives RDX and HMX. *Chemosphere* 2001, 44, 1259–1264.

Bose, B.; Srivastava, H. S. Absorption and accumulation of nitrate in plants: influence of environmental factors. *Ind J Exp Biol* 2001, 39(2), 101–110.

Burken, J. G.; Schnoor, J. L. Uptake and metabolism of atrazine by poplar trees. *Env Sci Technol* 1997, 31, 1399–1406.

Calheiros, C. S. C.; Rangel, A. O. S. S.; Castro, P. M. L. Constructed wetland systems veg-etated with different plants applied to the treatment of tannery wastewater. *Water Res* 2007, 41(8), 1790–1798.

Chang, Y. Y.; Kwon, Y. S.; Kim, S. Y.; Lee, I. N.; Bae, B. Enhanced degradation of 2,4,6-trini-trotoluene (TNT) in a soil column planted with Indian mallow (*Abutilon avicennae*). *J Biosci Bioeng* 2004, 97(2), 99–103.

Cherifi, M.; Raveton, M.; Picciocchi, A.; Ravanel, P.; Tissut, M. Atrazine metabolism in corn seedlings. *Plant Physiol Biochem* 2001, 39, 665–672.

Coniglio, M. S.; Busto, V. D.; González, P. S.; Medina, M. I.; Milrad, S.; Agostini, E. Application of *Brassica napus* hairy root cultures for phenol removal from aqueous solutions. *Chemosphere* 2008, 72(7), 1035–1042.

Cunningham, S. D.; Anderson, T. A.; Schwab, A. P.; Hsu, F. C. Phytoremediation of soils contaminated with organic pollutants. *Adv Agron* 1996, 56, 55–114.

Cunningham, S. D.; Berti, W. R.; Huang, J. W. Phytoremediation of contaminated soils. *Trends Biotechnol* 1995, 13(9), 393–397.

Darryl, L.; Kui, T.; Todd, A.; George, P. C.; Jun, L.; Jackson, W. A. Treatment of RDX using down-flow constructed wetland mesocosms. *Ecol Engg* 2008, 32(1), 72–80.

Das, P.; Datta, R.; Markis, K. C.; Sarkar, D. Vetiver grass is capable of removing TNT from soil in the presence of urea. *Environ Pollut* 2010, 158(5), 1980–1983.

Dautpure, P. S. Microbial degradation of nitroexplosive HMX (high melting explosive). PhD Thesis, Pune University, India, 2007.

Davies, L. C.; Carias, C. C.; Novais, J. M.; Martins-Dias, S. Phytoremediation of textile effluents containing azo dye by using *Phragmites australis* in a vertical flow intermittent feeding constructed wetland. *Ecol Engg* 2005, 25(5), 594–605.

Davies, L. C.; Cabrita, G. J. M.; Ferreira, R. A.; Carias, C. C.; Novais, J. M.; Martins-Dias, S. Integrated study of the role of *Phragmites australis* in azo-dye treatment in a constructed wetland: From pilot to molecular scale. *Ecol Engg* 2009, 35(6), 961–970.

De Prado, R.; Romera, E.; Menedez, J. Atrazine detoxification in *Panicum dichotomiflorum* and target site *Polygonum lapathifolium*. *Pesticide Biochem Physiol* 1995, 52, 1–11.

Dixit, G. B.; Merle, S. K. Cytotoxic effects of industrial effluents on *Allium cepa* L. *Geobios* 1985, 12, 237–240.

Dordio, A. V.; Duarte, C.; Barreiros, M.; Carvalho, A. J. P.; Pinto, A. P.; da Costa, C. T. Toxicity and removal efficiency of pharmaceutical metabolite clofibric acid by *Typha* spp. – potential use for phytoremediation? *Biores Technol* 2009, 100(3), 1156–1161.

Greger, M.; Landberg, T. Using of willow in phytoextraction. *Int J Phytoremed* 1999, 1(2), 115–123.

Groom, C. A.; Halasz, A.; Paquet, L.; Morris, N.; Olivier, L.; Dubois, C.; Hawari, J. Accumulation of HMX (octahydro-1,3,5,7-tetranitro-1,3,5,7-tetrazocine) in indigenous and agricultural plants grown in HMX-contaminated anti-tank firing-range soil. *Environ Sci Technol* 2002, 36, 112–118.

Gujarathi, N. P. Phytoremediation of tetracycline and oxytetracycline, PhD Thesis, Colorado State University, Fort Collins, USA, 2005.

Hanano, A.; Almousally, I.; Shaban, M. Phytotoxicity effects and biological responses of *Arabidopsis thaliana* to 2,3,7,8-tetrachlorinated dibenzo-*p*-dioxin exposure. *Chemosphere* 2014, 104, 76–84.

Hannink, N.; Rosser, S. J.; French, C. E.; Basran, A.; Murray, J. A.; Nicklin, S.; Bruce, N. C. Phytodetoxification of TNT by transgenic plants expressing a bacterial nitroreductase. *Nat Biotechnol* 2001, 19(12), 1168–1172.

Jayanthi, V.; Geetha, R.; Rajendran, R.; Prabhavathi, P.; Sundaram, S. K.; Dinesh Kumar, S.; Santhanam, P. Phytoremediation of dye contaminated soil by *Leucaena leucocephala* (subabul) seed and growth assessment of *Vigna radiata* in the remediated soil. *Saudi J Biol Sci* 2014, 21(4), 324–333.

Jha, P.; Jobby, R.; Kudale, S.; Modi, N.; Dhaneshwar, A.; Desai, N. Biodegradation of phenol using hairy roots of *Helianthus annuus* L. *Int Biodeter Biodegr* 2013, 77, 106–113.

Kagalkar, A. N.; Jadhav, M. U.; Bapat, V. A.; Govindwar, S. P. Phytodegradation of the triphenylmethane dye malachite Green mediated by cell suspension cultures of *Blumea malcolmii* Hook. *Biores Technol* 2011, 102(22), 10312–10318.

Kanekar, P.; Kumbhojkar, M. S.; Ghate, V.; Sarnaik, S. *Wolffia arrhiza* (L.) Wimmer and *Spirodella polyrrhiza* (L.) Schleiden as test plant systems for toxicity assay of microbially treated dyestuff wastewater. *J Environ Biol* 1993a, 14(2), 129–135.

Kanekar, P.; Kumbhojkar, M. S.; Ghate, V.; Sarnaik, S.; Kelkar, A. Evaluation of *Acacia nilotica* (L.) Del and *Casuarina equisetifolia* Forst. for tolerance and growth on microbially treated dyestuff waste water. *Environ Pollut* 1993b, 81, 47–50.

Kanekar, P.; Fanse, K.; Kumbhojkar, M. S.; Ghate, V.; Kelkar, A.; Sarnaik, S. Effect of mencozeb pesticide effluent with high sulfate content on growth of selected plant species. *J Environ Biol* 1998, 19(3), 205–209.

Kanekar, P.; Dautpure, P.; Sarnaik, S. Biodegradation of nitroexplosives. *Ind J Expt Biol* 2003, 41, 991–1001.

Kanekar, P. P.; Sarnaik, S. S.; Dautpure, P. S.; Patil, V. P.; Kanekar, S. P. Bioremediation of nitroexplosive waste wasters. In *Biological remediation of explosive residues*; Singh, S. N., Ed.; Springer International Publishing: Switzerland, 2014; pp 67–86.

Kirk, J. L.; Klironomos, J. N.; Lee· H.; Trevors, J. T. The effects of perennial ryegrass and alfalfa on microbial abundance and diversity in petroleum contaminated soil. *Environ Pollut* 2005, 133(3), 455–465.

Kwan, K. H. M.; Smith, S. The effect of Thallium on the growth of *Lemna minor* and plant tissue concentrations in relation to both exposure and toxicity. *Environ Pollut* 1988, 52, 203–220.

Lamoureux, G. L.; Stucki, G.; Shimabukuro, R. H.; Zaylskie R. G. Atrazine metabolism in sorghum: Catabolism of glutathione conjugate of atrazine. *J Agri Food Chem* 1973, 21, 1020–1030.

Lao, S. H.; Loutre, C.; Brazier, M.; Coleman, J. O. D.; Cole, D. J.; Edwards, R.; Theodoulou, F. L. 3,4-Dichloroaniline is detoxified and exported via different pathways in *Arabidopsis* and soybean. *Phytochem* 2003, 63(6), 653–661.

Low, D.; Tan, K.; Anderson, T.; Cobb, G. B.; Liu, J.; Jackson, W. D. Treatment of RDX using downflow constructed wetland mecosomes. *Ecol Engg* 2008, 32(1), 72–80.

Macek, T.; Macková, M.; Káš, J. Exploitation of plants for the removal of organics in environmental remediation. *Biotechnol Adv* 2000, 18(1), 23–34.

Maqbool, F.; Wang, Z.; Xu, Y.; Zhao, J.; Gao, D.; Zhao, Y. G.; Bhatti, Z. A.; Xing, B. Rhizodegradation of petroleum hydrocarbons by *Sesbania cannabina* in bioaugmented soil with free and immobilized consortium. *J Hazard Mater* 2012, 262-269, 237–238.

McCutcheon, S. C.; Jørgensen, S. E. Phytoremediation. In *Encyclopedia of Ecology,* Jorgensen, S. E., Fath, B., eds., Elsevier Science: Amsterdam, Netherlands, 2008, pp 2751–2766.

Meagher, R. B. Phytoremediation of toxic elemental and organic pollutants. *Curr Opin Plant Biol* 2000, 3(2), 153–162.

Medina, V. F.; Larson, S. L.; Bergstedt, A. E.; McCutcheon, S. C. Phyto-removal of trinitrotoluene from water with batch kinetic studies. *Water Res* 2000, 34(10), 2713–2722.

Megharaj, M.; Ramakrishnan, B.; Venkateswarlu, K.; Sethunathan, N.; Naidu, R. Bioremediation approaches for organic pollutants: A critical perspective. *Environ Int* 2011, 37(8), 1362–1375.

Mendez, M. O.; Maier, R. M. Phytostabilization of mine tailings in arid and semiarid environments - an emerging remediation technology. *Environ Health Persp* 2008, 116(3), 278–283.

Merini, L. J.; Bobillo, C.; Cuadrado, V.; Corach, D.; Giulietti, A. M. Phytoremediation potential of the novel atrazine tolerant *Lolium multiflorum* and studies on the mechanisms involved. *Environ Pollut* 2009, 157(11), 3059–3063.

Mimmo, T.; Bartucca, M. L.; Del Buono, D.; Cesco, S. Italian ryegrass for the phytoremediation of solutions polluted with terbuthylazine. *Chemosphere* 2015, 119, 31–36.

Nesterenko-Malkovskaya, A.; Kirzhner, F.; Zimmels, Y.; Armon, R. *Eichhornia crassipes* capability to remove naphthalene from wastewater in the absence of bacteria. *Chemosphere* 2012, 87(10), 1186–1191.

Perelo, L. W. *In situ* and bioremediation of organic pollutants in aquatic sediments. *J Hazard Mater* 2010, 177, 81–89.

Peterson, M. M.; Horst, G. L.; Shea, P. J.; Comfort, S. D. Germination and seedling development of switch grass and smooth brome grass exposed to 2,4,6-trinitrotoluene. *Environ Polln* 1998, 99(1), 53–59.

Piutti, S.; Hallet, S.; Rousseaux, L.; Philippot, L.; Soulas, G.; Martin-Laurent, F. Accelerated mineralization of atrazine in maize rhizoshere soil. *Biol Fertil Soils* 2002, 36, 434–441.

Podlipná, R.; Pospíšilová, B.; Vaněk, T. Biodegradation of 2,4-dinitrotoluene by different plant species. *Ecotox Environ Safe* 2015, 112, 54–59.

Rodgers, J. D.; Bunce, N. J. Treatment methods for the remediation of nitroaromatic explosives. *Water Res* 2001, 35, 2101–2111.

Rupassara, S. I.; Larson, R. A.; Sims, G. K.; Marley, K. A. Degradation of atrazine by hornwort in aquatic systems. *Bioremed J* 2002, 6(3), 217–224.

Rylott, E. L.; Neil, C.; Bruce, N. C. Plants disarm soil: Engineering plants for the phytoremediation of explosives. *Trends Biotechnol* 2009, 27(2), 73–81.

Sarbhoy, R. K.; Sharma, A.; Singh, R. M. The pisum test - an alternative in environmental studies, the relative toxicity of pesticides. *J Environ Biol* 1991, 12, 137–141.

Schwitzguebel, J. P.; van der Lelie, D.; Baker, A.; Glass, D. J.; Vangronsveld, J. Phytoremediation: European and American trends – successes, obstacles and needs. *J Soils Sediments* 2002, 2, 91–99.

Schwitzguebel, J. P. Potential of phytoremediation, an emerging green technology: European trends and outlook. *Proc Indian Nat Sci Acad* 2004, B70, 109–130.

Shi, X.; Leng, H.; Hu, Y.; Liu, Y.; Duan, H.; Sun, H.; Chen, Y. Removal of 2,4-dichlorophenol in hydroponic solution by four *Salix matsudana* clones. *Ecotox Environ Safe* 2012, 86, 125–131.

Shimabukuro, R. Atrazine metabolism in resistant corn and sorghum. *Plant Physiol* 1968, 43, 1925–1930.

Singh, O. V.; Jain, R. K. Phytoremediation of toxic aromatic pollutants from soil. *Appl Microbiol Biotechnol* 2003, 63, 128–135.

Stottmeister, U.; Wiessner, A.; Kuschk, P.; Kappelmeyer, U.; Kästner, M.; Bederski, O.; Müller, R. A.; Moormann, H. Effects of plants and microorganisms in constructed wetlands for wastewater treatment. *Biotech Adv* 2003, 22(1–2), 93–117.

Subramanian, M.; Oliver, D. J.; Shanks, J. V. TNT phytotransformation pathway characteristics in Arabidopsis: Role of aromatic hydroxylamines. *Biotechnol Prog* 2006, 22(1), 208–216.

Teixeira, J.; de Sousa, A.; Azenha, M.; Moreira, J. T.; Fidalgo, F.; Silva, A. F.; Faria, J. L.; Silva, A. M. T. *Solanum nigrum* L. weed plants as a remediation tool for metalaxyl-polluted effluents and soils. *Chemosphere* 2011, 85(5), 744–750.

Vanek, T.; Gerth, A.; Vakrikova, Z.; Podlipna, R.; Soudek, P. Phytoremediation of explosives. In *Advanced science technology for biological decontamination of sites affected by chemical and radiological nuclear agents, part III*; Marmiroli, N., Samotokin, B., Marmiroli, M., Eds.; Springer: Netherlands, 2007 pp 209–225.

Vaishampayan, P. A. Bioremediation of an herbicide, atrazine from soil. PhD Thesis, University of Pune, India, 2004.

Vaishampayan, P. A.; Kanekar, P. P. Use of atrazine sensitive leguminous plants as biological indicators to evaluate the atrazine degradation efficiency of bacterial inoculum. *World J Microbiol Biotechnol* 2007, 23, 447–449.

Vaishampayan, P. A.; Kanekar P. P.; Dhakephalkar, P. K. Isolation and characterization of *Arthrobacter* sp. strain MCM B-436, an atrazine-degrading bacterium from rhizospheric soil. *Int Biodet Biodeg* 2007, 60, 273–278.

Verlicchi, P.; Zambello, E. How efficient are constructed wetlands in removing pharmaceuticals from untreated and treated urban wastewaters? A review. *Sci Total Environ* 2014, 470–471, 1281–1306.

Vymazal, J. The use constructed wetlands with horizontal sub-surface flow for various types of wastewater. *Ecol Engg* 2009, 35(1), 1–17.

Wang, Y.; Zhu, H.; Tam, N. F. Y. Effect of a polybrominated diphenyl ether congener (BDE-47) on growth and antioxidative enzymes of two mangrove plant species, *Kandelia obovata* and *Avicennia marina* in South China. *Marine Poll Bull* 2014, 85(2), 376–384.

Williams, J. B. Phytoremediation in wetland ecosystems: Progress, problems and potential. *Crit Rev Plant Sci* 2002, 21(6), 607–635.

Zezulka, S.; Kummerová, M.; Babula, P.; Váňová, L. *Lemna minor* exposed to fluoranthene: Growth, biochemical, physiological and histochemical changes. *Aquat Toxicol* 2013, 140–141, 37–47.

Zhou, Q.; Diao, C.; Sun, Y.; Zhou, J. Tolerance, uptake and removal of nitrobenzene by a newly-found remediation species *Mirabilis jalapa* L. *Chemosphere* 2012, 86 (10), 994–1000.

CHAPTER 11

BIOCONVERSION OF PALM OIL AND SUGAR INDUSTRY WASTES INTO VALUE-ADDED POLYHYDROXYALKANOATE

NOOR FAZIELAWANIE MOHD RASHID[1], AIN FARHANA MOHD YATIM[1], AL-ASHRAF AMIRUL[2], and KESAVEN BHUBALAN[3]

[1]*School of Marine and Environmental Sciences, Universiti Malaysia Terengganu, Kuala Terengganu 21030, Terengganu, Malaysia*

[2]*School of Biological Sciences, Universiti Sains Malaysia, Pulau Pinang 11800, Malaysia*

[3]*Malaysian Institute of Pharmaceuticals and Nutraceuticals, MOSTI, Bayan Lepas 11700, Pulau Pinang, Malaysia*

CONTENTS

11.1   INTRODUCTION

Among the key features of tomorrow's plastic are eco-friendly bioplastics that can be degraded in shorter time and sustainability to ecosystem. Microbial polyhydroxyalkanoate (PHA) has attracted research and commercial interest because of their biodegradability, thermoplastic properties, and synthesis from renewable resources. The high PHA production cost is a major factor that hinders commercialization process. The production cost commonly involves fermentation set-up, carbon feedstocks of the bacteria, electricity as well as chemical usage (Lee, 2008). However, carbon feedstock is considered to make-up the major fraction of production cost. Therefore, efforts are taken to identify cheap and renewable carbon substrates. Agricultural industry wastes and by-products have been investigated in detail as potential carbon sources. The aim of reducing the total production cost of PHA is to increase the commercialization potential and demand in global market for this biodegradable polymer. In recent years, the use of organic wastes, such as palm oil mill effluent (POME) (Hassan, 1996), municipal wastewater (Coats et al., 2011), biodiesel wastewater (Dobroth et al., 2011), and food processing waste effluent (Venkateswar et al., 2012) provide double benefits because the wastes were converted into environmentally friendly biodegradable polymer. Several types of microbial-derived PHA have shown great potential to be commercialized in industrial scale, medical application, and utilized as substitutes to petrochemical-based plastics.

11.2   POME

By year 2014, the palm oil industries in Indonesia and Malaysia have grown rapidly and both nations have become the leading producers in the world, replacing Nigeria as a main producer in the last three decades. Figure 11.1 shows the palm oil producers in the world. From a humble source of the edible oil, palm oil has given us a very meaningful knowledge in the application from every part of its plant (Foo and Hameed, 2010). Oil palm currently accounts for more than a half of the total cultivated land in Malaysia, and its oil production is one of the highest among the producing countries. The contribution of the oil palm industry to Malaysia's economic development has indeed been impressive. Based on the statistics obtained from the Malaysian Palm Oil Council, Malaysia currently accounts 39% of world palm oil production, and 44% of world exports (MPOC, 2014). The palm oil industry generates large quantities of by-products composed of triglycerides and fatty acids, suitable for microbial utilization. Other than that, Malaysia also accounts for 12 and 27% of the world's production and exports of oils and fats. Being one of the biggest producers and exporters of palm oil and palm oil products, Malaysia has an important role to play in fulfilling the growing global need for oils and fats sustainably (MPOC, 2014). The high abundance of palm oil has shown the potential primary and secondary of resources for PHA biosynthesis.

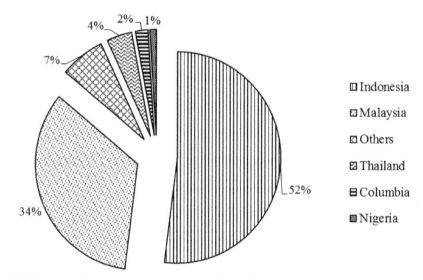

FIGURE 11.1   World palm oil producers (MPOC, 2014).

POME is considered as one of the most polluting agro-industrial effluent due to its high values of biochemical oxygen demand (BOD) and chemical oxygen demand (COD) (Ahmad et al., 2011). Based on Figure 11.2, the re-newable resources from palm oil extraction generate about 50 million tonnes of POME, 9 million tonnes of fiber, 16 million tonnes of empty fruit bunches, and 4 million tonnes of shells. POME is brownish in color with a discharge temperature of between 80 and 90 °C, acidic, viscous, and colloidal suspension that contain 95–96% of water, 0.6–0.7% of oil, and 2–4% suspension solids (Foo and Hameed, 2010). The wastewater of POME cannot be discharged directly into the drainage due its high levels of BOD and COD that is harmful to environment. The wastewater has to be removed in order to prevent interfaces in water treatment units and stability of aquatic biodiversity and also to avoid problems in biological treatment stages (Ahmad et al., 2005). POME is a strong wastewater from palm oil mills and has been identified as a potential source to generate renewable bioenergy through anaerobic digestion. The raw treated POME has an extremely high content of degradable organic matter, which is due in part to the present of unrecovered palm oil (Ahmad et al., 2003). Application of the renewable palm oil biomass for PHA production will help to reduce the overall production cost and address the waste management issue of these materials. Table 11.1 summarizes the information on palm oil and oil palm industry in Malaysia.

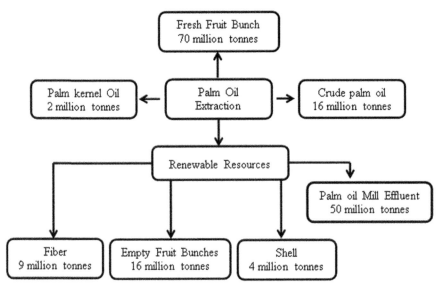

FIGURE 11.2 Palm oil derivatives and biomass generated from palm oil extraction processes (Hassan et al., 2013).

TABLE 11.1 Information on Palm Oil and Oil Palm Industry in Malaysia.

Derivative	Description	Value	References
Oil palm	Total planted area (million hectares)	4.92	MPOB (2014)
	Plantation density (palms/hectare)	135	USDA (2012)
	Economical life span (years)	30	USDA (2012)
Fresh fruit bunch (FBB)	Weight (kg)	15–25	MPOB (2014)
	Number of fruits/FFB	1000–3000	
	Average number FFB/palm	10	
Palm oil	Production volume (million tonnes/annum)	4.1	MPOB (2014)
	Average yield (million tonnes/hectare/annum)	4.0	
	Average market price (US$/tonne)		

TABLE 11.1 *(Continued)*

Derivative	Description	Value	References
	Edible use fraction (%)	65	
	Amount of by-product/waste generate by palm oil mills (million tonnes/annum)		
	Empty fruit bunch	16	
	Mesocarp fiber	9	
	Trunk	9	
	Shell	4	
	Palm oil mill effluent	50	Hassan et al. (2013)
	Cost of production (US$/tonne of palm products)		Sime Darby (2009)
	Estate cost	256.36	
	Mill cost	62.81	

11.3 MOLASSES

Molasses is a sugar-rich co-product stream produced during the extraction of sugar from sugar cane and sugar beets (Du et al., 2012). The global sugar production is exceeds 165 million tonnes per year. In 2012, 80% of global sugar production is produced from sugar cane, while remaining 20% is produced from sugar beet. Currently, Brazil is the world's largest producer of sugarcane, accounting for one-third of world production. Asian production, which includes India, China, and Thailand, accounts for another one-third of the world production. India is the world's second largest producer of sugarcane and China is the fourth largest. A vast global market for sugarcane derivatives keeps the industry booming. Sugar is prevalent in the modern diet and increasingly a source of biofuels and bioplastics.

Sugar cane industry generates some by-products namely molasses, bagasse, and press mud (Fig. 11.3). Molasses is generated during the centrifuging of sugar crystals, whereby the yield per ton of cane is approximately 4–4.5%. Bagasse is residual woody fiber of the cane, which used for several purposes, especially used as fuel for the boilers in the generation of process steam. Solid mud contained high amount of potassium, sodium, phosphorous, and organic matters produced during sugar cane processing.

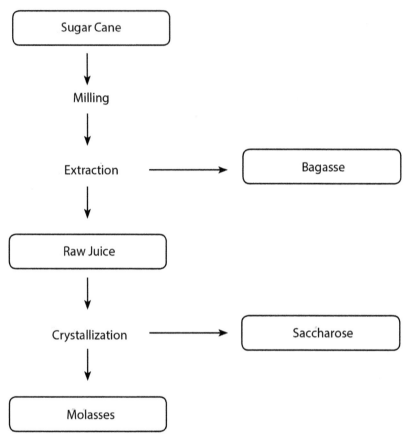

FIGURE 11.3   Simplified process flow of refined sugar production (Koller et al., 2010).

Molasses from both sugarcane and sugar beets is a major component of animal feed. Molasses composed of very high sugar content of mainly sucrose, glucose, and fructose, 41, 6.7, and 3.2% w/w, respectively, and other growth promoters, such as vitamins and biotin (Koller et al., 2010). Molasses and saccharose obtained will be used for the production of PHA-rich biomass and production of catalytically active biomass. Meanwhile, bagasse will be hydrolyzing into convertible sugars, namely glucose, xylose, and arabinose, which directly used for PHA production. Zhang et al. (1994) noted that molasses normally sells approximately 30–55% of the cost of glucose.

11.4 POLYHYDROXYALKANOTE

Polyhydroxyalkanote (PHA) is among the most widely investigated group of biodegradable polymer due to its natural, renewable, and biocompatible nature (Verlinden et al., 2011). It is considered to be a potential substitute for some commercially available petro-chemical derived plastics. PHA is produced intracellularly by various types of bacteria including *Cupriavidus* sp., *Aeromonas* sp., *Acinobacter* sp., *Azobacter* sp., *Bacillus* sp., *Klebsiella* sp., *Pseudomonas* sp., and recombinant *Escherichia coli* (Steinbüchel et al., 1992; Koller et al., 2010). Among the different strains tested, *Cupriavidus necator* (previously known as *Alcaligenes eutrophus*, *Ralstonia eutropha*, or *Wautersia eutropha*) has been studied most extensively due to its ability to accumulate large amount of PHA from simple carbon sources, for example, glucose, fructose, and acetic acid (Khanna and Srivastava, 2005).

PHAs are reported as biodegradable (100% biodegradable), insoluble in water, biocompatible, non-toxic, piezoelectric, thermoplastics, and/or elestomeric (Shivakumar, 2012). Their characteristic feature makes them as possible alternatives to replace petroleum-based polymers (Andreeßen and Steinbüchel, 2010). PHA is degraded aerobically by microorganism to CO_2 and H_2O upon disposal, thus reduce the issue of greenhouse gas emissions. PHA can be produced from renewable carbon source, thus allow the sustainable production process (Albuquerque et al., 2007). PHA assembles various types of monomers with approximately 150 structures. It can be classified according to their branching monomers. Short-chain length PHAs (scl-PHAs) consist of 3–5 carbon atoms, while medium-chain-length PHAs (mcl-PHAs) and long-chain-length PHAs (lcl-PHAs) composed of 6–14 and more than 14 carbon atoms, respectively (Anderson and Dawes, 1990).

Poly(3-hydroxybutyrate) [(P3HB)], a scl-PHA is the most common type of PHA produced by bacteria in the natural environment. The presence of P(3HB) was first identified in *Bacillus megaterium* by Lemoigne at the Pasteur Institute in Paris in the mid-1920s (Lemoigne, 1926). P(3HB) is the intracellular granule, which synthesized by bacteria under unbalanced growth conditions, such as high carbon concentration and limited concentration of N, P, S, or some trace elements, such as Mg, Fe, and Ca (Shivakumar, 2012). Under normal condition, N sources will be used by the bacteria to synthesize protein and is an essential element for growth. Limitation of N sources will inhibit the TCA cycles enzyme, such as isocitrate dehydrogenase, which eventually slows down the TCA cycle. As a result, acetyl-CoA is channeled for the synthesis of P(3HB) (Shivakumar, 2012). In bacteria,

P(3HB) can be accumulated up to 80% of dry cell weight (DCW) as membrane enclosed inclusion (Khanna and Srivastava, 2005). P(3HB) has a wide range of application in packaging, medical, insecticides, herbicides, cosmetic world, and disposable personal hygiene due to its physical properties, which remarkably similar to those conventional plastics (Ojumu et al., 2004; Kulpreecha et al., 2009; Chee et al., 2010). Advancement of PHA application in medical industry for bone replacement, wound dressings, and surgical pins were progressing studies.

P(3HB) are commercially produced by various types of cheap substrate, namely methanol (Suzuki et al., 1986), ethanol (Alderet et al., 1993), starch and whey (Kim, 2000; Ghaly, 2003), and beet molasses (Page, 1992a,b). Nevertheless, Page and Knosp (1989) proved that unrefined carbon source, such as cane molasses and beet molasses also promote biopolymer production comparable to, or even better than refined sugars. Shivakumar (2012) noted that polyhydroxybutyrate (PHB) production using fermentation resulted in a very high production cost, thus making their use unattractive. The cost of carbon source, fermentation strategy, and recovery process/downstream processing contributes to the high manufacturing cost for biopolymer production. Halami (2008) noted that carbon source for P (3HB) production can reach approximately 50% of the total production cost. Research currently focuses on the use of waste agriculture residue, namely starch, whey, molasses, CSL, bagasse, and soy meal as well as POME to downsize the production cost. Currently, the application of biopolymer for industrial production is restricted due to its extremely high-cost. Hence, the viable solution strategy to overcome this problem is by utilizing the waste materials generated from agro-based industry. Choi and Lee (1999) and Kim (2000) noted that the use of industrial waste and byproducts as sole carbon source for PHA production can reduce approximately 40–50% of total production cost. The utilization of waste materials provides a viable strategy for the production of biopolymer and overcoming the disposal problem as well. Other parameters which contribute significantly to the total production cost are bacterial strains, fermentation strategy, and recovery process as well.

Various types of wastes have been tested for the production of PHA by numbers of bacteria, namely waste lipids (waste cooking oil), glycerol (byproducts of biodiesel production), molasses (by products of sugar cane industry), whey (by-products in the manufactures of cheese), wastewater (municipal wastewater), and lignocellulosic raw materials. Table 11.2 summarizes the PHA fermentation by different cheap substrate as a carbon source.

TABLE 11.2 PHA Production Using Sustainable Raw Materials.

Carbon Sources	Strains	PHA Content (%)	References
Lipids			
Residual oil	*C. necator* H16	41.3	Füchtenbusch et al. (2000)
	P. oleovorans	38.9	
Tallow	*P. resinovorans*	15	Cromwick et al. (1996)
Olive oils	*Aeromonas caviae*	6–12	Doi et al. (1995)
Palm kernel oil, crude palm oil, palm acid oil	*C. necator*	5	Loo et al. (2005)
Crude palm oil	*Erwinia* sp. USMI-20	46	Majid et al. (1999)
Vegetable oil	*Pseudomonas* sp. strain DR2	23.5	Song et al. (2008)
Whey and hydrolyzed whey			
Whey	Recombinant *Escherichia coli* CGSC 4401	70	Park et al. (2002)
Hydrolysates whey	*C. necator*	37	Marangoni et al. (2002)
Molasses			
Sugar cane molasses	*Bacillus megaterium* BA-019	61.62	Kulpreecha et al. (2009)
Beet molasses	*Azobacter vinelandii* UWD	76	Chen and Page (1997)
Soy molasses	*Pseudomonas corrugata* 388	5–17	Solaiman et al. (2006)
Lignocellulosic wastes			
Xylose	*P. cepacia* ATCC 17759	60	Ramsay et al. (1995)
Hemicellulosic hydrolysates	*Burkholderia cepacia* ATCC 17759	1–61	Keenan et al. (2006)

11.5 PRODUCTION OF PHA FROM POME

POME is organic waste water, which has high carbon content and low nitrogen content. With these characteristics, the POME becomes a suitable carbon replacement for PHA biosynthesis (Hassan et al., 2013). This enables the bioconversion of environment polluting waste into value added material. The abundant supply of agro-industrial waste has all so far resulted in the successful accumulation of PHA, and there is a growing knowledge base in the use of these feedstocks (Chanprateep, 2010). Anaerobic digestion of POME produced a series of volatile fatty acid (VFA), namely formic acid, acetic acid, propionic acid, and butyric acids, which became a potential platform chemical for biofuel and material production (Cheng et al., 2010). It is likely to improve POME conversion efficiency and energy utilization further if bacterial PHA production can be integrated into the existing mill VFA-producing units (Mumtaz et al., 2010). Several studies proved that polymers from POME-derived VFAs are comparable to those obtained from commercially available organic acids. POME has been used as a carbon sources by numbers of bacteria, such as *Rhodobacter sphaeroides* (Hassan et al., 1996, 1997a, 1997b), *Comamonas* sp. (Mumtaz et al., 2010; Zakaria, 2010), and *C. necator* (Din, 2012). Those studies from POME cultivation have been reported for two-stage cultivation or pure culture systems in a simple and direct way. Both organic removal and PHA-producing microbial organisms are cultivated in the same system, which is called a hybrid fed-batch system (Din, 2012).

Mainly, PHA productions focus on the scaling-up of production and improvement of material properties. Based on Junpadit et al. (2014), POME produced high DCW and PHA content compared to others carbon sources. It might due to the high fatty acid content (VFA and LCFA). Fatty acid was degraded by β-oxidation, which is the main metabolic route for the PHA synthesis. In most cases, the organic carbon sources are converted into VFA in aerobic activated sludge in the first step, and then converted into the PHAs by mixing cell cultures in the second step (Du, 2012). PHA-producing bacteria from Junpadit and Boonsawang (2010) showed that POME gave the highest PHA monomer composition containing the 3-hydroxybutyrate (3HB), 3-hydroxyvalerate (3HV), 3-hydroxyhexanoate (3HHx), and 3-hydroxydecanoate (3HD) of 534, 552, 488, and 41 mg/L, respectively. Although the final PHA concentrations are still low at the current investigated conditions, PHAs could accumulate to approximately 50% of the cell dry weight in some cases (Liu, 2011).

The production of PHA from *Comamonas* sp. EB172 has shown a positive result to POME as a carbon sources. The combination of POME treatment and PHA production can provide a zero discharge system for palm oil mills (Hassan, 2002). The fed-batch cultivation of *Comamonas* sp. was cultivated in 2 L fermentation, the development of cell biomass and PHA accumulation recorded were 9.8 g/L and 59 (wt %) of PHA after 60 h of incubation (Mumtaz, 2009; Zakaria, 2010). The incorporation of P(3HV) monomer unit (21 mol%) in the poly(3-hydroxybutyrate-co-3-hydroxyvalerate) [P(3HB-*co*-3HV)] copolymers was obtained in this study due to the presence of propionic acid in the mixture of organic acids from anaerobically treated POME. Zakaria (2010) proved that high PHA accumulation was achieved from mixed microbial cultures by feeding them a high concentration of VFAs from anaerobically fermented POME. The maximum PHA content was 40% of the DCW and more than 80% of COD was removed. Based on the result, it has shown that *Comamonas* sp. with POME is ongoing to meet the market potentials and reduction of PHA production cost.

POME can be used as a carbon sources for PHA production, but the pH in the anaerobic treatment of POME has to be maintain at seven, so that no formic acid and biogas but only acetic and propionic acids will be produced (Hassan, 1996). As reported, the highest yield of PHA with 50% and PHA content with over 65% was obtained at pH 7. The control of agitation, pH, and time in the cultivation of *R. sphaeroides* by using POME as a carbon sources was required in order to achieve maximum yield and PHA content. *C. necator* are known to be able to accumulate a large amount of PHA when nitrogen and phosphorus limited also able to use residual oil from POME as the source of carbon to produce PHA (Hassan, 2013). By using a fed-batch production of PHA, *C. necator* produced 18–76% of PHA contents but the overall PHA productivity obtained was less than PHA obtained by other researchers (Hassan, 1997a, 1997b). This might due to the unstable cell concentration when concentrated acetic acid separated from POME was incorporated into the standard medium. Hence, further research and development to explore advanced technology to produce PHAs with consistent quality from wastewater treatment are required to ensure that target applications are achieved. The proposed strategy should not only reduce the overall PHA production costs, but also help to solve the waste management problems in the palm oil industry. Table 11.3 summarizes the PHA production using POME as carbon source.

TABLE 11.3   Biosynthesis of PHA Using POME as Carbon Source.

Microorganism	PHA Yield (g/L)	PHA Content (% cell dry weight)	References
Comamonas sp. EB172	0.31	85.8	Mumtaz et al. (2010); Zakaria et al. (2010)
Rhodobacter sphaeroides	0.5	67.0	Hassan et al. (1996), Hassan et al. (1997a, 1997b)
C. necator	0.32	45.0	Hassan et al. (1997a, 1997b)
Anaerobic bacteria			Junpadit et al. (2014)

11.6 PRODUCTION OF PHA FROM MOLASSES

11.6.1 SUGAR CANE MOLASSES

The use of sugar cane molasses, which is by-products of sugar cane industries, may help to reduce the cost production of biopolymer (Mona et al., 2001). It contained high amount of sugar (over 50% of dry weight) and widely used for production of PHA using various cultivation techniques (Albuquerque, 2007). Wu et al. (2001) have isolated *Bacillus* sp. JMa5 from molasses contaminated soil and further investigates the use of molasses as a sole carbon source for PHB production. *Bacillus* sp. JMa5 grew up to 30 g/L of CDW after 8 h of incubation in fed-batch fermentation with the initial molasses concentration of 210 g/L. However, after 30 h of incubations, cell reached approximately 70 g/L. Wu et al. (2001) concluded that aeration plays important roles in synthesizing P(3HB). High P(3HB) content was detected when culturing in non-baffled flask compared to baffled flask, which were 42 and 22%, respectively. This can be concluded that aeration allows the cell growth and PHB production was parallel.

Chaijamrus and Udpuay (2008) further investigated the use of sugar cane molasses and corn steep liquor (CSL) for P(3HB) production using *B. megaterium* ATCC 6748. Sugar cane molasses and CSL were used, as they provide sufficient amount of carbon and nitrogen source for bacteria growth. For the first parameter, sugar cane molasses at a level of 1–6% w/w and 0.5% of NH_4Cl were used, meanwhile for the second parameter, the experiment was conducted using 0–6% of CSL instead of NH_4Cl and 4% of molasses. The highest production of DCW obtained was 7.2 g/L using 4% of molasses and 6% of CSL. For P(3HB) production, the highest concentration was obtained after 48 h of incubation (43% w/w, dry matter) using 4% of molasses and 4% of CSL as C and N source, respectively. However, DCW produced is only 5 g/L, thus can be concluded that bacterial growth increased with the increasing of CSL, whereas P(3HB) accumulation decreased.

Kulpreecha (2009) reported the production of PHB by *B. megaterium* BA-019 using shake flask and fed-batch culture. The bacterium was isolated from soil in Thailand. In batch culture, approximately 8.78 g/L of DCW and 42.1% of P(3HB) content were achieved, respectively, after 12 h of incubation. Meanwhile, the DCW was increased up to 72.6 g/L when culturing using PH-stat-batch culture at 24 h of cultivation. Using PH-stat-batch culture, the P(3HB) content also increased to 61.62% with productivity of 0.45 (g/L/h).

Mona et al. (2001) have reported the use of different concentration of sugarcane molasses and CSL as sole carbon and nitrogen source for production of P(3HB) by *B. megaterium* strain. Different sugarcane molasses levels ranging from 1 to 5% were used. DCW obtained was ranging from 0.8 to 1.6 g/L after 48 h of incubation. Meanwhile, PHA and P(3HB) contents were ranging from 20 to 50%.

11.6.2 SUGAR BEET MOLASSES

Sugar beet molasses contains 30–50% (w/w) of sucrose. After hydrolysis, the sucrose will decompose to glucose and fructose. Liu (1998) has investigated the production of P(3HB) using beet molasses by recombinant *Escherichia coli* (HMS174/Ptz18u-PHB). Beet molasses was successfully replaced glucose in synthesizing PHB. In a batch culture, approximately 20–60 g/L of beet molasses was used. The maximum DCW and P(3HB) production achieved were ranging from 6.5 to 16.7 g/L and 68 to 85% w/w, respectively. Meanwhile, using 5 L fed-batch culture, maximum DCW obtained was 39.5 g/L with 80%w/w of P(3HB) content after 19 h of incubation.

Chen and Page (1997) have reported P(3HB) production by *Azobacter vinelandii* UWD using two-stage fermentation process. Aeration was used to promote cell growth and suppressed the P(3HB) for the first stage, and lower aeration of raw sugar medium containing fish peptone was used to promote P (3HB) production in second stage. P(3HB) yield of 36 g/L and productivity of > 1 g/L/h were obtained using 5% of beet molasses. P(3HB) content was reported at 76 wt%, after 40 h of incubation. Table 11.4 summarizes the PHA production from sugar cane and sugar beet molasses by different bacteria using different cultivation strategies.

TABLE 11.4 PHA Production by Various Microorganisms Using Sugar Cane and Sugar Beet Molasses by Different Cultivation Techniques.

Strain	Carbon Source	Cell Density (g/L)	Cultivation Time (h)	PHA Type	PHA Content (%)	Cultivation Strategy	References
Bacillus sp. JMa5	Cane molasses	30	8	P (3HB)	25–35	Batch culture	Wu et al. (2001)
B. megate-rium ATCC 6748	Cane molasses	5	45	P (3HB)	43	Batch culture	Chaijamrus and Udpuay (2008)

TABLE 11.4   *(Continued)*

Strain	Carbon Source	Cell Density (g/L)	Cultivation Time (h)	PHA Type	PHA Content (%)	Cultivation Strategy	References
B. mega-terium BA-019	Cane molasses	8.78	12	P (3HB)	61.62	Batch culture	Kulpreecha et al. (2009)
B. megaterium	Cane molasses	40	48	P (3HB)	46.3	Batch culture	Mona et al. (2001)
Recom-binant E. coli	Beet molasses	6.5–16.7	80	P (3HB)	68–85	Batch culture	Liu et al. (1998)
Recom-binant E. coli	Beet molasses	39.5	19	P (3HB)	80	PH-DO-stat fed-batch culture	Liu et al. (1998)
Azobacter vinelandii UWD	Beet molasses	36	40	P (3HB)	76	Two-stage cultivation	Chen and Page (1997)

11.7   CONCLUSION AND FUTURE OUTLOOK

Bioconversion of agricultural industry wastes into value added products is indeed a brilliant way to solve waste accumulation in the environment. PHA, which holds much potential as an alternative to chemically synthesized plastics is a value added material that can be produced using agricultural industry wastes. Considering economic, environmental, and social issues, the ultimate goal is to obtain an economically viable PHA production system based on clean and safe processes. The final commercial PHA based products can be environmentally compatible, leading to a truly sustainable manufacturing process. This chapter has highlighted the use of POME and molasses from sugar industry as carbon feedstock for PHA production. It is evident that these waste materials can be converted into biodegradable plastic, a value added product. Some strains have shown their ability to convert these wastes into PHA; however, researchers are still trying to isolate better PHA producing strains from the environment. This effort is also complemented by designing suitable fermentation processes for high yield PHA production. Major producers of palm oil, cane, and beet sugars, such as Malaysia, Indonesia, Thailand, Brazil, and India may exploit the readily available waste products as carbon feedstock for PHA production. Research can be geared toward fully exploiting this renewable carbon feedstock to its maximum and developing sustainable PHA production systems. Much potential and great economic revenue can be seen in conversion of organic substrate from agricultural wastes to PHA. With technological and scientific developments in production of bio-based as well as biodegradable products, such as PHA, a positive attitude toward sustainable green technology and increased usage of green products in future is evident.

KEYWORDS

- **Agro-industrial wastes**
- **Biopolymer**
- **Microbial fermentation**
- **Oil palm**
- **Palm oil mill effluent**
- **Poly(3-hydroxybutyrate)**
- **Polyhydroxyalkanoate**
- **Renewable raw material**
- **Sugar beet molasses**
- **Sugar cane molasses**

REFERENCES

Ahmad, A.; Ghufran, R.; Wahid, Z. A. Bioenergy from anaerobic degradation of lipids in palm oil mill effluent. *Rev Environ Sci Bio/Technol* 2011, 10, 353–376.

Ahmad, A.; Sumathi, S.; Hameed, B. Adsorption of residue oil from palm oil mill effluent using powder and flake chitosan: Equilibrium and kinetic studies. *Water Res* 2005, 39, 2483–2494.

Ahmad, A. L.; Ismail, S.; Bhatia, S. Water recycling from palm oil mill effluent (POME) using membrane technology. *Desalination* 2003, 157, 87–95.

Albuquerque, M. G. E.; Eiroa, M.; Torres, C.; Nunes, B. R.; Reis, M. A. M. Strategies for the development of a side stream process for polyhydroxyalkanoate (PHA) production from sugar cane molasses. *J Biotechnol* 2007, 130, 411–421.

Alderet, J. E.; Karl, D. W.; Park, C. H. Production of poly(hydroxybutyrate) homopolymer and copolymer from ethanol and propanol in a fed-batch culture. *Biotechnol Prog* 1993, 9, 520–525.

Anderson, A. J.; Dawes, E. A. Occurrence, metabolism, metabolic role and industrial uses of bacteria polyhydroxyalkanoates. *Microbiol Rev* 1990, 54 (4), 450–472.

Andreeßen, B.; Steinbüchel, A. Biosynthesis and biodegradation of 3-hydroxypropionate-containing polyesters. *Appl Environ Microbiol* 2010, 76 (15), 4919–4925.

Chaijamrus, S.; Udpuay, N. Production and characterization of polyhydroxybutyrate from molasses and corn steep liquor produced by *Bacillus megaterium* ATCC 6748. *Agric Eng Int: CIGR J* 2008, 10, FP07030.

Chanprateep, S. Current trends in biodegradable polyhydroxyalkanoates. *J Biosci Bioeng* 2010, 110 (6), 621–632.

Chee, J. Y.; Yoga, S. S.; Lau, S. W.; Abed, R. M. M.; Sudesh, K. Bacterially produced polyhydroxyalkanoate (PHA): Converting renewable resources into bioplastics. In *Current research, technology and education topics in applied microbiology and microbial biotechnology*; Méndez-Vilas, A., Ed.; Formatex Research Center: Spain, 2010, pp 1395–1404.

Chen, G. Q.; Page, W. J. Production of poly-β-hydroxybutyrate by *Azobacter vinelandii* in a two-stage fermentation process. *Biotechnol Tech* 1997, 11 (5), 347–350.

Cheng, J.; Zhu, X.; Ni, J.; Borthwick, A. Palm oil mill effluent treatment using a two-stage microbial fuel cells system integrated with immobilized biological aerated filters. *Bioresour Technol* 2010, 101, 2729–2734.

Coats, E. R.; Vandevoort, K. E.; Darby, J. L.; Loge, F. J. Toward polyhydroxyalkanoate production concurrent with municipal wastewater treatment in a sequencing batch reactor system. *J Environ Eng* 2011, 137(1), 46–54.

Cromwick, A. M.; Foglia, T.; Lenz, R. W. The microbial production of poly(hydroxyalkanoates) from tallow. *Appl Microbiol Biotechnol* 1996, 46, 464–469.

Din, M. F.; Mohanadoss, P.; Ujang, Z.; Loosdrecht, M. V.; Yunus, S. M.; Chelliapan, S. Development of Bio-PORec® system for polyhydroxyalkanoates (PHA) production and its storage in mixed cultures of palm oil mill effluent (POME). *Bioresour Technol* 2012, 124, 208–216.

Dobroth, Z. T.; Hu, S.; Coats, E. R.; McDonald, A. G. Polyhydroxybutyrate synthesis on biodiesel wastewater using mixed microbial consortia. *Bioresour Technol* 2011, 102, 3352–3359.

Doi, Y.; Kitamura, S.; Abe, H. Microbial synthesis and characterization of poly(3-hydroxybutyrate-co-3-hydroxyhexanoate). *Macromolecules* 1995, 28(14), 4822–4828.

Du, C.; Sabirova, J.; Soetaert, W.; Lin, S. K. C. Polyhydroxyalkanoate production from low-cost sustainable raw materials. *Curr Chem Biol* 2012, 6, 14–25.

Foo, K.; Hameed, B. Insight into the applications of palm oil mill effluent: A renewable utilization of the industrial agricultural waste. *Renew Sust Energ Rev* 2010, 14, 1445–1452.

Füchtenbusch, B.; Wullbrandt, D.; Steinbüchel, A. Production of polyhydroxyalkanoic acids by *Ralstonia eutropha* and *Pseudomonas oleovorans* from an oil remaining from biotechnological rhamnose production. *Appl Microbiol Biotechnol* 2000, 53 (2), 167–172.

Ghaly, E. A.; Tango, S. A.; Adams, M. A. Enhanced lactic acid production from cheese whey with nutrient supplement addition. *Int Comm Agric Eng* 2003, 5, FP02009.

Halami, P. M. Production of polyhydroxyalkanoate from starch by native isolate *Bacillus cereus* CFR06. *World J Microbiol Biotechnol* 2008, 24, 805–812.

Hassan, M. A.; Nawata, O.; Shirai, Y.; Rahman, N.; Yee, P. L.; Ariff, A. A proposal for zero emission from palm oil industry incorporating the production of polyhydroxyalkanoates from palm oil mill effluent. *J Chem Eng* 2002, 35(1), 9–14.

Hassan, M. A.; Shirai, Y.; Kusubayashi, N.; Karim, M. I.; Nakanishi, K.; Hashimoto, K. Effect of organic acid profiles during anaerobic treatment of palm oil mill effluent on the production of polyhydroxyalkanoates by *Rhodobacter sphaeroides*. *J Ferment Bioeng* 1996, 82(2), 151–156.

Hassan, M. A.; Shirai, Y.; Kusubayashi, N.; Karim, M. I.; Nakanishi, K.; Hashimoto, K. The production of polyhydroxyalkanoate from anaerobically treated palm oil mill effluent by *Rhodobacter sphaeroides*. *J Ferment Bioeng* 1997a, 83 (5), 485–488.

Hassan, M. A.; Shirai, Y.; Umeki, H.; Yamazumi, H.; Jin, S.; Yamamoto, S. Acetic acid separation from anaerobically treated palm oil mill effluent by ion exchange resins for the production of polyhydroxyalkanoate by *Alcaligenes eutrophus*. *Biosci Biotechnol Biochem* 1997b, 61 (9), 1465–1468.

Hassan, M. A.; Yee, L. N.; Yee, P. L.; Ariffin, H.; Raha, A. R.; Shirai, Y. Sustainable production of polyhydroxyalkanoates from renewable oil-palm biomass. *Biomass Bioenerg* 2013, 50, 1–9.

Junpadit, P.; Boonsawang, P.; Suksaroj, T. T. *Polyhydroxyalkanoate production from palm oil factory wastes and its application for 3-hydroxyalkanoate methyl esters as biofuels.* International conference on sustainable energy and environment (SEE 2014): Science, technology and innovation for ASEAN green growth, Bangkok, Thailand, 2014, pp 67–70.

Junpadit, P.; Boonsawang, P. *Isolation and selection of bacteria for polyhydroxyalkanoate (PHA) production from palm oil effluent (POME).* International conference on biotechnology for healthy living, Thailand, 2010, pp 787–793.

Keenan, T. M.; Nakes, J. P.; Tanenbaum, S. W. Polyhydroxyalkanoate copolymers from forest biomass. *Indian J Microbiol* 2006, 33, 616–626.

Khanna, S.; Srivastava, A. K. Recent advances in microbial polyhydroxyalkanoates. *Process Biochem* 2005, 40, 607–619.

Kim, B. S. Production of poly-(3-hydroxybutyrate) from inexpensive substrate. *Enzyme Microbe Technol* 2000, 27, 774–777.

Koller, M.; Atlic, A.; Dias, M.; Reiterer, A.; Braunegg, G. Microbial PHA production from waste raw materials. *Microbiol Monogr* 2010, 14, 85–119.

Kulpreecha, S.; Boonruangthavorn, A.; Meksiriporn, B.; Thongchul, N. Inexpensive fed-batch cultivation for poly(3-hydroxybutyrate) production by a new isolate of *Bacillus megaterium*. *J Biosci Bioeng* 2009, 107 (3), 240–245.

Lee, S. Y.; Choi, J. I. Production and degradation of polyhydroxyalkanoates in waste environment. *Waste Manage* 1999, 19(2), 133–139.

Lee, W. H.; Loo, C. Y.; Nomura, C. T.; Sudesh, K. Biosynthesis of polyhydroxyalkanoate co-polymers from mixtures of plant oils and 3-hydroxyvalerate precursors. *Bioresour Technol* 2008, 99, 6844–6851.

Lemoigne, M. Products of degradation and of polymerization of β-hydroxybutyric acid. *Bull. Soc. Chem. Biol.* 1926, 8, 770–782.

Liu, F.; Li, W.; Ridgway, D.; Gu, T. Production of poly-β-hydroxybutyrate on molasses by recombinant *Escherichia coli*. *Biotechnol Lett* 1998, 20 (4), 345–348.

Liu, Q.; Luo, G.; Zhou, X. R.; Chen, G. Q. Biosynthesis of poly(3-hydroxydecanoate) and 3-hydroxydodecanoate dominating polyhydroxyalkanoates by β-oxidation pathway inhibited *Pseudomonas putida*. *Metab Eng* 2011, 13, 11–17.

Loo, C. Y.; Lee, W. H.; Tsuge, T.; Doi, Y.; Sudesh, K. Biosynthesis and characterization of poly(3-hydroxybutyrate-*co*-3-hydroxyhexanoate) from palm oil products in a *Wautersia eutropha* mutant. *Biotechnology Letters* 2005, 27(18), 1405–1410.

Majid, M. I. A.; Akmal, D. H.; Few, L. L.; Agustien, A.; Toh, M. S.; Samian, M. R.; Najimudin, N.; Azizan, M. N. Production of poly(3-hydroxybutyrate) and its copolymer poly(3-hydroxybutyrate-*co*-3-hydroxyvalerate) by *Erwinia* sp. USMI-20. *International Journal of Biological Macromolecules* 1999, 25(1), 95–104.

Marangoni, C.; Furigo, Jr. A.; de Aragão, G. M. F. Production of poly(3-hydroxybutyrate-co-hydroxyvalerate) by *Ralstonia eutropha* in whey and inverted sugar with propionic acid feeding. *Process Biochem* 2002, 38, 137–141.

Mona, K. G.; Swellam, A. E.; Omar, S. H. Production of PHB by a *Bacillus megaterium* strain using sugarcane molasses and corn steep liquor as sole carbon and nitrogen source. *Microbiol Res* 2001, 156, 201–207.

MPOB. Malaysian Palm Oil Board. 2014, http://www.mpob.gov.my/ (Dec 22, 2014).

MPOC. Malaysian Palm Oil Council. 2014, http://www.mpoc.org.my/ (accessed Dec 23, 2014).

Mumtaz, T.; Suriani, A. A.; Rahman, N. A.; Shirai, Y.; Hassan, M. A. Fed-batch production of P(3HB-*co*-3HV) copolymer by *Comonas* sp. EB 172 using mixed organic acids under dual nutrient limitation. *Eur J Sci Res* 2009, 33 (3), 374–384.

Mumtaz, T.; Yahaya, N. A.; Suraini, A. A.; Rahman, N. A.; Yee, P. L.; Shirai, Y. Turning waste to wealth-biodegradable plastics polyhydroxyalkanoates from palm oil mill effluent - a malaysian perspective. *J Clean Prod* 2010, 18, 1393–1402.

Ojumu, T. V.; Yu, J.; Solomon, B. O. Production of polyhydroxyalkanoate a bacteria biodegradable polymer. *Afr J Biotechnol* 2004, 3, 18–24.

Page, W. J. Production of polyhydroxyalkanoates by *Azobacter vinelandii* UWD in beet molasses culture. *FEMS Microbiol Rev* 1992a, 103, 149–158.

Page, W. J. Suitability of commercial beet molasses fractions as substrate for polyhydroxyalkanoate production by *Azobacter vinelandii* UWD. *Biotechnol Lett* 1992b, 14, 385–390.

Page, W. J.; Knosp, O. Hyperproduction of poly-β-hydroxybutyrate during exponential growth of *Azobacter vinelandii* UWD. *Appl Environ Microbiol* 1989, 55, 1334–1339.

Park, S. J.; Park, J. P.; Lee, S. Y. Production of poly(3-hydroxybutyrate) from whey by fed-batch culture of recombinant *Escherichia coli* in a pilot-scale fermenter. *Biotechnol Lett* 2002, 24, 185–189.

Ramsay, J. A.; Hassan, M. C. A.; Ramsay, B. A. Hemicellulose as a potential substrate for production of poly(β-hydroxyalkanoates). *Can J Microbiol* 1995, 41, 262–266.

Shivakumar, S. Polyhydroxybutyrate (PHB) production using agro-industrial residue as substrate by *Bacillus thuringiensis* IAM 12077. *Int J Chem Tech Res* 2012, 4 (3), 1158–1162.

Sime Darby. *Annual Report 2009*. Sime Darby Berhad: Malaysia, 2009.

Solaiman, D. K. Y.; Ashby, R. D.; Hotchkissn, Jr. A. T.; Fogha, T. A. Biosynthesis of medium-chain-length poly(hydroxyalkanoate) from soy molasses. *Biotechnol Lett* 2006, 28, 157–162.

Song, J. H.; Jeon, C. O.; Choi, M. H.; Yoon, S. C.; Park, W. Polyhydroxyalkanoate (PHA) production using waste vegetable oil by *Pseudomonas* sp. strain DR2. *J Microbiol Biotechnol* 2008, 18, 1408–1415.

Steinbüchel, A.; Hustede, E.; Liebergesell, M.; Pieper, U.; Timm, A.; Valentin, H. Molecular basis for biosynthesis and accumulation of polyhydroxyalkanoic acids in bacteria. *FEMS Microbiol Lett* 1992, 103, 217–230.

Suzuki, T.; Yamane, T.; Shimizu, S. Mass production of poly-β-hydroxybutyric acid by fully automatic fed-batch culture of methylotroph. *Appl Microbiol Biotechnol* 1986, 23, 322–329.

USDA. *Malaysia: Stagnating palm oil yields impede growth.* Foreign Agricultural Service: Malaysia, 2012.

Venkateswar, R. M.; Venkata, M. S. Influence of aerobic and anoxic microenvironments on polyhydroxyalkanoates (PHA) production from food waste and acidogenic effluents using aerobic consortia. *Bioresour Technol* 2012, 103, 313–321.

Verlinden, R. A. J.; Hill, D. J.; Kenward, M. A.; Williams, C. D.; Piotrowska, S. Z.; Radecka, I. Z. Production of polyhydroxyalkanoates from waste frying oil by *Cupriavidus necator*. *AMB Express* 2011, 1, 1–11.

Wu, Q.; Huang, H.; Hu, G. H.; Chen, J.; Ho, K. P.; Chen, G. Q. Production of poly-3-hydroxy-butyrate by *Bacillus* sp. JMA5 cultivated in molasses media. *Antonie Van Leeuwenhoek* 2001, 80, 111–118.

Zakaria, M. R.; Tabatabaei, M.; Ghazali, F. M.; Suraini, A. A.; Shirai, Y.; Hassan, M. A. Polyhydroxyalkanoate production from anaerobically treated palm oil mill effluent by new bacterial strain *Comamonas* sp. EB172. *World J Microbiol Biotechnol* 2010, 26, 767–774.

Zhang, H.; Obias, V.; Gonyer, K.; Dennis, D. Production of polyhydroxyalkanoate in sucrose-utilizing recombinant *Escherichia coli* and *Klebsiella* strains. *Appl Environ Microbiol* 1994, 60, 1198–1205.

CHAPTER 12

INFLUENCE OF ENVIRONMENTAL FACTORS ON THE PREVALENCE OF POSTHARVEST DETERIORATION OF *RAPHIA* AND SHEA FRUITS IN NIGERIA

OKUNGBOWA FRANCISCA IZIEGBE[1] and
ESIEGBUYA OFEORITSE DANIEL[2]

[1]*Department of Plant Biology and Biotechnology, University of Benin, Benin City, Nigeria*

[2]*Plant Pathology Division, Nigerian Institute for Oil Palm Research (NIFOR), Benin City, Nigeria*

CONTENTS

12.1   INTRODUCTION

Coates and Johnson (1997) stated that losses due to postharvest disease may occur at any time during postharvest handling from harvest to consumption. When estimating postharvest disease losses, it is important to consider reductions in fruit quantity and quality, as some diseases may not render produce unmarketable yet reduce product value. Apart from direct economic considerations, diseased produce poses a potential health risk by mycotoxigenic fungi belonging to some genera such as *Penicillium*, *Alternaria* and *Fusarium*, which are known to produce mycotoxins under certain conditions. Losses due to postharvest disease are affected by a great number of factors including: commodity type, cultivar susceptibility to the causal agents of postharvest disease, postharvest storage environment, maturity and ripeness stage, disease control methods, and handling methods.

Virtually all postharvest diseases are caused by fungi and bacteria. In some root crops and brassicas, viral infections present before harvest can sometimes develop more rapidly after harvest. In general, however, viruses are not an important cause of postharvest disease. The so-called 'quiescent' or 'latent' infections are those where the pathogen initiates infection of the host at some point in time (usually before harvest) but then enters a period of inactivity or dormancy until the physiological status of the host tissue changes in such a way that infection can proceed (Coates and Johnson, 1997). The dramatic physiological changes which occur during fruit ripening are often the trigger for reactivation of quiescent infections. Examples of postharvest diseases arising from quiescent infections include anthracnose of various tropical fruits caused by *Colletotrichum* spp. and grey mould of strawberry caused by *Botrytis cinerea*.

The other major groups of postharvest diseases are those which arise from infections initiated during and after harvest. Often these infections occur through surface wounds created by mechanical or insect injury. Wounds need not be large for infection to take place and in many cases may be microscopic in size. Common postharvest diseases resulting from wound infections include blue and green mould (caused by *Penicillium* spp.) and transit rot (caused by *Rhizopus stolonifer*). Bacteria such as *Drutnia carotovora* (soft rot) are also common wound invaders. Many pathogens, such as the banana crown rot fungi, also gain entry through the injury created by severing the crop from the plant.

12.2   *RAPHIA* PALM

In Nigeria, *Raphia* palms grow wild in the lowland forest region and swamps of the South as well as river courses of the Savannah region of the North (Ndon, 2003). *Raphia* palms are peculiar for their hepaxanthic flowering and so a trunk usually flowers and fruits only once and dies after 3–35 years of vegetative growth (Otedoh, 1985). Economic products of *Raphia* palm include building materials such as bamboos, *Raphia* fiber, thatch, pissava and palm wine. Most of the species are tapped for wine by tapping the young terminal inflorescence (Otedoh, 1982). The wine is now successfully bottled for commercial purposes at the Nigerian Institute for Oil Palm Research (NIFOR), Benin City and other places in Nigeria such as Federal Institute of Industrial Research, Oshodi (FIIRO), Lagos. The palm wine is also used in distilling local dry gin. The trunk of the *Raphia* palm has been strongly recommended for paper making as well as for producing soft tissue paper because of its good quality (Odeyemi, 1984). According to Ndon (2003), the *R. hookeri* palm is the most popular among the twenty species of *Raphia* palm that have been identified. Its advantage over the other species includes its ability to mature between 3 and 6 years and also being able to yield 115–1145 L of palm wine within its life time when compared to the other species of the palm.

Other economic importance of *R. hookeri* fruits as reported by Ndon (2003) includes its oil which can be used for cooking and making of confectionery, the mature and ripe fruit which serves as food for coastal people of Akwa Ibom State, Nigeria. Ndon (2003) also reported that the fruit contains plant growth hormones which can be applied in tissue culture and saponin that can be used to stupefy fish. Literature shows that some saponins are toxic to cold-blooded organisms and insects at particular concentrations. Most saponins, which readily dissolve in water, are poisonous to fish.

The healthy ripe mesocarp of *R. hookeri* fruits have also been reported to posses some phytochemical agents such as phenols, flavonoid, alkaloids saponin oxalate, quinones, and other nutritional components such as moisture, minerals, fat, protein, carbohydrates, and other mineral components which are of beneficial and nutritional purposes to man (Murray et al., 2000; Esiegbuya, 2012, 2013).

12.3   CAUSES OF POSTHARVEST DISEASES OF *R. HOOKERI* FRUITS

There are two major types of postharvest diseases of *R. hookeri* fruits includes the black seed rot (Fig. 12.1) and dry seed rot (Fig. 12.2) diseases.

These postharvest diseases are of importance because they have the ability to affects the scale, mesocarp, and endocarp of the *Raphia* fruits, thus destroying the embryo and thereby making the seed unsuitable for planting (Esiegbuya et al., 2013a, 2013b, 2013c, 2013d). These diseases are of economic importance because the seed is the only means of propagation of the palm. The causes of these postharvest diseases on the *R. hookeri* fruit are as a result of the following:

FIGURE 12.1   Black seed rot (arrow) of *R. hookeri* fruits.

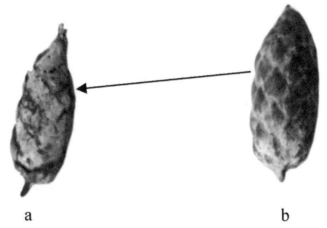

a   b

FIGURE 12.2   Dry rot seedrot disease affecting the scale (a) and mesocarp (b) of *R. hookeri* fruits.

12.3.1 SURFACE WOUND

Surface wound of *R. hookeri* fruits are mainly caused during harvesting when the fruit bunches are allowed to fall to the ground thereby creating wound on the body of the fruit. This allows the entrance of postharvest pathogens. *Chalara paradoxa* causing the black rot postharvest disease of *R. hookeri* palm was isolated from the spaces between the scales of the fruits, from the mesocarp and occasionally from the testa and embryo. According to Oruade and Ekundayo (1992) the pathogen can only penetrate the fruit through wound. The authors also stipulated that the inability of the *C. paradoxa* to penetrate the hard part of the scale and unwounded fruits reveals it to be a weak pathogen. The quick transformation of the *C. paradoxa* from the microconidia to macroconidia with thick walls within 48 h allows the pathogen to survive under adverse conditions. *Chalara paradoxa* and other fungi associated with the postharvest disease of the *R. hookeri* fruits include *Aspergillus niger*, *Fusarium* sp., *Botryodiplodia theobromae*, *Penicilium,* and *Trichoderma*. These are common moulds found in the air and soil and easily contaminate fruits and affect the mesocarp and endocarp of the *Raphia* fruits, thus destroying the embryo and thereby making the seed unsuitable for planting (Esiegbuya et al., 2013a, 2013b, 2013c, 2013d).

Xylaria feejensis the causal agent of the dry seed rot on the other hand posses the ability to penetrate intact fruit but its pathogenicity is also enhanced when the surface of the fruit is compromised (Esiegbuya et al., 2013a, 2013b, 2013c, 2013d). *Xylaria feejeensis* is an ascomycete of Class Sordariomycetes, Family Xylariaceae and Order Xylariales. The ecophysiological features of the Xylariaceae indicate a xerophilous lifestyle of their ancestors (Rogers, 2000) which partly explains the ability of *X. feejeensis* to cause postharvest dry seed rot on the fruit because it has the ability to thrive well on the all parts of the fruit (scale, mesocarp, testa, and embryo) and also forming a mycelia weft around the fruit thus preventing the invasion by other pathogens (Esiegbuya et al., 2013a, 2013b, 2013c, 2013d). As pointed out by Whalley (1996) the Xylariaceae have long been considered to be wood-destroying saprobes, aside from a few facultative tree parasites (Ostry and Anderson, 2009). Saprophytic Xylariaceae are considered to be white-rot fungi, owing to their ability to degrade lignin, but they even can degrade cellulose very effectively (Wei et al., 1992).

12.3.2   VIRULENCE ABILITY OF THE PATHOGEN

The virulence of these pathogens has been shown to be enhanced by environmental factors such as relative humidity and temperature under storage conditions. Locally, harvested bunches of *R. hookeri* fruits are usually left exposed on farmland or along corridors of farmhouse. This unwholesome practice allows conditions for creation of mechanical injuries to the fruits and also proliferation of the storage pathogens. Esiegbuya et al. (2014) reveals that environmental factors such as relative humidity and temperature have been found to significantly favor the prevalence and incidence of the black rot and dry rot diseases of *Raphia* palm caused by *C. paradoxa* and *X. feejensis* in storage. After two week of storage of 500 samples of *R. hookeri* fruits under room temperature, the percentage occurrence of black rot, dry rot, and the uninfected fruit was 48.54, 17.48, and 33.98%, respectively. However, after one and three months of storage the disease incidence of the black rot, dry rot, and healthy increased from 65.37 to 72.81%, 23.95 to 27.18%, and 10.68 to 0%, respectively, for the two storage periods (Figs. 12.1–12.3). The high intensity of these diseases in storage showed that environmental factors (relative humidity and temperature) played a significant role on the disease intensity and severity. The study also showed that the intensity of the black rot disease was more when compared to the dry rot disease. Fungi associated with postharvest diseases are important in that they can cause, among others, the following: (i) loss of viability of seeds; (ii) tainting, leading to decrease in market value of seeds and fruits; (iii) Deterioration of seed in storage; (iv) disease which result in rots of fruits and seeds; and, (v) introduction of new pathogen through fruit and seeds into areas where such pathogen are unknown.

FIGURE 12.3   Healthy *R. hookeri* fruits showing exposed mesocarp (arrow) (Esiegbuya, 2013e).

12.3.3 NUTRITIONAL STATUS OF THE FRUIT

As earlier stated, the ripe mesocarp of *R. hookeri* fruits serve as delicacies in the Southern part of the Nigeria. Comparative study of the proximate and mineral composition of healthy and black and dry rot infected *R. hookeri* fruits caused by the agents of these postharvest diseases showed a significant decrease in all the food components analyzed for, except for ash which showed a slight increase. The decrease in proximate and mineral composition of the fruit, indicate utilization of the food components by the postharvest pathogens and also ensure its survival in the pathogenicity of the fruit (Esiegbuya et al., 2013a, 2013b, 2013c, 2013d). Other authors have also reported on the changes in the mineral contents of fruits as a result of infection caused by microbes (Pathak, 1997). The Ascomycetes are also reported to be able to degrade mineral elements on substrates in which they grow (Lawal, 2011; Lawal and Fagbohun, 2012).

12.4 TRADITIONAL METHODS FOR MANAGEMENT OF POSTHARVEST DISEASE OF *R. HOOKERI* FRUIT

Two methods are commonly used which include fungicide application and storage of fruit endocarp in water. Fungicides such as benlate, dithane M45 and captan have been proposed by Oruade-Dimaro (1990) for the control of the mycelia growth of *C. paradoxa* causing the black seed rot disease of *R. hookeri* fruits. However, due to the risks of fungicides application to man and the environment, the traditional method of storing the seed of the fruits on a basin of water was adopted. This is done by totally removing the scale and the mesocarp from the fruit and the seed is then stored on plastic rubber containing water till the next planting season. The water is changed at intervals. This method is not environmentally friendly because the water serves as a breeding ground for mosquito larva.

12.5 EMERGING TECHNOLOGIES FOR POSTHARVEST DISEASE CONTROL OF *R. HOOKERI* FRUIT

12.5.1 BIOLOGICAL CONTROL

Increasing consumer concerns over the presence of chemical residues in food have prompted the search for non-chemical disease control measures.

Fungicides used before and after harvests are of particular concern because they are applied close to the time of consumption. However due to the high risk of fungicides to humans, animals, environment, and the increasing rate at which pathogens develop resistance against fungicides, new approaches to control the black rot postharvest diseases are currently under investigation. Okogbenin et al. (2014) have proposed the use of biocontrol agent such as *Aframomum sceptrum* for the control of the postharvest black rot disease of *R. hookeri* fruit caused by *C. paradoxa*.

Aframomum species have been reported to be fungitoxic against fungi such as *A. niger, P. digitatum, Helminthosporium solani,* and *Mucor piriformis*, against *E. coli, Klebsiella* spp., and *Salmonella* spp. (Doherty et al., 2010; Chiejina and Ukeh, 2012). The antimicrobial properties of *Aframomum* species reported by some authors are attributed to the phytochemical constituents such as flavonoids, phenolics, tannins, saponin, terpernoids, cardiac glycosides, and alkaloids present in the seeds. The ability of the different extracting solvent of *A. sceptrum* to inhibit the mycelia growth of *C. paradoxa* was attributed to the presence of phytochemical agents such as phenols, reducing sugar, steroids, oxalate, and alkaloids present in the different extracts. These extracts have been reported by Matasyoh et al. (2007) to have the ability to diffuse through the cell membranous structures of fungal cells and cause damage to the cell thereby altering or lowering the physiological activities of the cell.

The presence of phenolic compounds in these extracts indicates that the seed extract of the plants can serve as antimicrobial agents. Phenols and phenolic compounds have been extensively used in disinfection and remain the standard with which other fungicides are compared (Doherty et al., 2010). According to Doherty et al. (2010) alkaloids rank as the most efficient therapeutically significant plant substance. Pure isolated plant alkaloids and their synthetic derivatives are used as basic medicinal agents for their analgesic, antispasmodic, and bactericidal effects (Stary, 1998). They exhibit marked physiological activity when administered to animals. While the potential for biological control of postharvest diseases clearly exists, future success relies on the ability to achieve consistent results in the field and after harvest. It will be necessary to enhance the efficacy of biological control agents against postharvest disease and commercialize the technology involved.

12.5.2 MANIPULATION OF ENVIRONMENTAL STORAGE CONDITIONS

Studies have shown that the causal agents of the black rot of *R. hookeri* can be controlled by manipulation of environmental storage conditions such as temperature, lighting, relative humidity, and pH. The temperature response of the *C. paradoxa* was found to be similar to what is found in most fungi whose temperature optimum lies between 25 and 35 °C (Cochrane, 1958). The optimum temperature for the germination of *C. paradoxa* was found to be between 20 and 30 °C (Esiegbuya et al., 2013a, 2013b, 2013c, 2013d). Dede and Okungbowa (2007) have also reported on the growth of *C. paradoxa* at different temperatures on various growth media, surviving under a temperature range of 15–28 ± 2 °C. San-Juan (1997) and Bachiller (1998) also observed a temperature range of 25–30 °C as the optimum for *C. paradoxa* (date palm and coconut isolates, respectively). According to Esiegbuya et al. (2014) continuous light regime for 24 h was found to slow down the mycelia growth of *C. paradoxa* isolated from *R. hookeri* under ambient temperature. Belli et al. (2005) reported that illumination will sometimes increase or more commonly reduce the rate at which fungi spread across an agar surface. Such effects are sometimes due to the photochemical destruction of components of the medium but in other instances a direct effect on metabolism seems likely. Other isolates of *C. paradoxa* from date palm and coconut palm was also found to grow well at all the three conditions of light but sporulation was much less or nil, under continuous darkness (San-Juan, 1997; Bachiller, 1998). The survival of *C. paradoxa* under a wide range of relative humidity varying from 55–100%, with the relative humidity of 100% producing the highest mycelial growth and 55% slowing down the growth of the pathogen reveals that storing the *R. hookeri* fruits under environmental conditions with low relative humidity, high temperature, and complete lighting system will reduce the postharvest fungal deterioration of the *R. hookeri* fruits (Oruade-Dimaro and Ekundayo, 1992; Esiegbuya et al., 2013a, 2013b, 2013c, 2013d). There is however need to investigate the influence of these factors on the viability of the seed (Figs. 12.4–12.6).

FIGURE 12.4   Incidence of black rot and dry rot on *R. hookeri* fruits two weeks after harvest at room temperature condition.

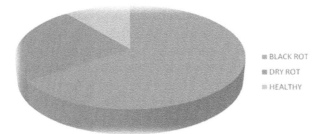

FIGURE 12.5   Incidence of black rot and dry rot on *R. hookeri* fruits one month after harvest at room temperature condition.

FIGURE 12.6   Incidence of black rot and dry rot on *R. hookeri* fruits three months after harvest at room temperature condition (Esiegbuya, 2012).

12.6 SHEA TREE

Shea butter tree (*Vitellaria paradoxa*) is indigenous to Sub-Saharan Africa and belongs to the family *Sapotaceae*. It grows in the wild and has a huge economic and ecological potential. Shea butter is naturally rich in Vitamins A, E, and F (Okullo et al., 2010). Shea butter is widely utilized for domestic purposes such as cooking, skin moisturizer, and commercially as an ingredient in cosmetic, pharmaceutical, and edible products (Alander, 2004). The fruit when very ripe can be eaten raw. Traditionally, Shea butter are used as cream for dressing hair, protecting skin from extreme weather and sun, relieving rheumatic and joint pains, healing wounds/swelling/bruising, and massaging pregnant women and children. It is also used in treatments of eczema, rashes, burns, ulcers, and dermatitis. Lovett (2004) concluded that Shea butter is a high–value export to Europe and the United States, where it is considered a luxury. Maranz and Wiesman (2003) stated that at least 500 million production trees are accessible in West Africa, which equates to a total of 2.5 million tons of dry kernel per annum (based on 5 kg dry kernel per tree).

It is also asserted that over two million people in 13 African countries process the commodity for cash and consumption. Shea butter is mostly processed manually in small villages in Nigeria. Shea butter processing is done by village women, and the method which they use is one passed down through generations. Moreover, there is no estimate of the overall balance between cost of input and the economic output of Shea butter, as the processing is not only arduous, labor-intensive, and time consuming, it also requires large amounts of water and firewood.

Nigeria is the largest producer of Shea nuts in Africa. According to USAID (2010) Nigerian production of Shea nuts in 2002 was 57% of the total Shea nuts production among some West African countries (Ghana, Cote d' Voire, Togo, Benin, Mali, and Burkina Faso). In that same year, Nigerian production was six times more than Ghana production. But the export rate of the Nigerian Shea butter in 2005 was 0% while that of Ghana was 15,000 million tons (USAID, 2004). According to the USAID report, one of the challenges facing the Shea industry in Nigeria is lack of certification of the product and weak organization of the processors. This has lead to the low penetration level of the Nigeria Shea butter in the international markets. The inefficiency of the processing techniques lowers the quality of Shea butter available in the market. Shea butter processing in West Africa involves minimum mechanical input, heavy drudgery and high input of firewood, which has a direct effect on the quality of Shea butter (Carette et al., 2009). The low quality of Shea butter is thus a concern, as it falls below international standard. Consequently, demand is decreasing and the potentials of Shea butter in alleviating rural poverty especially the women and children involved in the processing is dwindling, necessitating an assessment of the processing techniques. In Nigeria, lack of quality control system during the processing of the kernel and Shea butter at the rural end of the chain is one the major challenges of the Shea industry. The quality control of the Shea industry is affected mainly by poor postharvest practices. This is as a result of favorable environmental conditions. Recent research supported by USAID and commissioned by TechnoServe-Ghana, has shown that the first three steps in the post-harvest processing (accumulation of fresh Shea nuts, heating the fresh nuts and drying the kernel) are the critical determinants of kernel quality, for example Free Fatty Acids (FFA), Peroxide Value (PV), and fungal levels. Subsequent steps during extraction, can only 'maintain' quality, which if low, will almost certainly necessitate the need for refining before use in the Western marketplace (USAID, 2004).

12.7   INFLUENCE OF ENVIRONMENTAL CONDITIONS ON QUALITY OF PROCESSED SHEA BUTTER

The quality of fats and oils is dictated by several physical and chemical parameters that are dependent on the source of oil, geographic, climatic, and agronomic variables of growth in the case of plant oils as well as processing and storage conditions. The quality of processed Shea kernels and butter are also affected by some postharvest conditions which are as a result of environmental influence.

12.7.1   TIMELY PICKING OF SHEA FRUITS

The first factor affecting the quality of the Nigeria Shea industry is the delay in picking of the Shea fruits. This results in the fruits being colonized by aflatoxin producing fungi (Atehnkeng et al., 2013). The colonization of Shea fruits by fungi is as a result of the favorable environmental conditions such as temperature and moisture. Since Shea fruits are harvested during rainy season, the temperature and moisture conditions are favorable for fungi to thrive. In order to overcome this challenge, timely picking of Shea fruits from the bush is recommended. However this requires contingent arrangement because of the tedious and clumsy nature of the work in addition to being faced with some challenges such as snakebite, poor visibility, covering long distances, harassment by monkeys, and other reptiles (Nahm, 2011). As a result of the difficulties associated with Shea fruit collection, the processors insist on processing every kernel picked irrespective of its quality status that is a major determinant of Shea butter quality.

Moisture and temperature influence the initiation, development of infectious diseases and succulence of host to diseases in many interrelated ways. It may exist as rain or irrigation water on the plant surface or around the roots, as relative humidity in the air, and as dew. Moisture is indispensable for the germination of fungal spores and penetration of the host by the germ tube. It is also indispensable for the activation of bacterial, fungal, and nematode pathogens before they can infect the plant. Moisture, in such forms as splashing rain and running water, also plays an important role in the distribution and spread of many of these pathogens on the same plant and on their spread from one plant to another.

12.7.2 SHEA FRUIT GERMINATION

The Shea fruit germinates quickly when it falls to the ground. This is because the fruit has no dormancy period; it germinates within few days of dropping from the tree (Jøker, 2000). Environmental conditions such as moisture and temperature are thought to play a role in the quick germination of the fruit. Temperature affects cellular metabolic and growth rates. Seeds of different species and seed from same plant germinate over a wide range of temperatures. Seeds have temperature range within which they will germinate and they will not germinate at temperature above or below this range. Many seeds germinate at temperatures slightly above room temperature, others germinate only in the response to alternation in temperature between warm and cool temperature.

Germination process in seed sometimes may be a nuisance due to the depletion of the target reserves or a strategy to enhance some nutritive constituents of the seed. According to Munshi et al. (2007) the quantity of phospholipids, glycolipids, and sterols in cotyledons and embryonic axes in fast germinating seeds increased progressively between the 1 and 6 days after sowing (DAS) compared with the slow growing seeds. The fatty acid composition in cotyledons of fast growing seeds showed increased levels of palmitic and oleic acids 6 and 8 DAS, while a decline in palmitic and stearic acids as well as accumulation of oleic and linoleic acids were observed in slow growing seeds. On the other hand, Urbano et al. (2004) studied the effects of germination of *Pisum sativum*, for 2, 4 or 6 days, with and without light, on the proteolytic activity, the contents of soluble protein and non-protein nitrogen, and the amount of available starch of *P. sativum* L. as well as their nutritive utilization by growing rats, and concluded that germination of peas for 2 days would be sufficient to significantly improve the palatability and nutritive utilization of protein and carbohydrates from *P. sativum*.

According to Obubizor et al. (2013) depletion of butter in the germinated kernel, was due to the butter being mobilized and consumed during the germination process. This led to depletion in the value of the lipid content of Shea kernel, and elevated the FFA by seven-folds, the PV by 81, while iodine value decreased when compared to the ungerminated. Low FFA improves shelf life of butter. Such butter therefore attracts a premium price. The elevated free fatty value could be attributed to the germination process even though it is a known fact that hydrolysis could proceed via microbial, enzymatic and autocatalytic pathways. According to the author, for maximum butter yield, all the conditions necessary for Shea kernel germination must be eliminated or reduced to the barest minimum and prompt processing of the collected fruit must be considered as an important requirement.

12.7.3   SHEA NUT DRYING/STORAGE

The Shea nuts are usually sun dried for 1–2 weeks and dehusked to obtain the Shea kernel which is further sun dried for another 1–2 weeks. Although the Shea kernels can sometimes be baked to concentrate the oil in the kernel and lengthen the storage period, this has been discouraged because it is a limiting factor to quality of Shea butter. Methods of solar drying on polythene sheeting have been developed in some African countries, but they have limited durability (FAO and CFC, 2005). According to USAID (2004), the Shea kernels can be stored for several years without spoilage by maintaining its moisture content between 6 and 7%. This is so because the drying process inactivates enzymes responsible for the build-up of fatty acids in the seed kernel (USAID, 2004).

According to Esiegbuya et al. (2014) during the drying stage and storing of Shea nuts and kernels, they undergo various forms of postharvest deterioration, namely: nut cracks/holes, nut discoloration, kernel discoloration, and kernel deterioration. The possible source of the above problems in the Shea nuts and kernels is ascribed mainly to poor environmental storage conditions. The poor storage conditions lead to mechanical injury, fungal biodeterioration and discoloration. It also creates opportunity for insect larvae which have the ability to bore holes through the nuts and kernels, thus exposing the latter to microbial contamination.

It has been reported by Agboola (1992) and Mlambo et al. (1992) that species of insects that are associated with various stored seeds in Nigeria and other parts of Africa cause appreciable damage to stored rubber seeds by boring holes in the kernels, after gaining entrance through the micropyle, in the hard testa. Healthy Shea nuts and kernels are coffee brown in color, their discoloration to black/dark color is as a result of the activities of postharvest fungi. According to JICA (2007) Shea nuts will turn black if the nuts are not dry well or if they are wetted by rain or when direct sunlight is not available. Poorly dried or black nuts fetch lower prices in the market than well-dried kernels.

According to Igeleke and Ekpebor (1986) fungi such as *A. flavus* and *niger* growing on Palm kernels are known to impact various colors to the seeds and also increase the free fatty acid content. *Aspergillus niger, A. persii,* and *A. flavus* are common moulds associated with postharvest discoloration and deterioration of stored Shea kernels in Nigeria (Esiegbuya et al., 2014). Members of this group are common and widespread. They have been reported to produce mycotoxins including malformin and naphthopyrones and some strains are known to produce ochratoxin. Members of this species aggregate have been implicated in human and animal infections including superficial and local infections (cutaneous infections, otomycosis, and tracheobronchitis), infections associated with damaged tissue (aspergilloma, osteomyelitis), pulmonary infections, and clinical allergies (allergic bronchopulmonary aspergillosis, rhinitis, and Farmers's lung). However, the majority of infections relate to immunocompromised individuals while the *Aspergillus flavus* strain has been reported to produce both aflatoxins B and G (Klich, 2002).

12.8 TRADITIONAL STRATEGIES FOR CONTROL OF POSTHARVEST DISEASE OF SHEA NUTS AND KERNELS

12.8.1 USE OF STORAGE BAGS AND SACKS

Shea kernels are stored in sacks, woven baskets, and plastic buckets that are stored either in house, granary, or kitchen floors. Sometimes the kernels are hung in houses or kitchens instead of floors. In West Africa, jute bags from cocoa industry are widely used. Over the past decades, polythene bags or sacks have come into wide use for storage of Shea kernels. However, this has been reported to stimulate fungal growth important for quality because they do not allow air circulation (FAO and CFC, 2005). Moreover, because of the recalcitrant properties of Shea nuts, its storage is very difficult (Karin, 2004). The tightly woven plastic mesh also does not allow free circulation of air, and condensation of kernel moisture over a diurnal temperature gradient stimulates development of fungal spores leading to rapid contamination of the stored Shea nuts. The situation is made all the worse by the fact that the bags are often stored directly on the earthen floor of a house (USAID, 2004). Storage of Shea kernels in sack bags, also contributes to the high incidence of the postharvest fungal deterioration of the kernels such as kernel discoloration and kernel deterioration. The deteriorations were as a result of the favorable moisture content of the kernels which was above 10% (Esiegbuya et al., 2014).

12.8.2 PROPOSING BETTER ENVIRONMENTAL STORAGE CONDITIONS FOR SHEA NUTS AND KERNELS

The moisture content most favorable for the storage of Shea nut is 8–12% humidity/0.3–0.6 water activity. The adsorption is measured in conditions of ambient temperature (25 °C) and does not reach a certain value (superior to 0.75). It is advisable to store Shea kernels with a moisture content of $10 \pm 2\%$. This corresponds to an activity of water of 0.3–0.5 (Kapseu and Ngongang, 2002).

12.9 FUTURE PROSPECTS FOR MANAGEMENT OF POSTHARVEST DISEASES OF SEEDS

Due to the high risk of fungicides to humans, animals, environmental and the increasing rate at which pathogens develop resistance against fungicides, and also the difficulty in commercializing the technology of biological control agents, there is the need to develop an environmental growth storage models for pathogens causing postharvest diseases of fruits of high economic value. The environmental factors influencing the growth for a particular spoilage pathogens can be used to develop a storage screen house or box for against the spoilage pathogens. The internal environmental conditions such as humidity, temperature and different lighting conditions of the storage screen house/box can be designed to hinder the growth of spoilage pathogens and also maintaining the viability, physiological, and nutritional status of the fruits.

KEYWORDS

- **Biodeterioration**
- *Chalara*
- **Discoloration**
- **Germination**
- **Mycotoxins**
- **Postharvest quality**
- *Raphia* **palm**
- **Shea butter**
- **Shea tree**
- *Xylaria*

REFERENCES

Agboola, S. D. Technology for small storage of grains in Nigeria. In *Food storage, processing and utilization*. CODIR occasional paper, 1992, pp 22–37.

Alander, J. Shea butter - a multifunctional ingredient for food and cosmetic. *Lip Tech* 2004, 16(9), 202–205.

Atehnkeng, J.; Makun, H. A.; Osibo, O.; Bandyopadhyay, R. *Quality production of shea butter in Nigeria*. International Institute for Tropical Agriculture: Ibadan, 2013; pp 1–52.

Bachiller, N. C. S. J. Effect of environmental factors on the growth and sporulation of *Thielaviopsis paradoxa* (De synes) von Hohnel culture. *Phil J Crop Sci* 1998, 23(1), 37–43.

Bellí, N.; Ramos, A. J.; Sanchis, V.; Marin, S. Effect of photoperiod and day-night temperature stimulating field condition on growth and ochratoxin A production of *Aspergillus carbonarius* strain isolated from grapes. *Food Microb* 2005, 23(7), 622–627.

Carette, C.; Malotaux, M.; Van Leewen, M.; Tolkamp, M. *Shea nut and butter in Ghana, opportunities and constraints for local processing*, 2009, www.resilience-foundation.nl/docs/shea.pdf

Chiejina, N. V.; Ukeh, J. A. Antimicrobial properties and phytochemical analysis of methanolic extracts of *Aframomum melegueta* and *Zingiber officinale* on fungal diseases of tomato fruit. *J Nat Sci Res* 2012, 2(6), 10–15.

Coates, L.; Johnson, G. Postharvest diseases of fruits and vegetables. In *Plant pathogens and plant diseases*, Brown, J. F., Ogle, H. J., Eds.; Printey Armidale: Australia, 1997, pp 533–548.

Cochrane, V. W. *Physiology of fungi*. Wiley: New York, 1958.

Dede, A. P. O.; Okungbowa, F. I. *In vitro* growth of *Ceratocystis paradoxa* in oil palm (*Elaeis guineensis*) fruit extract media. *Micologia Aplicada Inter* 2007, 19(2), 51–55.

Doherty, V. F.; Olaniran, O. O.; Kanife, U. C. Antimicrobial activities of *Aframomum melegueta* (Alligator Pepper). *Inter J Bio* 2010, 2(2), 126–131.

Esiegbuya, O. D. Postharvest fungal deterioration and growth studies on the isolates. MSc Thesis, University of Benin: Nigeria, 2012.

Esiegbuya, D. O.; Osagie, J. I.; Okungbowa, F. I.; Ekhorutomwen, E. O. Fungi associated with the postharvest fungal deterioration of shea nuts and kernels. *Inter J Agric For* 2014, 4(5), 373–379.

Esiegbuya, D. O.; Okungbowa, F. I.; Oruade-Dimaro, E. A.; Airede, C. E. Comparative study of the proximate value and mineral composition of healthy and black rot infected *Raphia hookeri* fruits. *Nig J Mycol* 2013a, 5, 1–7.

Esiegbuya, D. O.; Okungbowa, F. I.; Oruade-Dimaro, E. A.; Airede, C. E. First report of postharvest dry rot of *Raphia hookeri* fruits caused by *Xylaria feejeensis*. *J Plant Path* 2013b, 95(2), 449.

Esiegbuya, D. O.; Okungbowa, F. I.; Oruade-Dimaro, E. A.; Airede, C. E. Dry rot of *Raphia hookeri* fruits and its effect on the mineral and proximate composition. *Nig J Biotech* 2013c, 26, 26–32.

Esiegbuya, D. O.; Okungbowa, F. I.; Oruade-Dimaro, E. A.; Airede, C. E. Mycelial growth response of *Chalara paradoxa Raphia* isolate to various environmental conditions. *Nig J Bot* 2013d, 26(1), 129–134.

Esiegbuya, O. D. Post harvest fungal deterioration and growth studies on the isolates. MSc Thesis, University of Benin: Nigeria, 2013e.

FAO and CFC. *International workshop on processing and marketing of shea products in Africa*. Food and Agriculture Organization of the United Nations, the Common Fund for Commodities and the Centre de Suivi Ecologique: Rome, Italy, 2005.

Igeleke, C. L.; Ekpebor, S. M. C. Fungal species associated with deterioration of rubber seeds during storage. In *Industrial utilization of natural rubber (Hevea brasiliensis) seed, latex and wood,* Enabor, E. E. Ed.; Research Institute of Nigeria: Benin, 1986, pp 135–145.

JICA. The Trial Program for the Shea Butter Industry in Northern Region, 2007.

Jøker, D. *Vitellaria paradoxa* Gaertn. f. *Seed Leaflet* 2000, 50, www.dfsc.dk

Kapseu, C.; Ngongang, D. Overview of post-harvest handling, processing and storage of the shea nut in African countries. *International workshop on processing and marketing of shea products in Africa*, Dakar, Senegal, 2002.

Karin, L. *Vitallaria paradoxa* and feasibility of shea butter project in the North of Cameroon. MS Thesis, University of Montana, USA, 2004.

Klich, M. A. Differentiation of *Aspergillus flavus* from *Aspergillus parasiticus* and other closely related species. *Tran Brit Myco Soc* 2002, 91, 99–108.

Lawal, U. O.; Fagbohun, E. D. Nutritive composition and mycoflora of sundried millet seeds (*Panicum miliacieum*) during storage. *Intern J Biosci* 2012, 2(2), 11–18.

Lawal, U. O. Effect of storage on the nutrient composition and the mycobiota of sundried water melon seeds (*Citrullus lanatus*). *J Micro Biotech Food Sci* 2011, 1(3), 267–276.

Lovett, P. N. The impact of certification on the sustainable use of shea butter (*Vitellaria paradoxa*) in Ghana, 2004.

Maranz, S.; Weisman, Z. Evidence of indigenous selection and distribution of the shea trees (*Vitellaria paradoxa*) and is important significance to prevailing parkland savanna tree patterns in sub saharan Africa north of equator. *J Biogeo* 2003, 30, 1505–1516.

Matasyoh, J. C.; Wagara, I. N.; Nakavuma, J. L.; Kiburai, A. M. Chemical composition of *Cymbopogon citratus* essential oil and its effect on mycotoxigenic *Aspergillus* species. *Afr J Food Sci* 2007, 5(3), 138–142.

Mlambo S. S.; Sithole, S. Z.; Tanyongana, R. Storage facilities in Zimbambwe and loss due to insects. In *Seed Pathology Technical Centre for Agricultural and Rural Cooperation (CTA) Nertherlands*, Mather, S. B., Jorgensen, J., Eds., Seed Pathology Technical Centre for Agricultural and Rural Cooperation (CTA): Netherlands, 1992, pp 109–116.

Munshi. S. K.; Sandhu, S.; Sharma, S. Lipid composition in fast and slow germinating sunflower (*Helianthus annuus* L.) seeds. *Gen App Plant Physiol* 2007, 33(3–4), 235–246.

Murray, R. K.; Granner, D. K.; Mayes, P. A.; Rodwell, V. W. *Harper's Biochemistry*, 25th Edn., McGraw-Hill, USA, 2000.

Nahm, H. S. Quality characteristics of West African shea butter (*Vitellaria paradoxa*) and approaches to extend shelf. MS thesis, The State University of New Jersey: USA, 2011.

Ndon, B. A. *The raphia palm, economic palm series*. Concept Publication: Lagos, Nigeria, 2003, pp 1–156.

Obubizor, J. U.; Abigor, R. D.; Omoriyekemwen, V.; Okogbenin, E. A.; Okunwaye, T. Effect of processing germinated shea kernels on the quality parameters of shea (*Vitellaria paradoxa*) butter. *J Cereals and Oilseeds* 2013, 4(2), 26–31.

Odeyemi, S. O. *Reaction of Raphia hookeri under some major chemical pulping processes*. Finnish Academy of Technical Sciences: Helsinki, 1984, pp 1–159.

Okogbenin, O. B.; Emoghene, A. O.; Okogbenin, E. A.; Esiegbuya, O. D.; Oruade-Dimaro, E. A. *In vitro* control of *Chalara paradoxa* isolated from *Raphia hookeri* (Mann and Wendel) fruits using diethyl ether, acetone and methanol extracts of the seeds of *Aframomum sceptrum*. *Nigeria J Pharma App Sci Res* 2014, 3(1), 15–23.

Okullo, J. B. L.; Omujal, F.; Agea, J. G.; Vuzi, P. C.; Namutebi, A.; Okello, J. B. A.; Nyanzi, S. A. Physico-chemical characteristics of shea butter (*Vitellaria paradoxa* C. F. Gaertn.) oil from the Shea districts of Uganda. *Afr J Food Agric Nut Develop* 2010, 10, 2070–2084.

Oruade-Dimaro, E. A. Preliminary investigation of diseases and disorder of *Raphia*. *N J Palms Oil Seeds* 1990, 10, 96–100.

Oruade-Dimaro, E. A.; Ekundayo, E. A. The biology of *Chalara paradoxa* (Desynes) Sacc. causing fruit rot of raphia palm in Nigeria. *Trop Sci* 1992, 33, 27–36

Ostry, M. E.; Anderson, N. A. Genetics and ecology of the *Entoleuca mammata-Populus* pathosystem: Implications for aspen improvement and management. *Forest Eco Manag* 2009, 257, 390–400.

Otedoh, M. O. A revision of the genus *Raphia* Beauv. *Palmae* 1982, 46, 145–204.

Otedoh, M. O. Flowering and fruiting in raphia palms and the terminology of reproductive parts. *J Nig Inst Oil Res* 1985, 7, 13–29.

Pathak, V. N. Post-harvest fruit pathology: Present status and future possibilities. *Indian Phytopathol* 1997, 50(2), 161–165.

Rogers, J. D. Thoughts and musings about tropical Xylariaceae. *Myco Res* 2000,104, 1412–1420.

San-Juan, N. C. *Etiology and dynamics of the stem bleeding disease coconut (Cocos nucifera L.) in the Philippines*. University of Los Banos, Laguna: Philippines, 1997, pp 1–134.

Stary, F. *The natural guide to medicinal herbs and plants*. Tiger Books International: London, 1998, pp 12–16.

Urbano, G.; Aranda, P.; Vílchez, A.; Aranda, C.; Cabrera, L.; Porres, J. M.; López-Jurado, M. Effects of germination on the composition and nutritive value of proteins in *Pisum sativum* L. *Food Chem* 2004, 93, 671–679.

USAID. Shea butter value chain, production transformation and marketing in West Africa. WATH technical report No. 2, 2004.

USAID. Investing in Shea in West Africa: A U.S. investor's perspective. West Africa Trade Hub: Accra, Ghana, 2010, pp 1–24.

Wei, D. L.; Chang, S. C.; Wei, Y. H.; Lin, Y. W.; Chuang, C. L.; Jong, S. C. Production of cellulolytic enzymes from the *Xylaria* and *Hypoxylon* species of Xylariaceae. *World J Microbiol Biotechnol* 1992, 8, 141–146.

Whalley, A. J. S. The xylariaceous way of life. *Mycol Res* 1996, 100, 897–922.

CHAPTER 13

SOIL REMEDIATION AND ECOLOGICAL RESTORATION FROM HEAVY METAL POLLUTION AND RADIOACTIVE WASTE MATERIALS USING FUNGAL GENETIC AND GENOMIC RESOURCES

JEYABALAN SANGEETHA[1], DEVARAJAN THANGADURAI[2], MUNISWAMY DAVID[3], JADHAV SHRINIVAS[3], ABHISHEK CHANNAYYA MUNDARAGI[2], PAIDI MURALI KRISHNA[3], ETIGEMANE RAMAPPA HARISH[3], PRATHIMA PURUSHOTHAM[2], and SWAPNA KISHOR DESHPANDE[2]

[1]Department of Environmental Science, Central University of Kerala, Kasaragod, Kerala 671316, India

[2]Department of Botany, Karnatak University, Dharwad 580003, Karnataka, India

[3]Department of Zoology, Karnatak University, Dharwad 580003, Karnataka, India

CONTENTS

13.1   INTRODUCTION

Soil executes pivotal role in our environment by providing an active atmosphere for existence of life on earth. A wide array of diverse organisms that are living on earth is haphazardly dependent on soil for growth, development, and survival. Soil is the niche habitat for numerous types of micro- and macroorganisms which assists the processes of biogeochemical cycling, degradation of organic/inorganic wastes and other natural activities that further helps in maintenance of ecological balance. Despite the fact that soil acquires major significance on earth, it is being polluted by number of hazardous activities by human beings. Urbanization and industrialization are the main agencies of environmental pollution. The exudates from industries includes several non-degradable chemicals, oils, toxic materials, synthetic dyes, heavy metals, acidic contents, and radioactive wastes from nuclear energy plants are the main sources that brings a threatening outcomes toward life on earth. In addition to this, the new advancements in the field of agriculture, pharmaceutical, and other sectors are also causing damage to the soil habitats. It was assumed that, if this dangerous practice continues for few more years, there will be a severe destruction may occur in the environment that eventually leads to an irreversible destructive impact on environment and the living creatures on earth.

Metals are said to be the major contaminants of nature. The indiscriminate use of these metallic elements by human beings made excessive release of certain harmful metals like cadmium, nickel, copper, zinc, lead, and other metals into the environment (Dixit et al., 2015). These heavy metals contaminate soil, aquatic environment, as well as food. By its long-term persistence in nature they may cause biomagnification. There is a possibility of occurrence of many harmful diseases in human, animals, as well as plants. In the past decades, many types of uncertain effects were seen by the metal accumulation.

Another major environmental pollution is radioactive wastes from nuclear atomic energy plants. The presence of heavy metals and radioactive wastes in the environment has to be removed in order to obtain healthy nature. In order to have contamination free environment restoration of heavy metal and radioactive is required. To preserve the soil from these kinds of pollutants, new and innovative physical, chemical, and biological methods are being developed which not only aims at removal of pollution, but also aims at long-term management of contaminated sites (Francis and Dodge, 1998). There are various methods which have been designed to reduce the amount of heavy metals and radioactive wastes from the soil. Acidic mixtures of many organic acids were used to treat contaminated soil and to extract the pollutants from it.

In order to overcome this unpredictable impact on living biota, number of experiments is performed. In this direction, some of the microorganisms that are beneficial having application in degrading the heavy metal contaminants have been used and this led the invention of new approach called bioremediation. Some beneficial microorganisms are capable in remediating the impurities of heavy metals and nuclear wastes from the environment. The bacterial species such as *Pseudomonas putida* are used in degradation of oil spills and some of the fungal species are also supports in bioremediation process. The spillage of oil on surface of earth can be decomposed by spreading a layer of mycelium on to the polluted soil. Mycelium in fungi facilitates the mycofiltration of chemicals thereby flowing of chemical pollutants into other habitats could be prevented. The saprophytic nature of these fungal species favors the elimination of organic wastes by decaying them. Use of microorganisms in general, fungi in particular, for pollution control is one of the beneficial approaches, because it is low-cost, low-maintenance biological solution to remediate toxin and heavy metals from the environment.

Biotechnological applications are helpful to obtain microbial strains with high efficiency toward pollution control. The genetic engineering approaches and strain improvement techniques are useful to develop a microbial system with good ability to degrade environmental pollutants. For the sake of bioremediation there requires several genetic systems to be exploited (Sayler et al., 1988; Menn et al., 2000). The technique of protein engineering also gives prominent results in bioremediation. Overall, this chapter provides a complete vision to understand the importance of fungal strains and their role in soil remediation, techniques involved in mycoremediation of environmental pollutants such as heavy metals and radioactive wastes, different sources of fungi to be used for soil remediation and applications of fungal strains for ecological restoration.

13.2   BIOMAGNIFICATION OF HEAVY METALS AND LOSS OF SOIL BIODIVERSITY

Pollution may be defined as an introduction of unwanted and unacceptable compounds in to the natural system which in turn may prevent natural processes in the environment and may have undesirable health effects. In that sense, heavy metal pollution is defined as pollution due to the metal whose elemental density is greater than 5–7 g/cm^3 and more than 23 atomic weight and number and poisonous even at low concentration (Duffus, 2002). Rapid urbanization, industrialization, mining and implementation of synthetic metal containing pesticides in a broad spectrum for pest management systems in developing countries, root up arsenic, cadmium, and lead metal ions in soil (Sherene, 2010). Heavy metals enter into the environment by natural phenomena like volcanoes, weathering, paedogenesis from rock surfaces and anthropogenic activities for human benefits (Jankaite et al., 2008; Mbah and Anikwe, 2010). In terms of soil pollution, anthropogenic causes are intense in terms of heavy metal pollution due to introduction of several contaminants like fertilizer, pesticides, and other xenobiotics. Ecotoxicologists have found in the past that natural resources such as air, soil, and water were polluted by heavy metals in and around of industrial and mining areas through discharging its wastes into water columns like water reservoirs, rivers, lakes, and canals (Natesan and Ranga Rama Seshan, 2011). The subsequent squeezing of heavy metals along with industrial wastes leads the soil and water pollution (Verma and Dwivedi, 2013). Particle matter in air may sometimes get deposited to soil by conveyance of particles from air to soil (Malizia et al., 2012). Geo-accumulation of heavy metal progressively increased in urban soil and road dust in cities (Wei and Yang, 2010).

However, the application of metal composed pesticides passively increased for wild pest's control throughout worldwide. Due to lack of awareness on pesticide compassion and its application limitations, the excessive applied pesticides fetch the toxic metals to non target organism through direct inhalation or water streams (Wuana and Okieimen, 2011). For example, cadmium containing pesticide used in a high range through the world, viz. *Excoecaria agallocha* mangrove species are in first line bio-indicator for heavy metal water pollution; the metal ions zinc, copper, and lead in the root, stem, and leaf of *Excoecaria agallocha* in north east coast of Bay of Bengal in and around Indian Sundarbans mangrove ecosystem (Chakraborty et al., 2014). Cadmium accumulation rising day by day anthropogenic activities such as tobacco smoking, mining, smelting, and refining of nonferrous metals, fossil fuel combustion, incineration of municipal waste (especially cadmium-containing batteries and plastics), manufacture of phosphate fertilizers, and recycling of cadmium-plated steel scrap and electric and electronic waste process magnify Cd in both aquatic and terrestrial ecosystems (WHO, 2010).

The modern urbanization and industrialization progressively increased automobile usage expel toxic metals particles into air in the form of dust, strengthen air pollution and increasing intake of arsenic, cadmium, copper, chromium and lead particles during inhalation. People who live in industrial waste and smoke discharged location are subsequently exposed to airborne lead from combustion of solid waste, coal, and oils, emissions from iron and steel production and lead smelters, and tobacco smoke (Wuana and Okieimen, 2011). Thermal power plants and metal smelters contribute majority of arsenic and mercury in to environment due to burning of arsenic rich coal (Sahu et al., 2012). The continuous intake of toxic air damage airways by surface mucus layer precipitation and stimulate inflammations (Fahy and Dickey, 2010).

Nevertheless, mining activities inflate the socioeconomic life of a country, but the resulted waste effluents are discharged in open area. Mining ore tailings are major source of toxic metals (Guan et al., 2014). The soil erosion is a natural phenomenon during higher rainfalls. The overburden ore tailing leachate runoff into water streams containing metal ions causes water with metal pollution (Reza and Singh, 2010) and accumulation in river sediments (Natesan and Ranga Rama Seshan, 2011; Nachiyunde et al., 2013). Gold mine ore dumps are the major source of toxic metals such as cadmium, chromium, lead, zinc, copper, arsenic, selenium, and mercury which can contaminate the environment through soil and water pollution (Cobbina et al., 2013).

Among developing countries, industrialization due to mining activities and exploitation of soil and water resources stands controversial because of socioeconomic status of the local people (Verma et al., 2012). In some extent mining operations and their waste disposal methods are considered the main sources for the environmental degradation. After mining, the leftover debris like rocky wastes is mainly dumped in waste dumping yards. Overburden deposits generated from mining activities constitute a potential risk to the environment fetching through leachate of potentially toxic elements hosted by a variety of minerals present in the mine-waste materials (Armienta et al., 2003). During metallurgy, ore tailings contain metal ions which are drastically increased while dumped (Jung, 2008). Mining activities change the geochemical nature of crop land, which induces the soil acidification, impairing the soil vegetation and degeneration of aquatic natural life (Christiana, 2012).

Historical mining background countries suffering from heavy metal contamination. For example in India, Gold Mine, Hutti and Greenstone Belt, Mangalur in Karnataka discharged dumps in open area causes heavy impact on agricultural lands and leads to the magnification of heavy metals in around mining areas since long many centuries (Chakraborti et al., 2013). Mine ore tailing show high nutritional quality and clay, hence which is advised in compost, school play ground preparations. While in some extent used in concrete composition mixture, enhance the flexible strength of concert (Skanda Kumar et al., 2014). Arsenic and cadmium content in school children toes and nails were reported in playing school ground prepared by mine ore tailings (http://www.healthyschools.org). The Environmental Protection Agency classified chromated copper arsenic (CCA) wood chips as hazardous wastes. This possessed an incredible 813–1654 ppm arsenic (www.health.usnews.com).

Industrial wastes contain good nutritional values and also high percentage of metal ions. Due to high nutritional quality farmers used these soils and liquid wastes as compost preparation and irrigation for cultivation of common vegetable (bulbs). The consequent application of these industrial wastes leads biomagnifications of metal ions in green vegetable *Coriandrum sativum*, Spinach (*Spinacia oleracea*) and *Amaranthus* sp. (Chiroma et al., 2014). Plants were grown in metal polluted sites escort the bioaccumulation of metal ion content in root, leaves and stem and also elevation of low molecular antioxidants such as ascorbate, glutathione, and flavonoids (Geneva et al., 2014).

Metal magnification in soil reduces the bacterial growth rate and diversity. Survival and colony size of several burrowing insects like ants is altered due to disturbances in ecological aspects of soil (Grześ, 2010). Toxic metal ions induce stress on cop plants by elevation of free radicals. Lead contamination in capsicum cultivation, despite a reduction in the plant growth and chlorophyll content and elevation of malondialdehyde (MDA), super oxide dismutase (SOD) and proline contents corresponding to the concentration of the metal ion (Britto et al., 2011). The aquatic metal pollution may alter the natural physiology of aquatic lives (Baby et al., 2010; Zaki et al., 2013). Metal toxicity to fresh water is considered to regulate chemical communication between freshwater habitants and this alteration in chemical communication could further devastate relationship between two species in an ecosystem (Boyd, 2010).

Heavy metal soil pollution reducing soil enzyme activity is a major cause of depletion of soil flora and fauna (Jose et al., 2011). The toxic metal ion transferred into food web in the form of biomagnification and through drinking water shows serious problems in vertebrates' growth and sexual maturations. It has been also reported that heavy metal toxicity could lead to endocrine disruption and DNA expression by molecular signaling. For example Zn fingers of the estrogen receptor, Zn can be replaced by several heavy metal molecules such as copper, cobalt, Ni, and Cd activities in mice (Georgescu et al., 2011). Cadmium chronic exposure induces free radical concentration, rennin angiotensin system abnormalities (RAS), microalbuminuria and Na^+

K^+ATPase depression in human being (Gulati et al., 2010). Heavy metal pollution could cause epigenetic modification in DNA mechanisms (Rzymski et al., 2015). According to the study of Benbrahim-Tallaa et al. (2007), Cd induced prostate epithelial malignant transformation through DNA hypermethylation at the global and gene specific levels. People who are living and predominantly exposing to cadmium, lead, and mercury metal contaminated location suffering from infertility (Joffe, 2003).

Aquatic toxic metal contamination may alter the fish physiology through elevation of free radicals, vital organelle cellular damage, and some extent, over exposure leads to death (Annabi et al., 2013). Metallothionein protein accumulation is a hall marker of metal exposure in fishes (Zaki et al., 2013) due to higher sensitivity of fishes and hence fishes are considered as bioindicator of effect of heavy metals (Authman, 2015).

Phytoinhalation is ubiquitous detoxification strategy in many plants at heavy metal stress conditions, elevation of glutathione (GHS) and phytochelatin synthase (PCS) enzymes activities are major indication of metal contamination which induces physiological stress. Metals like Al, Cu, Zn, and Fe abundance in soil influence on hyper activation of ascorbate oxidase (ASO), guaiacol peroxidase (GPX), ascorbate peroxidase (APX) and superoxide dismutase (SOD) activities and alter the plasma membrane phospholipids composition in *Zygophyllum* sp. (Morsy et al., 2012). Considerable reduction in chlorophyll, sugar, and protein contents were observed at road sites receiving higher toxic load is a significant bioindication of heavy metal pollution (Rai and Panda, 2015). The decline yielding capacity of common commercial vegetable cultivation in metal contaminated sites passively increases the accumulation of heavy metals. For example, plants (*Matricaria recutita*) grown in Cd, Pb, and Zn contaminated sites, the elevation account of APX in the above ground parts, glutathione peroxidase and guaiacol peroxidase in the leaves and dehydroascorbate reductase and glutathione-S-transferase in the flowers are for the toxic tolerance (Popova et al., 2012; Nadgorska-Socha et al., 2013).

The rapidly growing and uncontrolled population creates scarcity of essential natural resources which has become a major problem and shows significant impact on natural ecological degradation (Jahan, 2008; Anand, 2013). There are huge findings on heavy metal pollution and its toxicity throughout the world (Authman, 2015). Especially in developing countries lack of awareness on metal and its toxic potential in physiological life cycle, people are consequently exposing and applying excessively to the croplands. The anthropogenic activities are becoming essential for saturation of livelihood. In overall the above activities, our mother planet (Earth) is going to be contaminating with various pollutants. Hence, this is the time to raise the alarm on "pollution and toxicity" and early implementation of eco-friendly things and bioremediation, bioleaching and biodegradation on pollution affected locations.

13.3   RADIOACTIVE WASTE: POTENTIAL SOIL, AIR AND WATER POLLUTANT

Waste that emits harmful level of nuclear radiation is radioactive waste. None of the place on the Earth is free from natural radioactive background and every inhabitant on this earth is constantly exposed to both natural and artificially occurring ionizing and non-ionizing radiation. Cosmic rays from the celestial body, radioactive minerals widely distributed in rocks, sediments and soils, radionuclide's normally incorporated into our body's tissues, and radon and its products are the sources of natural radiation (Vogiannis and Nikolopoulos, 2015). Also exposed to ionizing radiation from man-made sources, mostly through medical procedures like X-ray diagnostics, civilian war, nuclear industries and nuclear explosion tests especially when carried out in the atmosphere, leakage of radiations from nuclear reactors, and other nuclear facilities (Rao, 2001).

Radioactivity was discovered over several years beginning with the discovery of X-rays in 1895 by Wilhelm Conrad Roentgen and continuing with such people as Henri Becquerel and the Curie family (http://www.nobelprize.org). After the Second World War, radioactivity was artificially and intentionally been introduced into nature through modern warfare and nuclear war head testing contaminating air, soil, and water bodies. Further, civil nuclear program for clean energy and biomedical research has its fare share in contamination (Rao, 2001; Shruti, 2010).

The nuclear residual waste from natural sources, from past mining, and its related operations, produced more than 1000 EBq/year of radioactive waste in the atmosphere and it is estimated to be around several million tons at many places and its radioactivity more or less equal to 0.001 EBq. Thousands of such sites are scattered all over the world. Exploiting nuclear energy in civil nuclear program has contributed major nuclear wastes in past 50 years, estimating 1000 EBq and is still on the growth by around 100 WBq/year (Gonzalez, 2000). Approximately 30 tons of high-level radioactive waste is generated by a mega nuclear energy plant with a capacity of 1000 MW electricity. A total of 440 tons of intermediate-level and 460 tons of low-level radioactive waste. While the coal fired power plant release nearly 400 tons of heavy metals and six million tons of greenhouse gases, 500,000 tons of mixture of sulfur and nitrogen oxides and about 320,000 tons of ashes to generate 1000 MW electric power (Rashad and Hammad, 2000). The ashes generated by thermal plants have the potential to pose humanity into the risk higher than that of collective dose of radiation due to nuclear waste from nuclear plants. Yet common perception among people is against those nuclear electric plants in several countries (Rao, 2001).

13.3.1 POTENTIAL CAUSES OF RADIOACTIVE WASTES

Radioactive soil pollution is commonly defined as excessive presence of radiation in soil system due to obvious anthropogenic or certain natural causes. Such solid waste dumping has become abundant landfill because of the indiscriminate discarding of solid waste on the earth. Nuclear industries, nuclear explosion tests, and nuclear bomb attack especially when carried out in the atmosphere are responsible for increasing the background level of radiation throughout the world and major cause of radiation pollution in soil. During atmospheric nuclear explosion tests, a number of long-lived radionuclides are released into the atmosphere (Simon et al., 2006). This radioactive dust (also known as radioactive fallout) gets suspended in air at a height of 6–7 km above the earth's surface and is dispersed over long distances by winds from the test site and some of the radioactive isotopes given off during nuclear test which affects the human body. The best example of fallout is the nuclear bomb attack on Hiroshima and Nagasaki, Japan in 1945 by United States of America during World War II. The radiation from the blast was so intense that it would leave shadows of the materials in way and as many as 2,25,000 people have lost their live in later five years of the blast due to radiation (John and Pastore, 1987; Simon et al., 2006).

Naturally occurring radioisotopes such as radon-222, potassium (K-40) and carbon (C-40), uranium, thorium, and radium are found in soil in small quantity is another source of radiation pollution in soil. Radiation due to Potassium-40 is the sole source of pollution for all potassium containing systems in the soil. Crops grown on such soil contain radioactive elements like carbon-14 and potassium-40 (http://mragheb.com). The unusable and unwanted waste products from nuclear industries, atmospheric fallout of the radioactive waste and radioisotopes from the natural sources may contain radionuclide's, often settle down by rain and get mixed with soil and water. Through bioaccumulation and biomagnification, it passes on from organism to organism finally reaching to human, where it can cause serious health hazards (Eisenbud and Gesell, 1997).

13.3.2 RADIATION POLLUTION AND ITS TOXIC EFFECTS

Whether or not, artificial and natural, the potential risk of radiation to human beings is noteworthy. The effect of these radiations was first reported in early Twentieth Century when person working in uranium mines suffered major health problems including cancer (Rao, 2001). The effects vary from organism to organism, level of radioactivity of nuclear isotopes and how much and how fast a radiation dose is received. On contradiction, it was fascinating to hear, some workers have reported low doses of radiations are in fact beneficial to organisms in terms of increased longevity, better disease resistance, and greater life span. But higher doses of radiation effect studies came only after Hiroshima attack in 1945 and people received major doses of radiation (Krishnan, 2014). About 12% of all the cancers that have developed among those survivors are estimated to be related to radiation effect (Jane and Orient, 2014). While, a chronic dose is a relatively small amount of radiation received over a long period of time may induce somatic and genetic effect which include the development of eye cataract, cancers, and genetic or heritable effects appear in the future generation. An acute radiation dose (a large dose delivered during a short period of time) may result in effects which are observable within a period of hours to weeks (Strom, 2004).

Radioactive particles in its most threatening form, forms ions which it reacts with biological molecules and has potential to damage genetic material of all organisms. Free radicals are then formed in the process which will start scavenging on biological materials like proteins, carbohydrates, and fats. A longer exposure to radioactive radiations can damage the DNA cells that results in cancer, genetic defects for the generations to come and even death (http://www.cna.ca).

13.3.3 BIOREMEDIATION OF RADIOACTIVE WASTES

Radioactive pollution in soil is defined as any site that has been exposed to naturally occurring or artificial radionuclides which may ultimately cause health and environmental hazards. It is imperative that radioactive waste is proven to be a severe threat toward environmental and human health (Kumar, 2004). Hence, there is an urgent need to develop efficacy management with respect to radioactive pollution management. According to Evrard (2012) there are three aspects in radioactivity management policies; site management according to the future perspective or anticipated uses; to record the past pollution hazards and remediation; and provide up to date information to the public with respect to concerning associated hazards.

Eight years after the publication of Controle Magazine devoted to the management of sites contaminated by radioactive substances, DGSNR (General Directorate for Nuclear Safety and Radiation Protection) came in force and ASN was tasked with the management of sites contaminated by radioactive materials and it became noticeable that an initial inventory of national and international practices was needed, in order to identify the major obstacles and the changes that were required. Hence, ASN organized its first national symposium on "Radioactive contamination: how to deal with polluted sites?" jointly with Ministry of Ecology on May 4, 2004 and also drafted the First National Radioactive Material and Waste Management Plan during 2006 for appropriate management of radioactive pollution. Again in 2007, ASN created National Commission for Assistance in Radioactive Field (CNAR) and recently launched operation radium diagnosis guide concerning the management of sites potentially polluted by radioactive substances (Evrard, 2012).

Management of polluted sites has gone from surveying the sites to more global aspects in past 20 years. This global approach allows faster and more sustainable management of the sites, by involving all the stakeholders as early as possible in the polluted site management process. Various measures were however initiated, to allow effective management of radioactive wastes which involves segregation, characterization, handling, treatment, conditioning, and monitoring prior to final storage or disposal even more transparent and efficient management of these forms of pollution (Wattal, 2013). However, mycoremediation is considered recently as an eco-friendly way of treating hazardous radioactive pollutants and is gaining attention nowadays and also it offers an efficient and cost effective way to treat contaminated ground water and soil.

13.4   FUNGAL GENETIC AND GENOMIC RESOURCES

Many fungi survive in extreme environmental conditions. These are recognized as extremotolerant fungi. Regrettably, the mechanisms involved in their ability to withstand harsh conditions against the abiotic and biotic stress have been poorly understood. Nevertheless, these fungi have evolved with exceptional resistance power to support their existence. Hence these may perhaps act as useful resources for developing novel compounds and enzymes for bioremediation. It is very important to analyze the genes responsible for the sorption and specificity to heavy metals. Stress-tolerant fungi such as extremophiles have evolved with a resistant power to withstand harsh environments and could be novel genetic resources for the decontamination of heavy metal polluted environment. The extremophiles can be further classified in to thermophiles (40–50 °C), psychrophiles (16–20 °C), acidophiles (pH 1.0–4.0), alkalophiles (pH>9.0), xerophiles or osmophiles (water activity (aw) of $0.85 \geq 0$ such as *Wallemia sebi*) and halophiles (tolerant to high salt concentrations such as *Wallemia ichthyophaga*) (Padamsee et al., 2012; Zajc et al., 2013). These extremotolerants indeed could be useful genetic resources for the detoxification or biosorption of heavy metals.

Certain fungi inhabiting soil are exposed to harmful radiations such as the ionizing and non ionizing radiations and these fungi are termed as pigmented fungi which have dark pigmentation due to the protective nature of fungi against radiations. According to Dadachova and Casadevall (2008), these fungi utilize the radiations as source of nutrition for their growth. Moreover, pigmented fungi such as *Cladosporium cladosporioides* and *Paecilomyces lilacinus* have been found at the site of Chernobyl Nuclear Power Plant (Zhdanova et al., 2004). These extremophiles can be hopeful of potential genetic resources. With powerful and novel functional screening strategies extremotolerant fungi could be a promising new opening for the identification of stress tolerant genes (Li et al., 1997; Trincone, 2011).

In the last two decades, with the advent of genomics and proteomics, an unprecedented growth has been achieved in understanding and characterizing fungal biodiversity and fungal biotechnology (Giaever and Nislow, 2014; Bianco and Perrotta, 2015). The power of fungal genetics and genomics has been transformed drastically due to the availability of mutations mapped and defined by available advanced genetic tools (Magee et al., 2003; Nagy et al., 2003). *Saccharomyces cerevisiae* is a best experimental model for eukaryotic studies. Genomic studies help us to better understand the complex biotrophic interactions and also in sustenance of natural environment. This was accomplished with the advent of two pioneering sequencing technologies. Maxam–Gilbert method and the Sanger method are the two sequencing tools developed in the 1970s. Sanger sequencing be the most followed standard method for genome sequencing in the past and future few decades too. Now, next generation sequencing technologies is being employed in DNA sequencing such as the Pyrosequencing (454), Illumina sequencing (Solexa), ABI SOLiD sequencing and the single molecule sequence (Helicos

HeliScope) (Zhang et al., 2011). Goffeau et al. (1996) decoded genome sequence of the yeast *S. cerevisiae* in their classical publication entitled "Life with 6000 genes", which was the first work to be reported on fungal genomics. Furthermore, the first ectomycorrhizal genome to be sequenced was *Laccaria bicolor* in 2008 (Martin and Selosse 2008). The genome of fungus *Tuber melanosporum* is the largest fungal genome (125 Mbp) decoded till date (Martin et al., 2010). Genome annotations of model fungi (*S. cerevisiae*, *Aspergillus nidulans*, *Neurospora crassa*, and *Schizosaccharomyces pombe*) lead to gene discovery and identification of novel gene activities (Seiler and Plamann, 2003; Galagan et al., 2005).

Though genomics era has begun in last decade, in upcoming years, it will pave the way in creating and understanding complex fungal pathways and bioprocesses for producing biofuels and biochemicals (Nevoigt, 2008; Dellomonaco et al., 2010). Comparative genomics is one such approach which deals with the study of two or multiple genomes, comparing them either through nucleic acid sequence or protein sequence to map gene positions (Koonin and Galperin, 2003). Until now more than 100 fungal genomes have been sequenced and many more are in progress with the target to achieve 1000 fungal genomes in near future; particularly, those fungi that are very important in industrial processes, bioenergy, and medical sciences. Major fungal genomic research institutes such as the Broad Institute, the National Human Genome Research Institute (NHGRI) and the U.S. Department of Energy's (DOE) Joint Genome Institute (JGI) are concurrently involved in the decoding of genomes of many important fungi. JGI's fungal genome project (FGP) is the major program which is involved in the extensive research on fungal genomics. This project has collaborated with major research institutes across the globe. In contrast, one of the key initiatives of FGP, the Genome Encyclopedia of Fungi is working to track down the genomes of fungi that are considered to be important in three thrust areas, plant feedstock health, biorefinery and fungal diversity. The sequenced genomes are deposited and maintained in the database MycoCosm (http://www.jgi. doe.gov/fungi) which is a fungal genomics resource portal that promotes users in submission, annotation and analysis of vast fungal genome data (Grigoriev et al., 2014). The following fungi have already been sequenced by NHGRI in the last decade: *Aspergillus nidulans, Batrachochytrium dendrobatidis, Candida albicans, Candida tropicalis, Candida guilliermondii, Candida lusitaniae, Chaetomium globosum, Coprinus cinereus, Coccidioides immitis, Cryptococcus neoformans* Serotype A, *Cryptococcus neoformans* Serotype B, *Geomyces destructans, Histoplasma capsulatum, Lodderomyces elongisporus, Myceliophthora thermophila, Pneumocystis carinii, Podospora anserine, Rhizopus oryzae, Schizosaccharomyces japonicus, Schizosaccharomyces octosporus, Saccharomyces cerevisiae* RM11 1A, *Ustilago maydis*, and *Unicinocarpus reesii*. High-throughput genomic assays, has enabled and resulted in understanding structure, functions, complex mechanisms, and variations among different genomes, genes of interest and their expression in model organisms and gene regulation in developmental processes.

Fungi are the most promising tools for biosorption as they have distinctive properties to sequester the toxic heavy metals from environment (Hossain et al., 2005; Ji et al., 2005; Romero et al., 2005; Baxter and Cummings, 2006; Monrroy et al., 2007). Yeast and filamentous fungi are able to bind heavy metals because of the special cell wall structure that they posses. Chitin is a major component in framework of cell wall composed of polysaccharides such as glycans, chitin, chitosan, mannans and phosphormannans. Though there are several established methods or chemical processes to reduce the heavy metal pollution and toxicity the fungal biosorption is quite accepted and popular as it is efficient, eco-friendly and cost effective (Rao and Bhargavi, 2013; Saraf and Vaidya, 2015). The mechanisms involved in heavy metal adsorption are methylation, reduction and dealkylation (Gadd, 1993). Recent studies indicate that the removal and recovery of toxic heavy metals has been successful by the innovations in the bioresource and bioprocessing technologies, many fungal strains have been used to recover these toxic heavy metals in lab scale bioreactors within short period of time. The approach is astonishing as the fungi used is the waste biomass from any food and pharmaceutical industries, the dead biomass absorb the heavy metals and recovery is also thought to be simple and effective. Many fungal strains and biomass has been tried out to assess their potentiality to sequester the heavy metals but still there are certain limitations such as the experiments have been carried out in the lab scale wherein still the *in situ* application is lacking and also more research has to be carried out in mathematical modeling, kinetics and suitability toward the environmental factors. These limitations could be overcome by adapting modern technologies and genomic techniques.

The enzyme Phytochelatin synthase (PCS) is a 95 kDa tetramer belongs to dipeptidyl transferases that was first characterized by Grill et al. (1989). PCS ((γ-Glu-Cys)n-Gly) catalyzes the synthesis of phytochelatin from GSE. Phytochelatin (earlier known as cadystins) are structurally similar to GSE. These are heavy metal binding peptides that protect cells against heavy metals toxicity. PCS genes from fungi have been characterized in species belonging to Ascomycota and these genes are sparsely distributed in most of fungi. Heavy metal cations such as Cd, Cu, Zn, Pb, and anions like nickel, mercury, and arsenate are the causes for activating PCS in cells and tissues. The molecular weights of various PCS enzymes deduced from DNA sequences range from 40–70 kDa and the genes have been characterized in the following fungi and other organisms: *Schizosaccharomyces pombe* (SpPCS), *Dictyostelium discoideum* (DdPCS1), *Arabidopsis thaliana* (AtPCS1), *Triticum aestivum* (TaPCS1) and *Caenorhabditis elegans* (CePCS1) (Cobbett, 2000a). Together, these observations indicate that PCS is primarily regulated by induction of heavy metal cations such as zinc and cadmium. The best activator tested was Cd, followed by Ag, Bi, Pb, Zn, Cu, Hg, and Au cations. PCS expressed in *E. coli* or in *S. cerevisiae* was activated to varying range of Cd, Cu, Ag, Hg, Zn, and Pb ions. Shen et al. (2015) reported PCS in *Sporobolomyces* sp. strain IAM 13481 belonging to *Pucciniomycotina* subphylum of the *Basidiomycota*.

Metallothioneins (MTs) are low molecular weight cysteine-rich polypeptides which bind heavy metals. MTs are widely distributed among living organisms and involved in heavy metal tolerance of many eukaryotes (Kagi and Nordberg, 1979; Butt and Ecker, 1987; Bellion et al., 2007). Margoshes and Vallee (1957) first reported purified MTs from horse kidney, who was working with bioregulation of cadmium and zinc. Pulido et al. (1966) isolated MTs from human liver. Since then, MTs have been isolated from different organisms including many fungi and several groups have reported copper-inducible protein from fungi. Depending upon their distribution in cell they have been characterized into three subfamilies such as cytosolic, microsomal, and mitochondrial GST or kappa-class GST subfamilies (Sheehan et al., 2001; Hayes et al., 2005; Frova, 2006; Shen et al., 2015). Based on amino acid/nucleotide sequences, immunological, kinetic, structural properties and other aspects, they have been further classified in to nine classes of cytosolic GSTs in fungi: GTT1, GTT2, Ure2p, MAK16, EFB1, GSTFuA (Mathieu et al., 2013; Thuillier et al., 2013), phi (Morel et al., 2013), omega, and glutathionyl hydroquinone reductase (GHR) (McGoldrick et al., 2005). Lindegren and co-workers for the first time demonstrated that a single gene was responsible for copper sensitivity (cupl') and copper resistance (CUP1"), in which MTs play key role in Copper (Cu) absorption, storage and homeostasis. *Saccharomyces cerevisiae* contain a single Cu-MT gene present in a CUP1 locus (Fogel et al., 1982) that encodes a Cys-rich protein which results Cu sorption in the fungi. This ability to bind might be due to Cu ions chelation, which is determined. Hence, high CUP1 expression levels result in increased Cu-binding capacity (Butt et al., 1984). More recently, MT-encoding gene from the AM fungi such as *Pisolithus tinctorius*, *Gigaspora rosea* and *Gigaspora margarita* have also been reported (Stommel et al., 2001; Voiblet et al., 2001; Lanfranco et al., 2002). Bellion et al. (2007) reported Pimt1 gene coding for a metallothionein from the ectomycorrhizal fungus *Paxillus involutus*.

Glutathione S-transferases (GSTs) earlier known to be ligandins are the metabolic isozymes that are involved in detoxification of xenobiotics and heavy metals by catalyzing their conjugation to GSE (sulfur-containing tripeptide thiol). Earlier reports suggested that the formation of the GSH–Cd conjugate seems to regulate cadmium absorption in *S. cerevisiae* cells (Gomes et al., 2002). In *Saccharomyces cerevisiae* Yeast Cadmium Factor 1 (YCF1) and in *S. pombe* Heavy-Metal Tolerance 1 (Hmt1) genes encode vacuolar ATP-binding cassette membrane proteins that transport bis-(glutathionato)-cadmium and PC–Cd conjugates in to vacuole, play significant role in cadmium detoxification (Ortiz et al., 1995; Li et al., 1997; Cobbett, 2000b). Yeast such as *S. cerevisiae* protects itself from non-essential elements such as cadmium would help to understand the mechanisms of metal ion detoxification. Adamis et al. (2004) investigated that two genes, GTT1 and GTT2 of *S. cerevisiae* conferred tolerance to cadmium absorption than the control strain, and proposed that the formation of the cadmium-glutathione complex is dependent on that transferase. The technique reverse Northern blot to assess molecular response of GSTs in arbuscular mycorrhizal fungi *Glomus intraradices* has been well-demonstrated (Waschke et al., 2006). Fraser et al. (2002) reported gene from *Aspergillus nidulans* that is similar to URE2 gene of *Saccharomyces cerevisiae* encodes a GST's that accounted in heavy metal and xenobiotic resistance. In a similar study twenty-four GST genes from the transcriptome of a metal-tolerant *Exophiala pisciphila* a dematiaceous fungus also known as dark septate endophyte (DSE) closely related to the heavy metal tolerance has been documented (Shen et al., 2015). Vallino et al. (2005) investigated the genetic basis of heavy metal tolerant ericoid mycorrhizal species *Oidiodendron maius*, isolated from roots of *Vaccinium myrtillus* growing in soil heavily contaminated with high zinc concentrations. Martino et al. (2007) studied genetic transformation of metal tolerant *Oidiodendron maius* and proposed that genetic transformation could be useful in understanding responsible genes and potential application in bioremediation strategies.

The potential use of fungi for the detoxification or biosorption of heavy metals in polluted environments is being increasingly exploited. Understanding the phytochelatins (PTs), metallothioneins (MTs), and glutathione (GSE) genes and their biosynthetic pathway on metal tolerance and sorption will soon lead to indications as to their advantageous in this attempt.

The mechanisms by which cells defend themselves against the heavy metals involve complex biosynthetic pathways. The structural and functional similarities of genes in microorganisms suggest that understanding molecular mechanisms and how fungi detoxify or biosorp heavy metals would aid to develop potential resource for bioremediation of heavy metal contaminated sites (Hossain et al., 2012).

13.5 POTENTIAL USE OF FUNGI FOR POLLUTION ABATEMENT

Fungi are unique group of microorganisms and occur ubiquitously in nature. From the ancient days to till today, fungi occupy a unique place in human life. They have a wide range of applications from cookery to industry. Fungi being used for multipurpose, they are considered as true and natural ecosystem engineers (Jones et al., 1994). Fungal cells have totipotent nature and hence they can reproduce for unlimited period. The concept of degrading unwanted, harmful pollutants by fungi was thought by observing decomposing nature of fungi by spreading and decaying materials such as wood, papers, textiles, leather, and other various wastes. Decomposition of polythene by the fungal species such as white rot fungi, enzymes from *Aspergillus flavus* and *Mucor rouxii* NRRL 1835 and *Penicillium simplicissimum* were used (Singh, 2006). In addition fungi are proved to be the well decomposers of effluents from dye industries, paper and pulp industries. In order to achieve complete degradation of pollutants by the use of fungi, it is necessary to understand the decomposing nature of fungi, fungal byproducts, effects after fungal decomposition and availability of fungi being used.

The process of absorption of heavy metals by fungi can be termed as mycosorption. This process has gained importance in environmental protection and metal recovery. Application of fungal strains to attain biosorption of polluted materials becomes one of the recent trends in area of environmental pollution control. Process of mycosorption is a pseudo-ion-exchange method, in which the metal ion exchanged for a counterion in biomass (Singh, 2006). The filamentous and aquatic fungi possess more affinity and capacity to absorb heavy metal contaminants from the environment. This approach of heavy metal uptake was described for the first time by Michelot and further demonstrated by cultivating mushrooms to achieve bioaccumulations. Some of the marine fungi namely, *Corollospora lacera* and *Monodictys pelagica* are identified to accumulate lead and cadmium extracellular in mycelia (Taboski et al., 2005). The process of biosorption and metal recovery can be increased by providing a stirrer magnetic field externally (Gorobets et al., 2004). Ion exchange, adsorption, chelating, crystallization, precipitation, and entrapment are few mechanisms involved in biosorption of heavy metals. A detailed account on the aspect of biosorption of heavy metals using fungi is discussed in this chapter.

In addition to heavy metals, the radioactive wastes from nuclear energy plants became a major threat to the environment. The flow of radioactive pollutants into the environment is a dangerous criterion which affects the human health and other living organisms on earth. Not only nuclear atomic energy plants, but environment also polluted by the use of nuclear and atomic weapons during wars. U, Th, Pu, ^{90}Sr, ^{135}Cs, and Tc are the major radioactive contaminants of environment. Some fungal strains have ability to colonize and remove these radioactive contaminants from the soil. Most interestingly, mushrooms are useful in detaching the radioactive wastes from the environment.

13.6 MYCOACCUMULATION OF XENOBIOTICS

Traditionally, chemical recovery processes are followed to extract heavy metals from environments, but the process is considered expensive and ineffective (Anahid et al., 2010). However, most common biological methods of remediating recalcitrant pollutants from environment are by the use of soil microbes including symbiotic fungi and/or arbuscular mycorrhiza alone or in association with plants. These fungal species have a greater potential in increasing plant uptake of heavy metals. Gonzaga et al. (2006) have reported mycorrhiza as a potential candidate in bioremediation strategies for heavy metals as they act by elevating the plant's ability to accumulate phosphorous and other static elements. It is a well-established fact that mycorrhiza are the integral and functional part of plant root systems. These fungi either act by attenuating plant heavy metal toxicity or by increasing the metal uptake. Weissenhorn and Leyval (1995) observed and reported toxic levels and higher accumulation of heavy metals in plants as a result of arbuscular mycorrhizal colonization. On the contrary, Heggo et al. (1990) reported less or no metal accumulation in plants as a result of mycorrhizal colonization.

Anahid et al. (2010) studied different species of fungi in bioleaching of ores and spent wash containing toxic metal pollutants (Mo, V, Mn, W, and Zn). The study revealed that *Penicillium simplicissimum* was highly resistant capable of withstanding up to 8000 ppm concentration as compared with different isolated species. Abdel-Aty et al. (2013) investigated biosorption behavior of Pb and Cd to blue green *A. sphaerica*. They concluded stating mechanism of biosorption was chemisorptions as from the experimental data obtained from Freundlish and Langmuir isotherms and amino-, carboxyl-, hydroxyl-, and carbonyl groups were responsible for the biosorption as from the data obtained by FTIR analysis. *A. sphaerica* could demonstrate maximum biosorption of 111.1 mg/g and 121.9 mg/g for Pb and Cd at optimum operating conditions, respectively.

There exist two ways of metal uptake by living and/or dead cells. The first way is by surface binding of metal ions to cell surface or to extra cellular materials. The second way involves active transport of metal ion across the cell membrane which is metabolism dependent and termed as active uptake or bioaccumulation (Volesky, 1990). Metal uptake by surface adsorption is common to both live and dead cells, whereas the active uptake is a measure exclusive to live cells since it requires metabolic processes. This may include production of certain metal-binding proteins within the cell. Thus, metal uptake may happen in different ways, based on whether or not the cell is alive or dead.

Further, certain physical factors like cell age, composition of growth media, pH of solution and temperature determine bioaccumulation of heavy metals. Cell age is a factor for fungi by which bioaccumulation of metals can occur. It has been observed during lag phase, there is an increase in bioaccumulation and accumulation comes to halt as culture reaches stationary phase. Volesky and May (1995) observed that bioaccumulation of uranium by baker's yeast was 2.6 times higher at 12 h of incubation than with 24 h incubation. Bioaccumulation of heavy metals may also differ with respect to culture conditions and growth media composition. Avery and Tobin (1992) investigated Sr^{2+} uptake by culture of *S. cerevisiae*. They concluded that uptake of Sr^{2+} in presence of glucose (2% w/v) was more as compared without the glucose amendment in the media. Further, they reported that growth media controls the composition and structure of cell wall, which in turn affects the bioaccumulation.

The pH is another most important factor determining metal uptake by fungi. Various fungal strains are pH sensitive. Cd biosorption on the other hand, is pH dependent too. Barros et al. (2003) observed four different organisms, *A. oryzae*, *A. niger*, *F. solani*, and *Candida utilis* which showed maximum biosorption rate at acidic range. This pH dependent biosorption can be explained on the basis of proton-competitive adsorption reaction (Huang et al., 1991).

It has been known for decades that fungi can interact with heavy metal present in environment by secreting their extracellular enzymes. This eventually causes a physiological response through uptake of heavy metals by fungi. One of the species involved is white rot fungi, which can uptake and accumulates heavy metals in their mycelia. Favero et al. (1991) have reported that *Pleurotus ostreatus* was able to accumulate 20 mg g^{-1} dry weight Cd from liquid medium containing 150 ppm Cd with at least 20% of accumulated Cd deposited intracellularly. Sanglimsuwan et al. (1993) have reported a seven day growth study with *Daedalea quercina* in potato dextrose medium having 5 mM Cu and Zn. The fungus accumulated 10 μg g^{-1} dry weight Cu and 5 μg g^{-1} dry weight Zn.

Preferentially, heavy metal uptake is species specific. Gabriel et al. (1994) studied four different species of white rot fungi, *D. quercina, Ganoderma applanatum, Stereum hirsutum,* and *Schizophyllum commune* and their tolerance levels to Al, Cd, Pb, and Ca at 1 mM each. During an eight day study, they reported that most Pb was found in *S. hirsutum* (90.6 μM g^{-1}), most Cd, Al, and Ca in *G. applanatum* (272, 600, and 602 μM g^{-1}, respectively). From equimolar solution (1 mM), Pb was preferentially accumulated by all fungi except *G. applanatum* that accumulated more Al. On the contrary, although research data provide sufficient evidence regarding heavy metal accumulation by fungi, the same is not obvious in natural systems. Purkayastha and Mitra (1992) observed that Cu was taken up much preferably by *Volvariella volvacea* and Pb was taken up very low in liquid cultures. Whereas, when grown on wheat substrate, the fungus demonstrated high uptake of Pb and low uptake of Cd and Cu.

Through the course of hyperaccumulation, heavy metals interact with fungi affecting their physiological, enzymatic, and reproductive processes. This may cause changes in community structures in ecosystem. It is therefore obvious that different species of fungi differ in the degree of their heavy metal tolerance. Sanglimsuwan et al. (1993) studied minimum inhibitory concentrations (MIC) for 21 strains of 16 species. They reported MIC were lowest in case of Hg (0.05–0.2 mM), Cd (0.5–5 mM) and Co (1–5 mM) whereas higher in case of Ni (0.7–7 mM), Zn (5–1 5mM) and Cu (3–20 mM). Baldrian and Gabriel (1997) also studied 15 wood rotting species, in which *S. hirsutum* was the most Cd tolerant and slow growing *I. obliquus* was the most sensitive. This variation in heavy metal resistance can occur among strains in a single species as well. Major differences in Cu tolerance have been found among isolates of brown rot fungi. 40 mM Cu is the tolerance level for about 35 strains of *Antrodia vaillantii*, whereas, other could only resist 3 mM Cu (Collett, 1992).

In general, many factors determine bioaccumulation of heavy metals by fungal biomass with pH being a more sensitive variable. Cultivating fungal biomass is relatively easy and it can be used again and again. More precisely, metal scavenging by fungal biomass is considered better than any other means of conventional remediation. This mycoaccumulation processes could be a better replacement for existing metal-remediation processes or could be considered as supplementary for existing treatment technologies.

13.7 MYCOREMEDIATION, MYCOSORPTION, AND MYCOFILTERATION

Fungi are diverse group of organisms due to its unique nature in its morphology, physiology and genetic features; fungi having greater ability of decomposing several enzymes, heavy metals and toxic substances, which can be utilized as a nutritional source by them. Intense industrial activity and urbanization led to disposal of heavy metals which causes deleterious effect on animals and humankind. Fungi are capable of degrading hazardous chemical substances like chlorinated hydrocarbons, heavy metals such as chromium, cadmium, lead, gasoline, zinc, arsenic, and benzene through several applications ranging from biological, chemical, and engineering techniques from the contaminated sites.

Mycoremediation is a process of using fungi for the conversion of highly contaminated soils to less contaminated state. Due to the production of diverse enzymes like lipases, pectin and cellulose, the organic matters are easily degraded by fungi (Gupta and Shrivastava, 2014). A number of mushroom, *Pleurotus platypus*, *Agaricus bisporus*, *Calocybe indica*, *Calvatia excipuliformis*, *Hygrophorus virgineus*, *Boletus edulis*, *Lepiota rhacodes*, *Lepis tanuda*, *Pleurotus sajor-caju*, *Pleurotus ostreatus*, *Psalliota campestris* and *Russula delica* shows significant role in remediation of heavy metals (Nilanjana, 2005; Elekes and Busuioc, 2013). Fungal treatment utilizes low cost than using any other physical and chemical methods (Jalc, 2002). Mycosorption is a technique used to detoxify the environmental contaminants by fungi, and the organisms responsible for detoxification are considered as mycosorbents. A wide variety of living and dead biomass like bacteria, algae, fungi, and plants are known to tolerate and accumulate heavy metals (Singh, 2006). Gorobets et al. (2004) showed that mycosorption involves various external factors (type of metal, ionic form in solution and the functional site) and tends to be exothermic. Other factors like pH, temperature, biomass concentration, type of biomass preparation and initial metal ion concentration, are also important in evaluating mycosorption.

Mycoremediation involves the physico-chemical and biological method of remediating toxic substances; a physico-chemical process includes precipitation, coagulation, a reduction process, ion exchange, membrane processes such as ultrafiltration, electrodialysis; its implementation is costly and effectiveness is limited. Biological method is a novel approach in the mycoremediation due to its lack of adverse effect on environment and the low cost implementation (Prasad et al., 2012). Some fungal species like Basidiomycetes, Ascomycetes are capable of degrading lignocellulosic materials which are present on wood, paper and other effluents (Pozdnyakova, 2012). White rot fungi degrade lignin, *Phanerochate chrysosporium* is one which fragments lignocellulosic macromolecules into simpler forms (Novotny et al., 2004). *Pleurotus ostreatus* is an edible mushroom, fruiting body of the organisms helps in degrading the organic polymers (Rhodes, 2014). Several soil residing fungi which shows significance in remediation of heavy metals as listed in Table 13.1.

TABLE 13.1 List of Fungi Exploited for Mycoremediation of Heavy Metals through Mycosorption and Mycofiltration.

Heavy Metals	Toxic Effects	Fungi Used in Mycoremediation	References
Antimony	Respiratory disorder, cardiovascular disorder, diarrhea, vomiting, ulcer, dermatitis, disturbance in menstruation, breakage in human leukocytes, cot death or sudden infant death syndrome (SIDS)	*Scopularioopsis brevicaulis, Phaeolus schweinitzii*	Jenkins et al. (1988), Paul et al. (2001)
Arsenic	Bronchitis, dermatitis, bone marrow depression, hemolysis, hepatomegaly	*Penicillium chrysogenum*	Mamisahebei et al. (2007)
Cadmium	Kidney damage, bronchitis, gastrointestinal disorder, cancer, hypertension, itai-itai disease, weight loss	*Aspergillus cristatus*	Barros et al. (2003)
Chromium	Carcinogenic, mutagenic, teratogenicity, epigastria pain nausea, vomiting, severe diarrhea, producing lung tumor	*Penicillium canescens, Pleurotus ostreatus*	Chhikara et al. (2010), Donghee et al. (2005)
Cobalt	Cardiomyopathy, pnemonoconiosis, goiter	*Saccharomyces cerevisiae*	Mapolelo and Torto (2004)
Copper	Neurotoxicity, acute toxicity, dizziness, diarrhea	*Pleurotus ostreatus*	Jha et al. (2012)
Iron	Nausea, vomiting, anemia, abdominal pain, nephropathy	*Neurospora crassa*	Rashmi et al. (2001)

TABLE 13.1 *(Continued)*

Heavy Metals	Toxic Effects	Fungi Used in Mycoremediation	References
Lead	Anemia, brain damage, anorexia, malaise, loss of appetite, liver, kidney, gastrointestinal damage, mental retardation in children	*Rhizopus nigricans, Trichoderma longibrachiatum*	Osman and Bandyopadhyay (1999)
Manganese	Parkinson-like syndrome, respiratory disorder, neuropsychiatric	*Penicillium spinulosum*	Townsley et al. (1986a), Ross and Townsley (1986)
Mercury	Damage to nervous system, protoplasm poisoning, corrosive to skin, eyes, muscle, dermatitis, kidney damage	*Aspergillus flavus, Aspergillus fumigatus*	Murugesan et al. (2006)
Nickel	Chronic bronchitis, reducing lung function, lung cancer	*Aspergillus niger*	Ahmad et al. (2006)
Selenium	Caustic burns, pneumonitis, hypotension, brittle hair and nails, red skin, paresthesia, hemiplegia	*Alternaria alternata*	Thompson-Eagle et al. (1991)
Silver	Hemorrhage, bonemarrow suppression, pulmonary edema, hepatorenal necrosis, argyria, blue-grey discolouration of skin, nails, mucosae	*Cladosporium cladosporiodes*	Pethkar et al. (2001)
Strontium	Lung cancer, skin inflammation	*Saccharomyces cerevisiae*	Avery and Tobin (1992)
Thallium	Vomiting, diarrhea, painful neuropathy, coma, instability, alopecia	*Mariannae* sp., *Trichoderma viride, Trichoderma asperellum*	Sun et al. (2015)
Uranium	Radiation-induced erythema, impairment of kidney	*Penicillium digitatum*	Galun et al. (1983)
Zinc	Short term metal-fume fever, gastrointestinal distress	*Aspergillus niger*	Kumar et al. (2011)

It is a similar process of remediating heavy metals by fungal mycelia to filter toxic substances and microorganisms from water in soil (Sagar, 2015). According to Taylor and Stamets (2014), mushrooms are ubiquitous in nature, where the mycelium network produces enzyme acids that breakdown hydrogen and carbon chains. Mycofiltration shows significant role in filtering pathogens (viruses, bacteria, and protozoa), chemical toxins, and silt/heavy metals from the contaminated soil and water. *Stropharia rugoso-annulata* is an ideal model for mycofiltration, which reduces fecal coliforms by 99% contaminant in 24–48 h using mycological filters.

Nutrient from soil is supplied to plant roots through mycorrhiza, which are symbionts (Smith and Read, 1997). Soil arbuscular mycorrhiza is a symbiotic association between higher plants and soil fungi, mycelium forms an bridge to plant and soils, which also translocate various elements to the plants. An arbuscular mycorrhizal fungus exerts potentiality on the soil metal contamination (Finlay, 2008). Heggo et al. (1990) demonstrated that arbuscular mycorrhiza reduces heavy metal concentrations in soil and the concentrations of Zn, Cd, and Mn in plant leaves; ecto-mycorrhizal and ericoid mycorrhizal fungi can increase the tolerance of their host plants to heavy metals when the metals are present at toxic levels. Several investigations are evident that mychorrhizal fungi can colonize plant root extensively even in heavy contaminated soil (Sambandan et al., 1992).

13.8   CONCLUSION AND FUTURE PERSPECTIVES

The heavy metals and radioactive wastes are long term persisting pollutants in nature; they cannot be degraded easily, thereby causing pollution and many major side effects. The main cause of environmental pollution is industrial effluents, automobiles, atomic energy plants, and other harmful human activities. In order to control the environmental pollution it is necessary to understand the root of pollutants, their side effects, and their persistency in the environment. Removal of harmful pollutants from the surroundings is an important criterion and many attempts with chemicals as well as biological mediators have been performed in order to attain pollution free environment. The application of fungal genetic resources to remove heavy metals and radioactive wastes are proven to be better when compared to other techniques. Fungi are prominent group of microorganisms which occurs ubiquitously in natural habitats, with their unique property of degrading organic and inorganic substances which are categorized as good pollution control agents. Using fungi as bio-remediating agent have much importance because it requires less cost, less work force, and least maintenance as well as negligible side effects.

Arrays of sources are available to study the application and technology of using fungi for bioremediation. The mycotechnologies such as mycoaccumulation, mycosorption, mycofiltration, and mycodegradation have got more importance and hence, these qualities facilitate the use of fungi in control of environmental pollution. The biosorption of heavy metals have got more importance since, it is possible to re-collect the metal from the fungi.

Many techniques with respect to the fungal applications in soil remediation are discussed above in this chapter. There are possibilities of removal of numerous toxicants from the environment by the implementation of the fungal resources. Furthermore, many advancements are possible by intermediate use of biotechnological approaches. There is a possibility of constructing a genetically modified fungal species by inserting a gene coding for particular type of protein which take part in enzymatic degradation of pollutants. Biotechnology also helpful to get a heavy metal and radioactive waste resistant strains of microorganisms. By repeated experimental approaches, it is possible to produce such type of fungal strains with better quality of soil remediation and ecological restoration.

KEYWORDS

- Acute somatic effects
- Bioindicators
- Bioleaching
- Biomagnifications
- Bioremediation
- Biosorption
- Comparative genomics
- Detoxification
- Ecological restoration
- Epigenetic modification
- Fungal genetic resources
- Fungal genomic resources
- Heavy metal tolerance
- Heavy metal toxicity
- Hyperaccumulation
- Metal remediation
- Metallothioneins
- Mycoaccumulation
- Mycodegradation
- Mycofiltration

- Mycoremediation
- Mycorrhizal colonization
- Mycosorbents
- Mycosorption
- Mycotechnologies
- Nuclear residual waste
- Phytochelatins
- Radioactive isotopes
- Radioactive nuclides
- Radioactive waste materials

REFERENCES

Abdel-Aty, A. M.; Ammar, N. S.; Abdel-Ghafar, H. H.; Ali, R. K. Biosorption of cadmium and lead from aqueous solution by fresh water alga *Anabaena sphaerica* biomass. *J Adv Res* 2013, 4, 367–374.

Adamis, P. D.; Gomes, D. S.; Pinto, M. L.; Panek, A. D.; Eleutherio, E. C. The role of glutathione transferases in cadmium stress. *Toxicol Lett* 2004, 154(1–2), 81–88.

Ahmad, I.; Ansari, M. I.; Aqil, F. Biosorption of Ni, Cr and Cd by metal tolerant *Aspergillus niger* and *Penicillium* sp. using single and multi-metal solution. *Ind J Exp Biol* 2006, 44, 73–76.

Anahid, S.; Yaghmaei, S.; Ghobadinejad, Z. Heavy metal tolerance of fungi. *Scientia Iranica* 2010, 18, 502–508.

Anand, S. V. Global environmental issues. *Open Access Scientific Reports* 2013, 2, 632, doi:10.4172/scientificreports.632.

Annabi, A.; Said, K.; Messaodi, I. Cadmium: Toxic effect and physiological impairment in fishes. *Int J Adv Res* 2013, 1, 77–82.

Armienta, M. A.; Talavera, O.; Morton-Bermea, O.; Barrera, M. Geochemistry of metals from mine tailings in Taxco Mexico. *Bull Environ Contam Toxicol* 2003, 71, 387–393.

Authman, M. M. Use of fish as bio-indicator of the effects of heavy metals pollution. *J Aquac Res Dev* 2015, doi: 10.4172/2155-9546.1000328.

Avery, S. V.; Tobin, J. M. Mechanism of strontium uptake by laboratory and brewing strains of *Sachharomyces cervisiae*. *Appl Environ Microbiol* 1992, 58, 3883–3889.

Baby, J.; Raj, J. S.; Biby, E. T.; Sankarganesh, P.; Jeevitha, M. V.; Ajisha, S. U.; Rajan, S. S. Toxic effect of heavy metals on aquatic environment. *Int J Biol Chem Sci* 2010, 4(4), 939–952.

Baldrian, P.; Gabriel, J. Effect of heavy metals on the growth of selected wood-rotting basidiomycetes. *Folia Microbiol* 1997, 42, 521–523.

Barros, L. M. Jr.; Macedo, G. R.; Duarte, M. M. L.; Silva, E. P.; Lobato, A. K. C. L. Biosorption of cadmium using the fungus *Aspergillus niger*. *Braz J Chem Eng* 2003, 20(3), 229–239.

Baxter, J.; Cummings, S. P. The current and future applications of microorganism in the bioremediation of cyanide contamination. *Antonie Van Leeuwenhoek* 2006, 90(1), 1–17.

Bellion, M.; Courbot, M.; Jacob, C.; Guinet, F.; Blaudez, D.; Chalot, M. Metal induction of a *Paxillus involutus* metallothionein and its heterologous expression in *Hebeloma cylindrosporum*. *New Phytol* 2007, 151, 158–174.

Benbrahim-Tallaa, L.; Waterland, R.; Dill, A.; Webbe, M.; Waalkes, M. Tumor suppressor gene inactivation during cadmium-induced malignant transformation of human prostate cells correlates with overexpression of *de novo* DNA methyltransferase. *Environ Health Perspect* 2007, 115(10), 1454–1459.

Bianco, L.; Perrotta, G. Methodologies and perspectives of proteomics applied to filamentous fungi: From sample preparation to secretome analysis. *Intl J Mol Sci* 2015, 16(3), 5803–5829.

Boyd, R. S. Heavy metal pollutants and chemical ecology: Exploring new frontiers. *J Chem Ecol* 2010, 36, 46–58.

Britto, J. D. A.; Sebastian, S. R.; Gracelin, D. H. S. Effect of lead on malondialdehyde, superoxide dismutase, proline activity and chlorophyll content in *Capsicum annum*. *Biores Bull* 2011, 1, 93–98.

Butt, T. R.; Ecker, D. J. Yeast metallothionein and applications in biotechnology. *Microbiol Rev* 1987, 51(3), 351–364.

Butt, T. R.; Sternberg, E. J.; Gorman, J. A.; Clark, P.; Hamer, D.; Rosenberg, M.; Crooke, S. T. Copper metallothionein of yeast, structure of the gene, and regulation of expression. *Proc Natl Acad Sci USA* 1984, 81(11), 3332–3336.

Chakraborti, D.; Rahman, M. M.; Murrill, M.; Das, R.; Siddayya.; Patil, S. G.; Sarkar, A.; Dadapeer, H. J.; Yendigeri, S.; Ahmed, R.; Das, K. K. Environmental arsenic contamination and its health effects in a historic gold mining area of the Mangalur greenstone belt of Northeastern Karnataka, India. *J Hazard Mater* 2013, 262, 1048–1055.

Chakraborty, S.; Zaman, S.; Mitra, A. *Excoecaria agallocha*: A potential bioindicator of heavy metal pollution. *Int J Eng Res Gen Sci* 2014, 2(6), 289–298.

Chhikara, S.; Hooda, A.; Rana, L.; Dhankhar, R. Chromium (VI) biosorption by immobilized *Aspergillus niger* in continuous flow system with special reference to FTIR analysis. *J Env Biol* 2010, 31(5), 561–566.

Chiroma, T. M.; Ebewele, R. O.; Hymore, F. K. Comparative assessement of heavy metal levels in soil, vegetables and urban grey waste water used for irrigation in Yola and Kano. *Int Ref J Eng Sci* 2014, 3, 1–9.

Christiana, O. Impact of mining and agriculture on heavy metal levels in environmental samples in Okehi local government area of Kogi state. *Int J Pure Appl Sci Technol* 2012, 12, 66–77.

Cobbett, C. S. Phytochelatin biosynthesis and function in heavy-metal detoxification. *Curr Opinion Plant Biol* 2000a, 3, 211–216.

Cobbett, C. S. Phytochelatins and their roles in heavy metal detoxification. *Plant Physiol* 2000b, 123(3), 825–832.

Cobbina, S. J.; Myilla, M.; Michael, K. Small scale gold mining and heavy metal pollution: Assessment of drinking water sources in Datuku in the Talensi-Nabdam District. *Int J Sci Technol Res* 2013, 2, 96–100.

Collett, O. Comparative tolerance of the brown-rot fungus *Anthrodia vaillantii* (DC.:Fr.) Ryv. isolates to copper. *Holzforschung* 1992, 46, 293–298.

Dadachova, E.; Casadevall, A. Ionizing radiation: How fungi cope, adapt, and exploit with the help of melanin. *Curr Opin Microbiol* 2008, 11, 525–531.

Dellomonaco, C.; Fava, F.; Gonzalez, R. The path to next generation biofuels: Successes and challenges in the era of synthetic biology. *Microb Cell Fact* 2010, 9, 3.

Dixit, R.; Wasiullah.; Malaviya, D.; Pandiyan, K.; Singh, U. B.; Sahu, A.; Shukla, R.; Singh, B. P.; Rai, J. P.; Sharma, P. K.; Lade, H.; Paul, D. Bioremediation of heavy metals from soil and aquatic environment: An overview of principles and criteria of fundamental processes. *Sustainability* 2015, 7, 2189–2212.

Donghee, P.; Yeoung-Sang, Y.; Jong, P. M. Use of dead fungal biomass for the detoxification of hexavalent chromium: Screen and kinetics. *Proc Biochem* 2005, 40, 2559–2565.

Duffus, J. H. Heavy metals - a meaningless term? *Pure Appl Chem* 2002 74, 793–807.

Eisenbud, M.; Gesell, T. *Environmental radioactivity from natural, industrial and military sources*, Academic Press Inc: New York, 1997.

Elekes, C. C.; Busuioc, G. Response of four *Russula* species under copper sulphate and lead acetate treatments. *Not Bot Horti Agrobo* 2013, 41(2), 538–545.

Evrard, L. Management of sites and soils polluted by radioactive substances. *Controle* 2012, 195, 1–9.

Fahy, J. V.; Dickey, B. F. Airway mucus function and dysfunction. *N Engl J Med* 2010, 363(23), 2233–2247.

Favero, N.; Costa, P.; Massimino, M. L. *In vitro* uptake of cadmium by basidiomycete *Pleurotus ostreatus. Biotechnol Lett* 1991, 10, 701–704.

Finlay, R. D. Ecological aspects of mycorrhizal symbiosis: With special emphasis on the functional diversity of interactions involving the extraradical mycelium. *J Exp Bot* 2008, 59(5), 1115–1126.

Fogel, S.; Welch, J. W. Tandem gene amplification mediates copper resistance in yeast. *Proc Natl Acad Sci USA* 1982, 79, 5342–5346.

Francis, A. J.; Dodge, C. J. Remediation of soils and wastes contaminated with uranium and toxic metals. *Environ Sci Technol* 1998, 32, 3993–3998.

Fraser, J. A.; Davis, M. A.; Hynes, M. J. A gene from *Aspergillus nidulans* with similarity to URE2 of *Saccharomyces cerevisiae* encodes a glutathione S-transferase which contributes to heavy metal and xenobiotic resistance. *Appl Environ Microbiol* 2002, 68(6), 2802–2808.

Frova, C. Glutathione transferases in the genomics era: New insights and perspectives. *Bio Mol Eng* 2006, 23, 149–169.

Gabriel, J.; Mokrejs, M.; Bily, J.; Rychlovsky, P. Accumulation of heavy metals by some wood-rotting fungi. *Folia Microbiol* 1994, 39, 115–118.

Gadd, G. M. Microbial formation and transformation of organometallic and organometalloid compounds. *FEMS Microbial Rev* 1993, 11, 297–316.

Galagan, J. E.; Henn, M. R.; Ma, L. J.; Cuomo, C. A.; Birren, B. Genomics of the fungal kingdom: Insights into eukaryotic biology. *Genome Res* 2005, 15(12), 1620–1631.

Galun, M.; Keller, P.; Feldstein, H.; Galun, E.; Siegel, S.; Siegel, B. Recovery of uranium(VI) from solution using fungi II release from uranium loaded *Penicillium* biomass. *Water Air Soil Poll* 1983, 20, 277–285.

Geneva, M.; Yu, M.; Todorov, I.; Stancheva, I. Accumulation of Cd, Pb, Zn and antioxidant response in Chamomile (*Matricaria recutita* L.) grown on industrially polluted soil. *Genet Plant Physiol* 2014, 4, 131–139.

Georgescu, B.; Georgescu, C.; Dărăban, S.; Bouaru, A.; Paşcalău, S. Heavy metals acting as endocrine disrupters. *Sci Pap Anim Sci Biotechnol* 2011, 44, 89–93.

Giaever, G.; Nislow, C. The yeast deletion collection: A decade of functional genomics. *Genetics* 2014, 197(2), 451–465.

Goffeau, A.; Barrell, B. G.; Bussey, H.; Davis, R. W.; Dujon, B.; Feldmann, H.; Galibert, F.; Hoheisel, J. D.; Jacq, C.; Johnston, M.; Louis, E. J.; Mewes, H. W.; Murakami, Y.; Philippsen, P.; Tettelin, H.; Oliver, S. G. Life with 6000 genes. *Science* 1996, 274, 546–567.

Gomes, D. S.; Fragoso, L. C.; Riger, C. J.; Panek, A. D.; Eleutherio, E. C. A. Regulation of cadmium uptake by *Saccharomyces cerevisiae*. *Biochim Biophys Acta* 2002, 1573, 21–25.

Gonzaga, M. I. S.; Santos, J. A. G.; Ma, L. Q. Arsenic phytoextraction and hyperaccumulation by fern species. *Sci Agric* 2006, 63(1), 90–101.

Gonzalez, A. J. The safety of radioactive waste management, achieving internationally acceptable solutions. *IAEA Bulletin* 2000, 42(3), 5–18.

Gorobets, S.; Gorobets, O.; Ukrainetz, A.; Kasatkina, K.; Goyko, I. Intensification of the process of copper ions by yeast of *Saccharomyces cerevisiae* 1968 by means of a permanent magnetic field. *J Magn Magn Mater* 2004, 272–276, 2413–2414.

Grigoriev, I. V.; Nikitin, R.; Haridas, S.; Kuo, A.; Ohm, R.; Otillar, R.; Riley, R.; Salamov, A.; Zhao, X.; Korzeniewski, F.; Smirnova, T.; Nordberg, H.; Dubchak, I.; Shabalov, I. MycoCosm portal: Gearing up for 1000 fungal genomes. *Nucl Acid Res* 2014, 42, 699–704.

Grill, E.; Loffler, S.; Winnacker, E. L.; Zenk, M. H. Phytochelatins, the heavy metal-binding peptides of plants are synthesized from glutathione by a specific-glutamylcysteine dipeptidyl transpeptidase (phytochelatin synthase). *Proc Natl Acad Sci USA* 1989, 86, 6838–6842.

Grześ, I. M. Ants and heavy metal pollution - A review. *Eur J Soil Biol* 2010, 46, 350–355.

Guan, Y.; Shao, C.; Ju, M. Heavy metal contamination assessment and partition for industrial and mining gathering areas. *Int J Environ Res Public Health* 2014, 11, 7286–7303.

Gulati, K.; Banerjee, B.; Lall, S. B.; Ray, A. Effects of diesel exhaust, heavy metals and pesticides on various organ systems: Possible mechanisms and strategies for prevention and treatment. *Indian J Exp Biol* 2010, 48, 710–721.

Gupta, M.; Shrivastava, S. Mycoremediation: A management tool for removal of pollutants from environment. *Ind J Appl Res* 2014, 4(8), 289–291.

Hayes, J. D.; Flanagan, J. U.; Jowsey, I. R. Glutathione transferases. *Annu Rev Pharmacol Toxicol* 2005, 45, 51–88.

Heggo, A.; Angle, A.; Chaney, R. L. Effects of vesicular arbuscular mycorrhizal fungi on heavy metal uptake by soybeans. *Soil Biol Biochem* 1990, 22, 865–869.

Hossain, M. A.; Pukclai, P.; Teixeira da Silva, J. A.; Fujita, M. Molecular mechanism of heavy metal toxicity and tolerance in plants: Central role of glutathione in detoxification of reactive oxygen species and methylglyoxal and in heavy metal chelation. *J Bot* 2012, doi:10.1155/2012/872875

Hossain, S. M.; Das, M.; Begum, K. M. M. S.; Anantharaman, N. Studies on biodegradation of cyanide (AgCN) using *Phanerochaete chrysosporium*. *J Institution Engineers* 2005, 85, 45–49.

Huang, C.; Huang, C. P.; Morehart, A. L. Proton competition in Cu(II) adsorption by fungal mycelia. *Water Res* 1991, 25, 1365–1375.

Jahan, M. The impact of environmental degradation on women in Bangladesh: An overview. *Asian Affairs* 2008, 30(2), 5–15.

Jalc, D. Straw enrichment for fodder production by fungi. In *The mycota XI, agricultural applications*. Kempken, F., Ed.; Springer-Verlag: Berlin, 2002, pp 19–38.

Jane, M.; Orient, M. D. Fukushima and reflections on radiation as a terror weapon. *J Americ Physician Sur* 2014, 19(2), 48–55.

Jankaite, A.; Baltrėnas, P.; Kazlauskienė, A. Heavy metal concentrations in roadside soils of Lithuania's highways. *Geologija* 2008, 50(4), 237–245.

Jenkins, R. O.; Craig, P. J.; Goessler, W.; Miller, D. P.; Ostah, N.; Ingolic, K. J. Biomethylation of inorganic antimony compounds by an aerobic fungus *Scopulariopsis brevicaulis*. *Env Sci Technol* 1998, 32, 882–885.

Jha, S.; Dikshit, S. N.; Pandey, G. Comparative study of growing/immobilized biomass verses resting biomass of *A. lentulus* for the effect of pH on Cu^{2+} metal removal. *Res J Pharm Bio Chem Sci* 2012, 3(3), 421–427.

Ji, J.; Wang, X.; Li, F.; Zeng, Y.; Qiao, Y. Isolation screen and application of highly-effective cyanide-degrading fungus. *Gongye Shuichuli* 2005, 25(7), 35–38.

Joffe, M. Infertility and environmental pollutants. *Br Med Bull* 2003, 68, 47–70.

John, O.; Pastore, M. D. The short-term effects of nuclear war: The medical legacy of Hiroshima and Nagasaki. *Prev Med* 1987, 16, 293–307.

Jones, C. G.; Lawton, J. H.; Shachak, M. Organisms as ecosystem engineers. *Oikos* 1994, 69, 373–386.

Jose, J.; Giridhar, R.; Anas, A.; Loka Bharathi, P. A.; Nair, S. Heavy metal pollution exerts reduction/adaptation in the diversity and enzyme expression profile of heterotrophic bacteria in Cochin Estuary, India. *Environ Pollut* 2011, 159, 2775–2780.

Jung, M. C. Contamination by Cd, Cu, Pb, and Zn in mine wastes from abandoned metal mines classified as mineralization types in Korea. *Environ Geochem Health* 2008 30, 205–217.

Kagi, J. H. R.; Nordberg, M. *Metallothionein*. Birkhauser Verlag: Basel, 1979, pp 1–378.

Koonin, E. V.; Galperin, M. Y. *Sequence - evolution - function: Computational approaches in comparative genomics*. Kluwer Academic Publishers: Netherlands, 2003.

Krishnan, G. Synthesis, characterisation and catalytic evaluation of mesoporous Mcm 41 molecular sieves for the removal of organic contaminants in aqueous solution. PhD Thesis, Anna University: Chennai, India, 2014.

Kumar, A.; Bisht, B. S.; Joshi, V. D. Zinc and cadmium removal by acclimated *Aspergilus niger*: Trained fungus for biosorption. *Int J Environ Sci Res* 2011, 1(1), 27–30.

Kumar, A. *A Text Book of Environmental Science*. APH Publishing: New Delhi, 2004.

Lanfranco, L.; Bolchi, A.; Ros, E. C.; Ottonello, S.; Bonfante, P. Differential expression of a metallothionein gene during the presymbiotic versus the symbiotic phase of an arbuscular mycorrhizal fungus. *Plant Physiol* 2002, 130(1), 58–67.

Li, Z. S.; Lu, Y. P.; Zhen, R. G.; Szczypka, M.; Thiele, D. J.; Rea, P. A. A new pathway for vacuolar cadmium sequestration in *Saccharomyces cerevisiae*: YCF1-catalyzed transport of bis (glutathionato) cadmium. *Proc Natl Acad Sci USA* 1997, 94, 42–47.

Magee, P. T.; Gale, C.; Berman, J.; Davis, D. Molecular genetic and genomic approaches to the study of medically important fungi. *Infect Immun* 2003, 71(5), 2299–2309.

Malizia, D.; Giuliano, A.; Ortaggi, G.; Masotti, A. Common plants as alternative analytical tools to monitor heavy metals in soil. *Chem Cent J* 2012, 6(S2), S6, doi:10.1186/1752-153X-6-S2-S6

Mamisahebei, S.; Khaniki, G. R. H.; Torabian, A.; Nasseri, S.; Naddafi, K. Removal of arsenic from an aqueous solution by pretreated waste tea fungal biomass. *Iran J Environ Health Sci Eng* 2007, 4(2), 85–92.

Mapolelo, M.; Torto, N. Trace enrichment of metal ions in aquatic environments by *Saccharomyces cerevisiae*. *Talanta* 2004, 64, 39–47.

Margoshes, M.; Vallee, B. L. A cadmium protein from equine kidney cortex. *J Am Chem Soc* 1957, 79, 4813–4814.

Martin, F.; Kohler, A.; Murat, C.; Balestrini, R.; Coutinho, P. M.; Jaillon, O.; Montanini, B.; Morin, E.; Noel, B.; Percudani, R. Périgord black truffle genome uncovers evolutionary origins and mechanisms of symbiosis. *Nature* 2010, 464, 1033–1038.

Martin, F.; Selosse, M. A. The laccaria genome: A symbiont blueprint decoded. *New Phytol* 2008, 180(2), 296–310.

Martino, E.; Murat, C.; Vallino, M.; Bena, A.; Perotto, S.; Spanu, P. Imaging mycorrhizal fungal transformants that express EGFP during ericoid endosymbiosis. *Curr Genet* 2007, 52, 65–75.

Mathieu, Y.; Prosper, P.; Favier, F.; Harvengt, L.; Didierjean, C.; Jacquot, J. P.; Morel-Rouhier, M.; Gelhaye, E. Diversification of fungal specific class A glutathione transferases in saprotrophic fungi. *PLoS One* 2013, 8(11), e80298.

Mbah, C.; Anikwe, M. Variation in heavy metal contents on roadside soils along a major express way in South East Nigeria. *New York Sci J* 2010 3(10), 103–107.

McGoldrick, S.; O'Sullivan, S. M.; Sheehan, D. Glutathione transferase-like proteins encoded in genomes of yeasts and fungi: Insights into evolution of a multifunctional protein superfamily. *FEMS Microbiol Lett* 2005, 242, 1–12.

Menn, F. M.; Easter, J. P.; Sayler, G. S. Genetically engineered microorganisms and bioremediation. *Curr Opin Biotechnol* 2000, 11, 286–289.

Monrroy, M.; Baeza, J.; Freer, J.; Rodriguez, J. Degradation of tribromophenol by wood-decaying fungi and the 1,2-dihydroxybenzene-assisted Fenton reaction. *Bioremediation J* 2007, 11(4), 195–200.

Morel, M.; Meux, E.; Mathieu, Y.; Thuillier, A.; Chibani, K.; Harvengt, L.; Jacquot, J. P.; Gelhaye, E. Xenomic networks variability and adaptation traits in wood decaying fungi. *Microb Biotechnol* 2013, 6, 248–263.

Morsy, A. A.; Salama, K. H. A.; Kamel, H. A.; Mansour, M. M. F. Effect of heavy metals on plasma membrane lipids and antioxidant enzymes of *Zygophyllum* species. *Eurasian J Biosci* 2012, 10, 1–10.

Murugesan, G. S.; Sathishkumar, M.; Swaminathan, K. Arsenic removal from groundwater by pretreated waste tea fungal biomass. *Biores Tech* 2006, 97, 483–487.

Nachiyunde, K.; Ikeda, H.; Okuda, T.; Nishijima, W. Assessment of dissolved heavymetal pollution in five provinces of Zambia. *J Environ Protect* 2013, 4, 80–85.

Nadgorska-Socha, A.; Ptasinski, B.; Kita, A. Heavy metal bioaccumulation and antioxidative responses in *Cardaminopsis arenosa* and *Plantago lanceolata* leaves from metalliferous and non-metalliferous sites: A field study. *Ecotoxicology* 2013, 22, 1422–1434.

Nagy, A.; Perrimon, N.; Sandmeyer, S.; Plasterk, R. Tailoring the genome: The power of genetic approaches. *Nat Gen* 2003, 33, 276–284.

Natesan, U.; Ranga Rama Seshan, B. Vertical profile of heavy metal concentration in core sediments of Buckingham canal, Ennore. *Indian J Mar Sci* 2011, 40, 83–97.

Nevoigt, E. Progress in metabolic engineering of *Saccharomyces cerevisiae*. *Microbiol Mole Biol Rev* 2008, 72(3), 379–412.

Nilanjana, D. Heavy metal biosorption by mushrooms. *Nat Prod Rad* 2005, 4, 454–459.

Novotny, C.; Svobodova, K.; Erbanova, P.; Cajthaml, T.; Kasinath, A.; Lang E, Saek, V. Ligninolytic fungi in bioremediation: Extracellular enzyme production and degradation rate. *Soil Biol Biochem* 2004, 36, 1545–1551.

Ortiz, D. F.; Ruscitti, T.; McCue, K. F.; Ow, D. W. Transport of metal-binding peptides by HMT1, a fission yeast ABC-type vacuolar membrane protein. *J Biol Chem* 1995, 270, 4721–4728.

Osman, M. S.; Bandyopadhyay, M. Biosorption of lead(II) ions from wastewater by using a fungus *P. ostreatus*. *J Civ Eng* 1999, 27(2), 193–196.

Padamsee, M.; Kumar, T. K.; Riley, R.; Binder, M.; Boyd, A.; Calvo, A. M.; Furukawa, K.; Hesse, C.; Hohmann, S.; James, T. Y.; LaButti, K.; Lapidus, A.; Lindquist, E.; Lucas, S.; Miller, K.; Shantappa, S.; Grigoriev, I. V.; Hibbett, D. S.; McLaughlin, D. J.; Spatafora, J. W.; Aime, M. C. The genome of the xerotolerant mold *Wallemia sebi* reveals adaptations to osmotic stress and suggests cryptic sexual reproduction. *Fungal Genet Biol* 2012, 49, 217–226.

Paul, A.; William, R. C.; Elena, P.; Ken, J. R. Antimony biomethylation by the wood rotting fungus *Phaeolus schweinitzii*. *App Org Chem* 2001, 16(6), 473–480.

Pethkar, A. V.; Kulkarni, S. K.; Paknikar, K. M. Comparative studies on metal biosorption by two strains of *Cladosporium cladosporioides*. *Bioresour Technol* 2001, 80, 211–215.

Popova, L. P.; Maslenkova, L. T.; Ivanova, A.; Stoinova, Z. Role of salicylic acid in alleviating heavy metal stress. In *Environmental adaptations and stress tolerance of plants in the era of climate change*. Ahmad, P., Prasad, M. N. V., Eds.; Springer: New York, 2012, pp 441–466.

Pozdnyakova, N. N. Involvement of the ligninolytic system of white-rot and litter-decomposing fungi in the degradation of polycyclic aromatic hydrocarbons. *Biotech Res Int* 2012, 243217, doi:10.1155/2012/243217

Prasad, M.; Ankita, G.; Maheshwari, R. Decontamination of polluted water employing bioremediation processes: A review. *Int J LifeSci Bt Pharm* 2012, 1(3), 11–21.

Pulido, P.; Kagi, J. H.; Vallee, B. L. Isolation and some properties of human metallothionein. *Biochemistry* 1966, 5, 1768–1777.

Purkayastha, R. P.; Mitra, A. K. Metal uptake by mycelia during submerged growth and by sporocarps of an edible fungus *Volvariella volvacea*. *Ind J Exp Biol* 1992, 30, 1184–1187.

Rai, P. K.; Panda, L. L. S. Roadside plants as bioindicators of air pollution in an industrial region, Rourkela, India. *Int J Adv Res Technol* 2015, 4(1), 14–36.

Rao, K. R. Radioactive waste: The problem and its management. *Curr Sci* 2001, 81(12), 1534–1546.

Rao, P. R.; Bhargavi, C. Studies on biosorption of heavy metals using pretreated biomass of fungal species. *Int J Chem Chem Eng* 2013, 3(3), 171–180.

Rashad, S. M.; Hammad, F. H. Nuclear power and the environment: Comparative assessment of environmental and health impacts of electricity-generating systems. *App Energy* 2000, 65, 211–229.

Rashmi, K.; Sowjanya, T. N.; Mohan, P. M.; Balaji, V.; Venkateswaran, G. Bioremediation of ^{60}Co from simulated spent decontamination solutions. *Sci Total Environ* 2001, 328, 1–14.

Reza, R.; Singh, G. Heavy metal contamination and its indexing approach for river water. *Int J Environ Sci Technol* 2010, 7, 785–792.

Rhodes, C. J. Mycoremediation (bioremediation with fungi) growing mushroom to clean the earth. *Chem Spec Bioavailab* 2014, 26(3), 196–198.

Romero, M. C.; Hammer, E.; Hanschke, R.; Arambarri, A. M.; Schauer, F. Biotransformation of biphenyl by the filamentous fungus *Talaromyces helicus*. *World J Microbiol Biotechnol* 2005, 21(2), 101–106.

Ross, I. S.; Townsley, C. C. The uptake of heavy metals by filamentous fungi. In *Immobilization of ions by biosorption*. Eccles, H., Hunt, S., Eds.; Ellis Horwood: Chichester, UK, 1986, pp 49–57.

Rzymski, P.; Tomczyk, K.; Rzymski, P.; Poniedziaek, B.; Opala, T.; Wilczak, M. Impact of heavy metals on the female reproductive system. *Ann Agric Environ Med* 2015 22, 259–264.

Sagar, R. B. Bio-remediation a secure and reverberation module for ground-water and soil. *Int J Innv Eme Res Eng* 2015, 2(2), 32–38.

Sahu, R.; Saxena, P.; Johnson, S.; Mathur, H. B.; Agarwal, H. C. Mercury pollution in the Sonbhadra district of Uttar Pradesh, India, and its health impacts. *Toxicol Environ Chem* 2012, 96(8), 1272–1283.

Sambandan, K.; Kannan, K.; Raman, N. Distribution of vesicular-arbuscular mycorrhizal fungi in heavy metal polluted soils of Tamil Nadu, India. *J Environ Biol* 1992, 13, 159–167.

Sanglimsuwan, S.; Yoshida, N.; Morinaga, T.; Murooka, Y. Resistance to and uptake of heavy metals in mushrooms. *J Ferment Bioeng* 1993, 75, 112–114.

Saraf, S.; Vaidya, V. K. Comparative study of biosorption of textile dyes using fungal biosorbents. *Int J Curr Microbiol App Sci* 2015, 2, 357–365.

Sayler, G. S.; Matrubutham, U.; Menn, F. M.; Johnston, W. H.; Stapleton, R. D. Molecular probes and biosensors in bioremediation and site assessment. In *Bioremediation: Principles and practice, Vol. 1*. Sikdar, S. K., Irvine, R. L., Lancaster, P. N., Eds.; Technomic Publishing Company: Lancaster, PA, 1998, pp 385–434.

Seiler, S.; Plamann, M. The genetic basis of cellular morphogenesis in the filamentous fungus *Neurospora crassa*. *Mole Biol Cell* 2003, 14 (11), 4352–4364.

Sheehan, D.; Meade, G.; Foley, V.; Dowd, C. Structure, function and evolution of glutathione transferases: Implications for classification of non-mammalian members of an ancient enzyme superfamily. *Biochem J* 2001, 360, 1–16.

Shen, M.; Zhao, D. K.; Qiao, Q.; Liu, L.; Wang, J. L.; Cao, G. H.; Li, T.; Zhao, Z. Identification of glutathione-S-transferase (GST) genes from a dark septate endophytic fungus (*Exophiala pisciphila*) and their expression patterns under varied metal stress. *PLoS One* 2015, 10(4), e0123418, doi: 10.1371/journal.pone.0123418

Sherene, T. Mobility and transport of heavy metals in polluted soil environment. *Biol Forum* 2010, 2, 112–121.

Shruti, A. Disposal of nuclear waste: A global scenario. *J Environ Research Develop* 2010, 5(1), 251–258.

Simon, S. L.; Bouville, A.; Land, C. E. Fallout from nuclear weapons tests and cancer risks. *Am Sci* 2006, 94, 48–57.

Singh, H. *Mycoremediation: Fungal bioremediation*. John Wiley and Sons, Inc.: Hoboken, New Jersey, 2006, pp 484–532.

Skanda Kumar, B. N.; Suhas, R.; Shet, S. U.; Srishaila, J. M. Utilization of iron ore tailings as fine aggregate in ultra-high performance concrete. *Int J Res Eng Technol* 2014, 3(7), 369–376.

Smith, S. E.; Read, D. J. *Mycorrhizal symbiosis*. Academic Press: London, 1997.

Stommel, M.; Mann, P.; Franken, P. EST-library construction using spore RNA of the arbuscular mycorrhizal fungus *Gigaspora rosea*. *Mycorrhiza* 2001, 10, 281–285.

Strom, D. J. Health impacts from acute radiation exposure. U.S. Department of Energy: USA, 2004, pp 1–41.

Sun, J.; Zou, X.; Xiao, T.; Jia, Y.; Ning, Z.; Sun, M.; Liu, Y.; Jiang, T. Biosorption and bioaccumulation of thallium-tolerant fungal isolates. *Environ Sci Pollut Res* 2015, 22, 16742–16748.

Taboski, M. A. S.; Rand, T. G.; Piorko, A. Lead and cadmium uptake in the marine fungi *Corollospora lacera* and *Monodictys pelagica*. *FEMS Microbiol Ecol* 2005, 53, 445–453.

Taylor, A. W.; Stamets, P. E. Implementing fungal cultivation in biofiltration systems - the past, present and future of mycofiltration. USDA Forest Service Proceedings 2014, RMRS-P-72, 23–28.

Thompson-Eagle, E. T.; Frankenberger, W. T.; Longley, K. E. Removal of selenium from agricultural drainage water through soil microbial transformation. In *The economics and management of water and drainage in agriculture*. Dinar, A., Zilberman, D., Eds.; Kluwer: New York, 1991, pp 169–186.

Thuillier, A.; Roret, T.; Favier, F.; Gelhaye, E.; Jacquot, J. P.; Didierjean, C.; Morel-Rouhier, M. A typical features of a Ure2p glutathione transferase from *Phanerochaete chrysosporium*. *FEBS Lett* 2013, 587, 2125–2130.

Townsley, C. C.; Ross, I. S.; Atkins, A. S. Biorecovery of metallic residues from various industrial effluents using filamentous fungi. In *Fundamental and applied biohydrometallurgy*, Lawrence, R. W., Branion, R. M. R., Enner, H. G., Eds.; Elsevier: Amsterdam, 1986, pp 279–289.

Trincone, A. Marine biocatalysts: Enzymatic features and applications. *Mar Drugs* 2011, 9(4), 478–499.

Vallino, M.; Drogo, V.; Abba, S.; Perotto, S. Gene expression of the ericoid mycorrhizal fungus *Oidiodendron maius* in the presence of high zinc concentrations. *Mycorrhiza* 2005, 15, 333–344.

Verma, S. R.; Chaudhari, P. R.; Satyanaranyan, S. Impact of leaching from iron ore mines on terrestrial and aquatic environment. *Int J Environ Sci* 2012, 2(4), 2378–2386.

Verma, R.; Dwivedi, P. Heavy metal water pollution - a case study. *Rec Res Sci Technol* 2013, 5(5), 98–99.

Vogiannis, E. G.; Nikolopoulos, D. Radon sources and associated risk in terms of exposure and dose. *Front Public Health* 2015, 2, 207.

Voiblet, C.; Duplessis, S.; Encelot, N.; Martin, F. Identification of symbiosis-regulated genes in *Eucalyptus globulus-Pisolithus tinctorius* ectomycorrhiza by differential hybridization of arrayed cDNA. *Plant J* 2001, 25, 181–191.

Volesky, B. *Biosorption of heavy metals*. CRC Press: Boca Raton, Florida, 1990, pp 3–43.

Volesky, B.; May, H. Biosorption of heavy metals by *Saccharomyces cerevisiae*. *Appl Microbiol Biotechnol* 1995, 42, 797–806.

Waschke, A.; Sieh, D.; Tamasloukht, M.; Fischer, K.; Mann, P.; Franken, P. Identification of heavy metal-induced genes encoding glutathione-S-transferases in the arbuscular mycorrhizal fungus *Glomus intraradices*. *Mycorrhiza* 2006, 17, 1–10.

Wattal, P. K. Indian programme on radioactive waste management. *Ind Acad Sci* 2013, 38(5), 849–857.

Wei, B.; Yang, L. A review of heavy metal contaminations in urban soils, urban road dusts and agricultural soils from China. *Microchem J* 2010, 94, 9–107.

Weissenhorn, I.; Leyval, C. Root colonization of maize by a Cd- sensitive and a Cd-tolerant *Glomus mosseae* and cadmium uptake in sand culture. *Plant Soil* 1995, 175, 233–238.

WHO. *Exposure to cadmium: A major public health concern*. World Health Organization: Geneva, 2010, pp 3–6.

Wuana, R. A.; Okieimen, F. E. Heavy metals in contaminated soils: A review of sources, chemistry, risks and best available strategies for remediation. *ISRN Ecol* 2011, 402647, doi:10.5402/2011/402647.

Zajc, J.; Liu, Y.; Dai, W.; Yang, Z.; Hu, J.; Gostinčar, C.; Gunde-Cimerman, N. Genome and transcriptome sequencing of the halophilic fungus *Wallemia ichthyophaga*: Haloadaptations present and absent. *BMC Genomics* 2013, 14, 617.

Zaki, M. S.; Shalaby, S. I.; Ata, N.; Noor El–Deen, A. I.; Abdelzaher, M. F. Effect of aquatic pollution on fish (review). *Life Sci J* 2013, 10(1), 637–642.

Zhang, J.; Chiodini, R.; Badr, A.; Zhang, G. The impact of next-generation sequencing on genomics. *J Gen Genomics*, 2011, 38(3), 95–109.

Zhdanova, N. N.; Tugay, T.; Dighton, J.; Zheltonozhsky, V.; McDermott, P. Ionizing radiation attracts soil fungi. *Mycol Res* 2004, 108, 1089–1096.

CHAPTER 14

A MULTIFACETED BIOREMEDIATION OF XENOBIOTICS USING FUNGI

DEVIPRIYA RABIN MAJUMDER

Department of Microbiology, Abeda Inamdar Senior College, 2390-KB Hidayatullah Road, Azam Campus, Camp, Pune 411001, Maharashtra, India

CONTENTS

14.1 INTRODUCTION

14.1.1 WHAT ARE XENOBIOTICS?

Xenobiotics (Greek – Xenos: Strange, foreign, foreigner) are chemically synthesized compounds that are not found in nature and are therefore branded as "Foreign to Biosphere". Microorganisms have not been exposed to such 'strange' structures during evolution. This leads to resistance to biodegradation of xenobiotics by microorganisms. These foreign compounds with unnatural structures are mostly toxic and are indeed harmful to life in general. Some of the core anthropogenic compounds (pollutants) present in the environment are petroleum hydrocarbons, halogenated solvents, endocrine disrupting drugs, explosives, agricultural chemicals, heavy metals, metalloids, and radionuclides. Pollutants with xenobiotic structural features like polycyclic aromatic hydrocarbons, halogenated aliphatic and aromatic hydrocarbons, nitroaromatic compounds, azo compounds, s-triazins, organic sulphonic acids, and synthetic polymers are also recalcitrant in the environment.

14.1.2 MYCOREMEDIATION

Fungi are chemoheterotrophic, ubiquitous in sub-aerial and subsoil environments, and are infallible decomposers. The study of fungi in fundamental geological processes can be termed 'geomycology'. These include organic and inorganic transformations, element cycling, mineral transformations, mycogenic bioweathering, metal-fungal interactions, and the significance and relevance of such processes in the environmental biotechnology such as bioremediation. A fungal role in biogeochemical cycling of the elements is interlinked with the ability to adopt a variety of growth, metabolic and morphological strategies, their adaptive capabilities to environmental extremes and, their mutualistic associations with animals, plants, algae, and cyanobacteria. Fungal polymorphism and reproduction by spores help successful colonization of many different environments. Most fungi exhibit filamentous growth, which has the ability to adopt both explorative or exploitative growth strategies, and the formation of aggregated hyphae for protected fungal translocation. Some fungi are polymorphic, occurring as both filamentous mycelium and unicellular yeasts or yeast-like cells, as in microcolonial fungi colonizing rocks. The capability of fungi to translocate nutrients through the mycelial network is an important feature for exploring heterogeneous environments. The ability of fungi to form extended mycelial networks, the low specificity of their catabolic enzymes, and their capacity to use pollutants as a growth substrate make fungi most coveted for bioremediation processes. Bioremediation and waste treatment by a variety of fungi through an array of processes have been discussed in this chapter.

Fungi are intimately involved in biogeochemical transformations at both aquatic and terrestrial habitats. The geochemical transformations by fungi influence plant productivity, mobility of toxic elements, and are therefore of considerable socio–economic relevance, specially the mutualistic symbioses, lichens, and mycorrhizas. Fungal transformations from bioremediation point of view have beneficial applications in environmental biotechnology, e.g., in metal leaching, recovery and detoxification, and xenobiotic and organic pollutant degradation. Thus, a multidisciplinary approach is essential to understand fully all the phenomena encompassed within mycoremediation of xenobiotics and it is hoped that this chapter will serve to stimulate interest in an area of mycoremediation of global significance (Gadd, 2007; Harms et al., 2011).

14.2   RECALCITRANT ORGANIC COMPOUNDS

Xenobiotics may persist in the environment for months and years, but most biogenic compounds are biodegraded rapidly by omnipotent microorganisms. Recalcitrance (i.e., the structure-immanent stability) of a xenobiotic compound is due to its 'unphysiological' chemical bonds and/or substituents, which resist the attack by microbial catabolic enzymes. Type, number and position of bonds, and substituents affect the xenobiotic character.

14.2.1   *CHARACTERISTICS OF RECALCITRANT XENOBIOTICS*

Typical features of recalcitrant organic compounds are as follows: high molecular mass, low solubility in water, condensed benzene and pyridine rings, especially polycyclic structures, three-fold substituted N atoms, quarternary C atoms, unphysiological bonds and substituents R–X, especially polysubstitution, where $-X = -O-R, -N=N-, -F, -Cl, -Br, -NO_2, -CF_3,$ or $-SO_3H$. Figure 14.1 gives the examples of relatively persistent xenobiotics which are dangerous recalcitrant environmental pollutants.

D.T. Gibson in 1980 had said, "Many of these compounds bear little relationship to the biological products from which they were originally derived. For example, soils and young sediments contain thousands of substituted polycyclic aromatic hydrocarbons. These molecules, formed by the thermal alteration of cellular material, have been in contact with living organisms throughout evolutionary periods of time. Consequently, one would predict the existence of microorganisms that will degrade them, and organisms that metabolize aromatic hydrocarbons ranging in size from benzene to benzo[a] pyrene have been described".

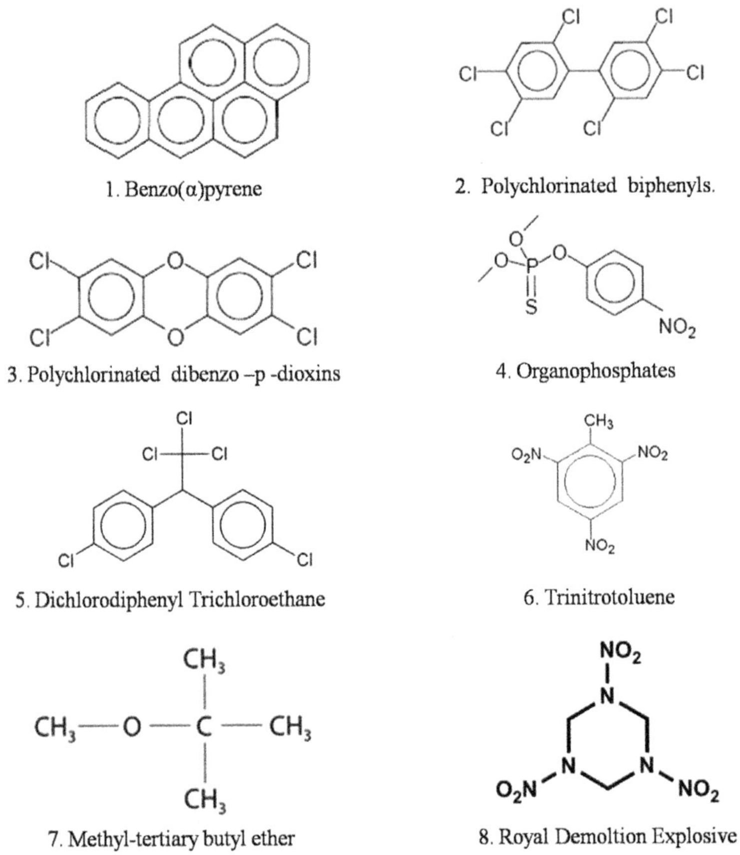

1. Benzo(α)pyrene

2. Polychlorinated biphenyls.

3. Polychlorinated dibenzo –p -dioxins

4. Organophosphates

5. Dichlorodiphenyl Trichloroethane

6. Trinitrotoluene

7. Methyl-tertiary butyl ether

8. Royal Demoltion Explosive

FIGURE 14.1   Examples of relatively persistent xenobiotics, which are dangerous recalcitrant environmental pollutants.

Organic chemicals of anthropogenic origin are not necessarily recalcitrant. There are several industrial products that are degraded by microorganisms. These compounds are readily recognized by microbial catabolic enzymes. Extensive research in biodegradation has demonstrated that a number of xenobiotic compounds such as polychlorinated biphenyls (PCBs) and nitroaromatics which once were thought to be recalcitrant are actually degraded by microbial enzymes. Moreover, microorganisms throughout geological time have been exposed to a variety of xenobiotic chemicals produced by abiotic natural processes (Fetzner, 2002).

14.3 FATE OF XENOBIOTICS

Xenobiotic compounds are present in the biosphere: air, water, soil, sediment. These pollutants are exposed to microorganisms in different environmental compartments. The most competent and efficient mechanism of bioremediation of xenobiotic compounds lies with microorganisms on the face of the Earth. The different physiochemical properties of the environment may affect the bioremediation efficiency.

14.3.1 ENVIRONMENTAL FRIENDLY BIOREMEDIATION BY NATURAL ATTENUATION

Today, bioremediation is regarded as the default method for the rehabilitation of pollutants due to its cost-effectiveness and environmental friendliness. But the extent to which such advantage can be exploited depends on the degree of technical expertise. The efficiency with which bioremediation is achieved ranges from the most intensive *ex situ* mechanism with specialized organisms, mechanical forces, heat, solvents, detergents through the stimulation of indigenous microbial communities with nutrients or oxygen to the completely passive *in situ* natural attenuation. Since 1990s, passive methods have been adopted by implementing risk-based remediation of targets, which lead to physical or chemical stabilization of pollutants instead of their elimination. Thus, to attain a better ecological status, risk reductions by stabilization become the only option. Depending on the fate of xenobiotics in air, water, soil, or sediment, they may become available to microorganisms in different environmental compartments.

14.3.2 MICROBIAL DEGRADATION OF XENOBIOTICS

Eco-friendly bioremediation involve four types of mechanisms, viz. bio-degradation, biotransformation, co-metabolism, and biosorption. The substances transformed or degraded by microorganisms are used as a source of energy, carbon, nitrogen, or other nutrient, or as final electron acceptor of a respiratory (ETC, Electron Transport Chain) process.

'Biodegradation' involves the breakdown of organic compounds, usually by microbial enzymes, into less complex compounds, like water, carbon dioxide, and the oxides or mineral salts of elements. 'Biotransformation' is the metabolic modification of the molecular structure of a compound, resulting in the loss or alteration of some characteristic properties (toxicity) of the original compound, with no loss of molecular complexity. 'Co-metabolism' is a process where a microbial population growing on one compound may fortuitously transform a contaminating chemical that cannot be used as carbon and energy source. This phenomenon has been designated 'co-oxidation' and 'gratuitous' or 'fortuitous' metabolism. Usually, the primary substrate induces production of (an) enzyme(s) that fortuitously alter(s) the molecular structure of another compound. The organisms do not benefit in any way from the co-metabolic process. Co-metabolic transformation may result in a minor modification of the molecule, or it may lead to incomplete or complete degradation. 'Biosorption' is bioengineering where metabolism independent adsorption of xenobiotics to living or dead cells takes place. Microorganisms dead or alive are successfully exploited for bioremediation of xenobiotics by biosorption.

In the biosphere, the products of bioconversion processes may be further transformed or degraded by other microorganisms and eventually leading to complete degradation by the microbial consortium. Co-metabolic processes and biodegradation by microbial consortia are of enormous ecological importance. However, persistent xenobiotics and metabolic dead-end products accumulating in the environment, become part of the soil humus, enter the food chain leading to biomagnification. Fungal bioremediation of xenobiotic compounds is depicted in Figure 14.2.

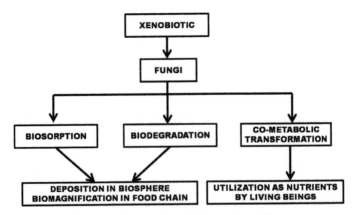

FIGURE 14.2   Fungal bioremediation of xenobiotic compounds.

14.4 FACTORS INFLUENCING RATE OF BIODEGRADATION

The extent and the rate of biodegradation depend on the chemical structure and concentration of the xenobiotic compound being degraded, the type and number of microorganisms present, and the physico-chemical properties of the environment. Bioavailability of the pollutant is controlled by several parameters such as the physical state of the pollutant (solid, liquid, gaseous), its solubility in water, and its tendency to adsorb or bind to soil or sediment particles. Sorption, immobilization, and micropore entrapment are major causes for the persistence of xenobiotics. 'Aging', is the length of time a soil or sediment has been exposed to contamination, thereby affecting bioavailability. Pollutants may form stable chemical bonds with soil and sediment material, becoming increasingly unavailable to microorganisms with the passage of time. Many xenobiotics, like the polycyclic aromatic hydrocarbons and the polychlorinated biphenyls, are poorly soluble in aqueous phase, and tend to adsorb to and be immobilized by the soil matrix and sediment material which impede or even prevent biodegradation. Concentrations of the xenobiotic compound affect biodegradation. In high concentrations, many xenobiotics are toxic to degradative bacteria. On the other hand, there is a minimum concentration below which a compound is not degraded as synthesis of catabolic enzymes may not occur when the concentration of a chemical is below a level that is effective for induction of the corresponding catabolic genes. The minimal threshold concentration depends mainly on the kinetic parameters of growth and metabolism, and on the thermodynamics of the overall transformation reaction. The substrate affinity constant is the important parameter with respect to the biodegradation of contaminants to very low concentrations. Minimal substrate concentrations for aerobic systems are in the range of 0.1 to 1.0 mg/L, and the desired end concentrations in environmental systems often are 1 μg/ L or less. Other factors that influence biodegradation are environmental conditions such as temperature, pH, water content and salinity, presence of inhibitory chemicals, availability of electron donors and nutrients, and availability of oxygen or other electron acceptors. In soil, oxygen availability is very often the limiting factor for aerobic biodegradation processes. An observed 'disappearance' of a xenobiotic from an ecosystem does not necessarily mean that it was biodegraded, since loss can occur by partial degradation, biotransformation, by volatilization, leaching, and chemical conversion (polymerization, modification, and breakdown). While the environmental fate of a chemical, one should monitor the products formed, not only the disappearance of the parent compound. The time required for xenobiotic biodegradation in the environment may range from days to weeks to years and decades. The organophosphate insecticide malathion disappears from soil within approximately one week, and the herbicide 2,4-D (2,4-dichlorophenoxyacetic acid) is degraded within four to six weeks in soil. Simple structural changes of a molecule, such as the addition of a chlorine substituent, can convert a readily biodegradable compound such as 2,4-D into a more persistent substance such as 2,4,5-T (2,4,5-trichlorophenoxyacetic acid) which is degraded in soil within approximately 6–12 months. A very persistent xenobiotic is the insecticide DDT (1,1,1-trichloro-2,2-bis[ρ-chlorophenyl]ethane), which was used extensively from the 1930s until its ban in 1979. DDT persists with an average half-life of 4.5 years in field soils, and a half-life in anoxic soils of about 700 d. Stable metabolites of DDT have been detected in soil, groundwater, and in the tissue of organisms (Fetzner, 2002).

14.5   ROLE OF FUNGI AS EFFICIENT BIOREMEDIATOR

This chapter focuses on mechanisms and pathways of fungal bioremediation of the previously mentioned xenobiotic pollutants. Fungi possess the most efficient ecological and biochemical capability to bioremediate environmental xenobiotic pollutants thereby decreasing the risk associated especially with metals, metalloids, and radionuclides by chemical or physical modification. Fungi form extended mycelial networks and have low specificity of their catabolic enzymes. Furthermore, a fungus does not always use the pollutants as their growth substrates thus making them ideal for bioremediation process. Fungi dominate the living biomass in soil and are also abundant in aqueous systems, thus it becomes the obvious choice for exploitation for bioremediation of xenobiotic pollutants.

Fungal action on either naturally-occurring or anthropogenically-derived organic and inorganic pollutants is depicted in Figure 14.3. There are aquatic and terrestrial habitats of fungi. They exist as aerial spores, yeasts, aquatic saprobes, symbiotic partners of lichens and mycorrhizal symbionts. They can form endophytic, ecto-mycorrhizal, and endo-mycorrhizal associations. They are exposed to anthropogenic chemicals which they act upon. We need to understand in detail the fungus as a unique bio-system for bioremediation (Table 14.1). Knowledge on fungal biogeochemistry in freshwater and marine systems, sediments, and the deep subsurface is required. Fungal roles have been categorized based on growth, organic and inorganic metabolism, physico-chemical attributes, and symbiotic relationships. However, many if not all of these are inter-linked, and almost all directly or indirectly depend on the mode of fungal growth (symbiotic relationships). Figure 14.4 shows the biogeochemical action of free-living and mycorrhizal fungi for metal mineral transformations.

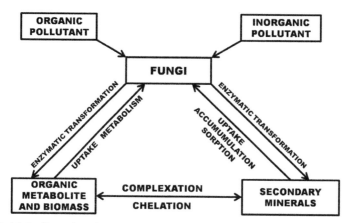

FIGURE 14.3 Fungal action on naturally-occurring and/or anthropogenically-derived organic and inorganic pollutants.

TABLE 14.1 Activities of Fungi in Biogeochemical Processes.

Fungal Activity	Biogeochemical Processes
Growth and mycelium development	Mineral tunneling; plant colonization (mycorrhizas); animal colonization (symbiotic); translocation of inorganic and organic nutrients; production of exopolymeric substances (nutrient resource for other organisms); mycelium acting as a reservoir of nitrogen and/or other elements (e.g. wood decay fungi)
Carbon and energy metabolism	Organic matter decomposition; cycling and transformations of organic compounds and biomass: carbon, hydrogen, oxygen, nitrogen, phosphorus, sulphur, metals, metalloids, radionuclides (natural and accumulated from anthropogenic sources); breakdown of polymers; changes in redox, oxygen, pH; production of inorganic and organic metabolites, for example, protons, carbon dioxide, organic acids; extracellular enzyme production; oxalate formation; metalloid methylation (e.g., arsenic, selenium); xenobiotic degradation (e.g. polynuclear aromatic hydrocarbons); organometal formation and degradation
Inorganic nutrition	Cycling of inorganic nutrient, for example, nitrogen, sulphur, phosphorus, essential and inessential metals, by transport and accumulation; transformation and incorporation of inorganic elements into macromolecules; alterations in oxidation state; metal(loid) oxido-reductions; heterotrophic nitrification; siderophore production for iron(III) capture; translocation of nitrogen, phosphorus, calcium, magnesium, sodium, potassium through mycelium to plant hosts; degradation of organic and inorganic sulphur compounds

TABLE 14.1   *(Continued)*

Fungal Activity	Biogeochemical Processes
Mineral dissolution	Bioweathering including carbonates, silicates, phosphates and sulphides; bioleaching of metals and other components; manganese dioxide (MnO_2) reduction; bioavailability of metals, phosphorus, sulphur, silicon, aluminum
Mineral formation	Element immobilization including metals, radionuclides, carbon, phosphorus, and sulphur; mycogenic carbonate formation; mycogenic metal oxalate formation; soil storage of carbon and other elements
Sorption of soluble and particulate metal	Metal distribution and bioavailability; leads to secondary mineral formation
Exo-polysaccharide production	Complexation of cations; chemical interactions of exopolysaccharide with mineral substrates
Mycorrhizas	Bioavailability of nutrient, nitrogen, phosphorus, sulphur; carbon flow and transfer between plant and fungus; mineral dissolution and metal and nutrient release from bound and mineral sources
Lichens	Bioweathering; mineral dissolution by 'lichen acids'; metal accumulation; metal sorption

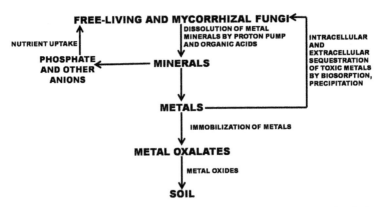

FIGURE 14.4   Biogeochemical action of free-living and mycorrhizal fungi for metal mineral transformations.

We should identify the unique characteristics of fungi that qualify them for pollutant degradation, bioremediation, the fungal way of life and its function in ecosystems where important groups of fungi act on environmental chemicals. One of the unique characteristic of fungi is their ability to attack organic compounds using a range of extracellular oxido-reductases with relatively nonspecific activities. Figure 14.5 explains the fungal degradation of organic pollutants. Detailed knowledge of enzymatic and genetic mechanisms would propose several fields of environmental biotechnology in which the use of fungi promises to be particularly effective. The list of fungal enzymes for catabolism of organic pollutants is given in Table 14.2.

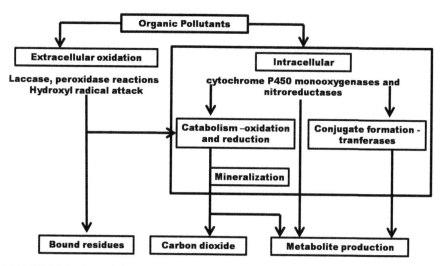

FIGURE 14.5   Fungal degradation of organic pollutants.

TABLE 14.2   Fungal Enzymes Involved in the Catabolism of Organic Pollutants.

Fungal Phylum	Enzymes	Reaction Mechanism
Ascomycota Basidiomycota	Laccases Extracellular	Direct oxidation of organic compounds like phenols, aromatic amines, and anthraquinone dyes; Activity in the acidic pH range
Ascomycota Basidiomycota Mucoromycotina	Tyrosinases extracellular/ intracellular	Hydroxylation of monophenols to o-diphenols (cresolase activity); Oxidation of o-diphenols to catechols (catecholase activity); Oxidation of various highly chlorinated phenols; Activity from the acidic to the alkaline pH range
Basiiomycota	Lignin peroxidases Extracellular	Direct oxidation of aromatic compounds like PAHs (polycyclic aromatic hydrocarbon), dyes, phenols; Activity in the acidic pH range
Basidiomycota	Manganese peroxidases Extracellular	H_2O_2-dependent one-electronoxidation of Mn^{2+} to Mn^{3+}, which subsequently oxidizes various phenols and aromatic amines; Extended substrate range in the presence of co-oxidants (organic SH-containing compounds, unsaturated fatty acids and their derivatives); Activity in the acidic pH range

TABLE 14.2 *(Continued)*

Fungal Phylum	Enzymes	Reaction Mechanism
Basidiomycota	*Coprinop-siscinerea* peroxidase Extracellular	H_2O_2-dependent one-electron oxidation of aromatic compounds; Direct oxidation of phenols and dyes; Activity from the acidic to the alkaline pH range
Basidiomycota	Versatile Peroxidases Extracellular	H_2O_2-dependent direct one-electron oxidation of phenols, dyes and aromatic compounds; H_2O_2-dependent one-electron oxidation of Mn^{2+} to Mn^{3+}, which subsequently oxidizes organic compounds; Mn^{3+}-dependent reactions as for manganese peroxidase; Activity in the acidic pH range
Basidiomycota	Dye-decolorizing Peroxidases Extracellular	H_2O_2-dependent one-electron oxidation of anthraquinone dyes (only rarely oxidized by other peroxidases); Additional hydrolysing activity; Highly stable at high pressure, high temperature and very low pH; Activity in the acidic pH range
Ascomycota	*Caldariomyces fumago* heme–thiolate chloroperoxidase Extracellular	H_2O_2-dependent halogenations of organic compounds in the presence of halides (one-electron transfer); H_2O_2-dependent one-electron oxidations of phenols and anilines in the absence of halides; H_2O_2-dependent peroxygenation (two-electron oxidation), leading to epoxidation of (cyclo)alkenes, hydroxylation of benzylic carbon and sulphoxidation of S-containing organic compounds; No activity on non-substituted aromatic rings and *n*-alkanes; Activity in the acidic pH range
Basidiomycota	Heme–thiolate peroxygenases Extracellular	H_2O_2-dependent peroxygenation of aromatic, aliphatic and heterocyclic compounds, leading to aromatic and alkylic carbon hydroxylation, double-bond epoxidation, ether cleavage, sulphoxidation or *N*-oxidation reactions (depending on the substrate); H_2O_2-dependent one-electron abstractions from phenols; H_2O_2-dependent bromination of organic substrates; Peroxygenation of various monoaromatic to polyaromatic pollutants, including PAHs, dibenzofuran, and mono-hydroxylated and polyhydroxylated products; Ether bond cleavage between aromatic and aliphatic parts of molecules and in alicyclic and aliphatic ethers (for example, MTBE: methyl-*tert*-butylether); Activity from the acidic to the alkaline pH range
Ascomycota Basidiomycota Mucoromycotina Chytridiomycota	Cytochrome-P450 Monooxygenases Cell bound	Incorporation of a single atom from O_2 into a substrate molecule, with concomitant reduction of the other atom to H_2O; Epoxidation and hydroxylation of aromatic or aliphatic structures of many pollutants, including PAHs, PCDDs (polychlorinated dibenzo-*p*-dioxin), alkanes and alkyl-substituted aromatics

TABLE 14.2   *(Continued)*

Fungal Phylum	Enzymes	Reaction Mechanism
Ascomycota Basidiomycota	Phenol-2-mono-oxygenases Cell bound	Incorporation of a single atom from O_2 into a substrate molecule, with concomitant reduction of theother atom to H_2O; *Ortho*-hydroxlyation of various (halo) phenols to the corresponding catechols
Ascomycota Basidiomycota Mucoromycotina	Nitroreductases Cell bound	NAD(P)H-dependent reduction of nitro-aromatics to hydroxyl-amino and amino(nitro) compounds, and of nitro functional groups of N-containing hetero-cycles; Reduction of TNT (2,4,6-trinitrotoluene) to hydroxylamino-dinitrotoluene and amino-dinitrotolu-enes; Formation of mononitroso derivatives and ring cleavage products from cyclic nitramine explosives; Widespread among fungi
Basidiomycota	Quinone reduc-tases Cell bound	NAD(P)H-dependent reduction of quinones; Func-tions in quinone detoxification, in the conversion of quinones arising from extracellular pollutant oxida-tion into substrates for extracellular and intracellular oxido-reductases, and in pollutant attack by hydroxyl radicals arising from quinine redox cycling; Occur-rence in white-rot and brown-rot basidiomycetes
Basidiomycota Ascomycota (perhaps)	Reductive deha-logenases Cell bound	Two-component system comprising a membrane-bound glutathione *S*-transferase that produces gluta-thionyl conjugates with concomitant chlorine removal, and a soluble glutathione conjugate reductase that releases reductively de-chlorinated compounds; Re-ductive dechlorination of chloro-hydroquinones arising from chlorophenol degradation and of diphenylether herbicides (basiodiomycetes); Perhaps responsible for reductive de-chlorination of chloro-catechols arising from PCDD degradation (ascomycetes)
Ascomycota Basidiomycota Mucoromycotina	Miscellaneous transferases Cell bound	Formation of glucoside, glucuronide, xyloside, sulphate or methyl conjugates from hydroxylated compounds; Phase II enzymes are prominent in fungal PAH metabolism but also act on other pollutants; Widespread among fungi

Bioremediation by fungi is due to their unique characteristics coupled with environmental circumstances, which make fungi particularly suitable for application in environmental biotechnology. Fungi should be considered for pollutant classes that are inefficiently degraded by bacteria. Bacteria might be disadvantaged if substrates contain rare structural elements, have a low bioavailability, contain little energy or occur permanently at min-ute concentrations. With disturbingly high biological activities, human and veterinary drugs (including synthetic hormones or antibiotics) and chemicals with low environmental concentration together with active ingredients are now found in environmental matrices (water, aquatic sediments and soil), as they are not retained in wastewater treatment plants. Ligninolytic basidiomycetes, mitosporic ascomycetes and aquatic fungi, are known to degrade endocrine disrupting chemical (EDC) (nonylphenol, bisphenol A and 17-α-ethinylestradiol); analgesic, anti-epileptic and non-steroidal anti-inflammatory drugs; X-ray contrast agents; polycyclic musk fragrances; and ingredients of personal care products. The use of filamentous fungi is advantageous for the long-range transport or translocation of essential factors (nutrients, water, and pollutants). They are required for the transformation or detoxification of environmental chemicals. Fungi in their natural habitat cope with resource heterogeneity by translocating resources between different parts of their mycelium. Time-lapse recordings demonstrate the immense traffic of resources in fungal hyphae, in the direction of growth, for example hyphal transport of autofluorescent polyaromatic hydrocarbons (PAHs) particularly. Resource translocation is the recycling of hyphal biomass located in substrate-depleted regions for the benefit of exploration for food in other regions. Owing to growing fungal hyphae they have wedge-shaped, hydrophobic tips which penetrate air–water interfaces and soil aggregates.

14.5.1 ECOLOGICAL FEATURES OF FUNGI

The kingdom fungi include moulds, mushrooms, lichens, rusts, smuts, and yeasts – eukaryotes with remarkably diverse life cycles. Fungi exist as saprobes, parasites, mutualists of plants (mycorrhizae), or algae (lichens). Fungi have been defined as eukaryotic, heterotrophic, absorptive organisms that typically develop a branched, tubular body called a mycelium and reproduce by means of sporulation. The microscopic diameters of these fungal hyphae are between 2 and 10 μm. Some of the largest living organisms on earth are fungi with networks extending over hundreds of hectares. Therefore, fungi can be regarded as 'macro-organisms packaged in microscopic units' that is, they exhibit a unique lifestyle that is adapted to heterogeneous environments. Table 14.3 lists the major organic pollutants degraded by diverse fungal phyla and subphyla.

TABLE 14.3 Organic Pollutants Degraded by Various Fungal Phyla and Subphyla.

Phylum or Subphylum	Organic Chemicals Degraded
Microsporidia, Kickxellomycotina, Zoopagomycotina, Entomophthoromycotina, Blastocladiomycota, Chytridiomycota	PAHs
Mucoromycotina, Neocallimastigomycota	Benzoquinoline, biphenyl, synthetic dyes, TNT, pesticides
Glomeromycota	PAHs, Pesticides
Ascomycetes	
Pezizomycotina	Alkanes, alkylbenzenes, biphenyl, chlorophenols, coal tar oil, crude oil, diesel, EDCs, fragrances, PAHs, PCDDS, pesticides, synthetic dyes, TNT, toluene
Saccharomycotina	Alkanes, alkylbenzenes, biphenyl, crude oil, EDCs, PAHs, TNT
Other ascomycetes	Alkanes, diesel, coal tar oil, crude oil, MTBE, PAHs, pesticides, RDX, toluene, synthetic dyes
Basidiomycota	
Agaricomycotina	Alkanes, BTEX compounds, chloroaliphatics, lignols, phenols, crude oil, coal tar, EDCs, PAHs, PCBs, PCDDs, PCDFs, personal care product ingredients, pesticides, pharmaceuticals, drugs, RDX, synthetic dyes, synthetic polymers, TNT, other nitroaromatics
Pucciniomycotina	Cresols, crude oil, dibenzothiophene, PAHs, RDX

14.5.2 CATABOLISM OF RECALCITRANT ORGANIC POLLUTANTS

Both polychlorinated dibenzo-ρ-dioxins (PCDDs) and polychlorinated dibenzofurans (PCDFs) are highly oxidized owing to the electronegativity of the chlorine atoms which withdraw electrons from the aromatic rings. These compounds are poor electron donors and always occur at very low environmental concentrations. The white-rot fungus *Phanerochaete sordida* transforms 2,3,7,8-tetrachlorodibenzo-ρ-dioxin (TCDD) into chlorocatechols in 10 d and *Phanerochaete chrysosporium* even mineralizes the compound. PCDDs with 6–8 chlorines and PCDFs with 4–8 chlorines were degraded by *Phanerochaete sordida*. White-rot and litter-decaying basidiomycetes mineralize TNT (2,4,6-trinitrotoluene), the nerve gases VX and sarin rapidly. Fungi of the Ascomycota, Basidiomycota and Mucoromycotina hydroxylate PAHs intracellularly and convert them to water-soluble products which are then excreted. This is a nonspecific mechanism for their detoxification. Various fungi use extracellular oxido-reductases in PAH degradation and mineralize even high-molecular-mass PAH, such as the highly carcinogenic benzo[*a*]pyrene. Benzo[*a*]pyrene is converted to quinones, carbon dioxides, gluconic acids, and sulphate conjugates. Figure 14.6 shows the fungal–metal interactions for bioremediation. Fungi can also precipitate a variety of uranium-containing minerals after growth on uranium oxides (Gadd, 2007; Harms et al., 2011). Several fungi and oomycotes like *Phaeosphaeria spartinicola*, *Mycosphaerella* sp., *Buergenerula spartinae*, *Phaeosphaeria alima*, *Passeriniella obiones*, *Phaeosphaeria neomaritima*, *Pleospora spartinae*, *Tremateia halophila*, *Spartina patens*, *Phoma* sp., and *Fusarium lateritum* have exhibited their bioremediation potential of Dimethylsulfoniopropionate (DMSP) from marine plants by DMSP lyase activity (Bacic et al., 1998). Bioremediation of other recalcitrant pollutants by fungi is listed in Table 14.4.

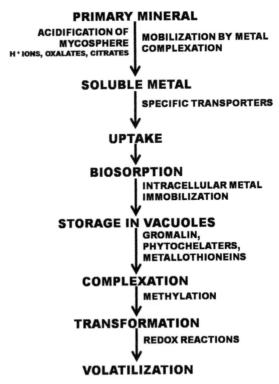

FIGURE 14.6   Fungal-metal interactions for bioremediation.

TABLE 14.4 Catabolism of Recalcitrant Pollutants.

Fungi	Pollutant	Reference
Fusarium oxysporium f. sp. *pisi*	Butyl benzyl phthalate	Kim et al. (2002)
Aspergillus sp.	Sulphur in coal	Acharya et al. (2005)
Aspergillus niger, *Phoma* sp., *Penicillium* sp., *Geomyces pannorum*, *Candida rugosa*	Polyurethane	Russell et al. (2011); Norton (2012)
Graphium sp., *Fusarium* sp., *Penicillium* sp., *Hypocrea* sp., *Discophaerina fagi*, *Aureobasidium pullulans*, *Botryodiplodia theobromae*, *Bipolaris* sp., *Cunnighamella* sp., *Drechslera* sp., *Helminthosporium* sp., *Macrophomina phaseolina*, *Mucor* sp., *Rhizopus* sp., *Talariomyces* sp., *Aspergillus flavus*	Hydrocarbon	Adekunle and Oluyode (2005); Rauch et al. (2006); Norton (2012)
Trichoderma viride, *Aspergillus niger*, *Aspergillus fumigatus*, *Curvularia* sp., *Fusarium* sp.	Municipal solid waste	Gautam et al. (2011)
Fusarium solani, *Penicillium variabile*	Rubber	Linos and Steinbüchel (2001)
Penicillium frequentans, *Kvehneromyces mutabilis*	TNT	Spain (1995)
Mortierella sp.	DCP	Nakagawa et al. (2006)
Nematoloma frowardii	Aromatic and Aliphatic compounds	Hofrichter et al. (1998)
Antrodia vaillantii FRLP-14G	Cu, Cr, As	Sierra-Alvarez (2009)
Aspergillus niger, *Fusarium solani*, *Fusarium oxysporium* f. sp. *melonis*	Nitrites	Martínková et al. (2009)
Rhodotorula, *Acremonium* spp.	RDX (Royal Demolition Explosive)	Harms et al. (2011)
Penicillum sp.	Atrazine	Singh et al. (2008)
Aspergillus fumigatus, *A. japonicas*, *A. niger*, *A. carbonarius*, *A. ellipticus*, *A. foetidus*, *A. alliaceus*, *A. clavatus*, *A. flavus*, *A. ochraceus*, *A. terreus*, *A. ustus*, *A. versicolor*, *A. wentii*, *Rhizopus homothallicus*, *Rhizopus japonicus*, *R. oryzae*, *R. stolonifer*, *Botrytis cinerea*, *Alternaria* sp., *Aureobasidium pullulans*, *Emericella nidulans*, *Mucor* sp., *Penicillium aurantiogriseum*, *Penicillium spinulosum*, *Trichoderma* sp., *Trichothecium roseum*	Ochratoxin A	Abrunhosa et al. (2002, 2010)

The following fungi were found in domestic sewage effluents treatment plant: nonpigmented yeasts, *Geotrichum candidum*, *Penicillium* spp. (including *P. lilacinum*), *Mucor* spp.,Yeast-like fungi, *Rhodotorula* spp., *Fusarium aquaeductuum*, *Fusarium* spp., *Mycelia sterile*, *Trichoderma viride*, *Phoma* spp., *Cladosporium cladosporioides*, *Margarinomyces heteromorphum*, *Gliomastix convolute*, *Aspergillus* spp., *Isoachlya* spp., *Moniliaceae*, *Rhizopus* spp., *Scopulariopsis brevicaulis*, *Pullularia pullulans*, *Cephalosporium* spp., *Mucorales*, *Saprolegnia* spp., *Syncephalastrum* spp., *Acrostalagmus cinnabarinus*, *Aspergillus flavus*, *Cunninghamella* sp., *Monilia sitophila*, *Myrothecium verrucaria*, *Spicaria violacea*, *Thamnidium* sp., *Trichoderma album* and *Verticillium* sp. (Becker et al., 1954). This proves that these indigenous fungal species are capable of bioremediating domestic sewage. Different fungal species isolated from endosulfan treated pine forest soil are *Alternaria alternate*, *A. humicola*, *Aspergillus candidus*, *A. flavus*, *A. fumigatus*, *A. niger*, *A. terreu*, *Cladosporium cladosporioides*, *Fusarium lateritium*, *F. solani*, *Gliocladium deliquescen*, *G. fimbriatum*, *Mammaria echinobotryoides*, *Mucor hiemalis*, *M. varians*, *Paecilomyces inflatus*, *P. lilacinus*, *Penicillium decumbens*, *P. frequentans*, *P. lilacinum*, *P. notatum*, *P. restrictum*, *P. spinulosum*, *P. striatum*, *P. tardum*, *Phoma fimeti*, *Trichoderma atroviride*, *T. aureoviride*, *T. harzianum*, *T. longibrachiatum*, *T. pseudokoningii*, and *Verticillium tenuissimum* (Bisht et al., 2014). These fungi are surviving in the soil contaminated with endosulfan, which indicates that they are degrading this pollutant.

14.5.3 BIOSORPTION BY LIVE AND DEAD FUNGI

Biosorption is considered to be an eco-friendly technique. This technique was first introduced by Ames Crosta Mills and Company, Ltd. in 1973. Biosorption is defined as the ability of biological materials to accumulate pollutants from waste water through physico-chemical interaction. It can be also defined as the process that utilizes inexpensive dead biomass to sequester toxic xenobiotics. Biosorption process involving the use of dead biomass is more faster in comparison to living cells as it is cell surface based binding and displays high affinity for metal ion removal from aqueous solution (Mali et al., 2014a, 2014b; Takey et al., 2014). Table 14.5 gives the biosorption capacity of different fungi with different heavy metals. According to investigations, the possibility of living organism to accumulate metallic elements could be toxic. In comparison to living cells the use of dead biomass is an easy and a non-destructive method for recovery of adsorbed metal ions which allows regeneration and reuse of biosorbents. The process of bioremediation by live organism has several limitations: (a) only some of the contaminants are biodegradable; (b) bioaccumulation and biomagnification; (c) not all contaminants are treated, some heavy metals are not absorbed by organisms; (d) biological process is highly specific; (e) requires suitable environmental growth conditions; (f) appropriate level of nutrients and contaminants; and (g) contaminants may be present in all the three phases (solid, liquid, and gas) which might not be adequately bioremediated. To overcome these problems the process known as biosorption is applied, which is becoming an excellent bioengineering process of heavy metal removal. The mechanism of biosorption is represented in Figure 14.7 which includes coplexation, precipitation, reduction, chelation, and ion exchange. Figure 14.8 gives the schematic representation of advantages of biosorption (Mali et al., 2014a).

TABLE 14.5   Biosorption Capacity of Various Fungi with Different Heavy Metals (Mali et al., 2014a).

Heavy Metal	Organism	% of Sorption
Cd	*Penicillium*	95.27
	Aspergillus amari	51.69
	Trichoderma species	89
Ni	*Aspergillus amari*	58.74
	Trichoderma species	77–89.41
Cu	*Amanita muscaria*	~90
	Spirulina platensis	~90.6
Zn	*Aspergillus tamarii*	54.3
Pb	Mycelial *Aspergillus tamarii*	74
Sb	*Agaricus campester*	~95
Al	*Agaricus campester*	~95
Cr	*Trichoderma* species	81.5
Mn	*Aspergillus tamarii*	46.99
Fe	Mycelial *Aspergillus tamarii*	46

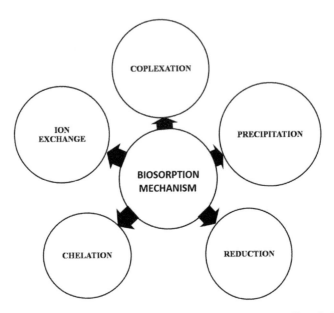

FIGURE 14.7   Schematic representation of biosorption mechanism (Mali et al., 2014a).

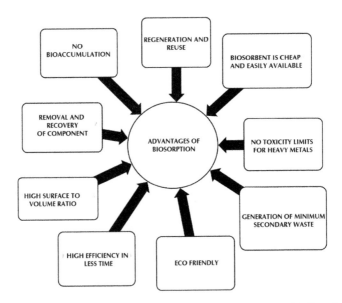

FIGURE 14.8   Schematic representation of advantages of biosorption (Mali et al., 2014a).

14.5.4 FUNGI AS BIOINDICATORS FOR BIOPROSPECTING

Extensive research has shown that a correlation exists between the quantities of metals in a growth substrate and the amounts subsequently found in fruit bodies of fungi. Therefore, they are bioindicators of metal and radionuclide contamination (Table 14.6). The concept of bioindicators is in terms of reaction indicators and accumulation indicators. Reaction indicators are individual organisms and/or communities that may decline or disappear (sensitive species) or show increases (tolerant species). In case of accumulation indicators, the indicator organism is analysed for the pollutant theoretically and technically. Some organisms can serve as both reaction and accumulation indicators. *Laccaria laccata* increased in frequency at more polluted locations. *Amanita muscaria* and several species of *Boletus* tolerate high metal pollution. *Hebeloma cylindrosporum* exhibited the highest uranium and thorium transfer factors, suggesting that this species was a good bioindicator of soil radioactive content. *H. cylindrosporum* and *Lycoperdon perlatum* exhibited plutonium-239 (^{239}Pu), plutonium-240 (^{240}Pu) and americium-241 (^{241}Am) transfer values. These species were therefore proposed as bioindicators for ^{239}Pu, ^{240}Pu and ^{241}Am. Metal concentrations are species-dependent, and the highest levels were found in *Calvatia utriformis* (235.5 mg Cu/kg), *Macrolepiota procera* (217.8 mg Cu/kg), *Agaricus macrosporus* (217.7 mg Cu/kg), *C. utrifomis* (265.8 mg Zn/kg), *Lactarius deliciosus* (231.0 mg Zn/kg), and *A. macrosporus* (221.3 mg Zn/kg) for copper and zinc, respectively. Mercury concentrations in fungi generally occur in the range 0.03–21.60 mg kg$_{D.W.}^{-1}$ (Dry Weight), although concentrations greater than 100 mg [kg DW]$^{-1}$ have been recorded from polluted sites. The concentrations of seven metals (lead, cadmium, manganese, copper, nickel, silver, and chromium) were determined in 32 species of wild mushrooms from Konya, an Inner Anatolian region of Turkey. The highest metal concentrations were given as 39 mg kg^{-1} Pb and 3.72 mg kg^{-1} Cd in *Trichaptum abietinum*, 467 mg kg^{-1} Mn in *Panaeolus sphinctrinus*, 326 mg kg^{-1} Cu in *Trametes versicolor*, 69.4 mg kg^{-1} Ni in *Helvella spadicea*, 6.97 mg kg^{-1} Ag in *Agaricus campestris*, and 84.5 mg kg^{-1} Cr in *Phellinus igniarius*. The maximum contents were 1.52, 2.22, and 60.2 mg kg^{-1} in *Pleurotus eryngii* (for lead), *Amanita vaginata* (for cadmium), and *Helvella leucomelana* (for copper), respectively. Environmental factors are of paramount importance in relation to metal accumulation by macrofungi, and include physico-chemical soil properties like moisture and temperature, all of which influence metal availability as well as the physiological activity of the fungus (Gadd, 2007).

TABLE 14.6   Bioindicators for Metal Pollution.

Metals	Fungi
Mercury, cadmium	*Agaricus arvensis, A. campestris, A. edulis, Mycena pura*
Mercury	*Amanita rubescens, A. strobiliformis, Coprinus comatus, Lycoperdon perlatum, A. haemorrhoidarius, A. xanthodermus, Marasmius oreades*
Lead, zinc, copper	*Agaricus* sp., *Lycoperdon* sp.

14.5.5   UNIQUE POTENTIAL OF WHITE-ROT FUNGI FOR MYCOREMEDIATION

White-rot fungi digest lignin by the secretion of ligninolytic enzymes and give a bleached appearance to wood, from undissolved cellulose, hence their name. White-rot fungi are excellent mycoremediators of toxins held together by hydrogen–carbon bonds. Table 14.7 gives the list of xenobiotics bioremediated by white-rot fungi (Hansen, 2012). The prospect of using principally white-rot fungi for mycoremediation is because of its effective degradation of a wide range of highly recalcitrant organopollutants due to their release of extra-cellular lignin-modifying enzymes, with a low substrate-specificity, which enables them to act upon various molecules that are broadly similar to lignin. The enzymes of the lignin degradation system of white-rot fungi being extracellular, they do not need to internalize the pollutants and thus enabling the fungi to tolerate a high concentration of pollutants. The enzymes present in the system employed for degrading lignin include lignin-peroxidase (LiP), manganese peroxidase (MnP), various H_2O_2 producing enzymes and laccase. White-rot fungi cannot utilize lignin as a source of energy for growth and instead require cosubstrates as a carbon source. They grow by hyphal extension and thus can reach pollutants in the soil in ways that other organisms cannot.

The white-rot fungus *Phanerochaete chrysosporium* is an ideal model for bioremediation by fungi, since it is more efficient than other fungi or microorganisms in degrading toxic or insoluble materials like chlorinated organic compound (chloroaliphatics, chlorolignols, chlorophenols, polychlorinated biphenyls, and PCDDs), aromatic hydrocarbon (benzene, toluene, ethylbenzene, and xylenes – BTEX compounds), polyaromatic hydrocarbons, synthetic dyes, lignocellulosic materials, nitro-substituted compounds (nitroaromatic and *N*-heterocyclic explosives – TNT and RDX, respectively), modified synthetic polymers and pesticides.

TABLE 14.7 Xenobiotics Bioremediated by White-Rot Fungi.

Type	Examples
Polycyclic aromatic hydrocarbon	Anthracene, 2-methyl anthracene, 9-methyl anthracene, benzo(α) pyrene, fluorene, napthalene, acenapthene, acenapthylene, phenanthrene, biphenylene, pyrene
Chlorinated aromatic compounds	Chlorophenols (e.g. pentachlorophenols [PCP], trichlorophenols [TCP], dichlorophenols [DCP]); Chlorolignols, 2,4-dichlorophenoxyacetic acid (2,4-D), 2,4,5-trichlorophenoxyacetic acid (2,4,5-T), Polychlorinated biphenyls (PCBs), dioxins, chlorobenzenes
Dyes	Azure B, Congo red, Disperse Yellow 3 (DY3), Orange II, Poly R, Reactive Violet 5, Reactive Black 5, Reactive Orange 96, Reactive Brilliant Blue R (RBBR), Solvent yellow 14, Tropaeolin
Nitroaromatics	TNT (2,4,6-trinitrotoluene); 2,4-dinitrotoluene; 2-amino-4,6-dinitrotoluene; 1-chloro-2,4-dinitrobenzene; 2,4-dichloro-1-nitrobenzene; 1,3-dinitrobenzene
Pesticides	Alachlor, Aldrin, Chlordane, Heptachlor, Lindane, Mirex, Atrazine, 1,1,1-trichloro-2,2-bis(4-chlorophenyl)ethane [DDT]
Other environmental pollutants	Benzene, Toluene, Ethylbenzene, o-, m-, p-xylenes (BTEX compounds), linear alkylbenzene sulfonate (LAS), trichloroethylene

It presents simultaneous oxidative and reductive mechanisms which permit its use in many different situations, regarding the type of contamination, its degree, and the nature of the site itself. A number of other white-rot fungi also can degrade persistent xenobiotic compounds, e.g., *Pleurotus ostreatus*, *Trametes versicolor*, *Bjerkandera adusta*, *Lentinula edodes*, *Irpex lacteus*, *Agaricus bisporus*, *Pleurotus tuberregium*, and *Pleurotus pulmonarius* (Rhodes, 2014).

Seven white-rot fungi, viz. *Pleurotus cystidiosus*, *Pleurotus sajor-caju*, *Pleurotus ostreatus*, *Polystictus sanguineaus*, *Trametes socotrana*, *Trametes versicolor*, and *Phanerochaete chrysosporium* degrade the pesticides like simazine, trifluralin, and dieldrin (Fragoeiro, 2005; Magan et al., 2010). Ten different genera of white-rot fungi are capable of phenanthrene degradation – *Coriolopsis gallica*, *Irpex lacteus*, *Phanerochaete chrysosporium*, *Pleurotus ostreatus*, *Phlebia lindtneri*, *Phlebia radiate*, *Trametes versicolor*, *Trichaptum biforme*, *Puntularia* sp., *Bjerkandera* sp. and *Phomopsis* sp. (B3) (Dai et al., 2010; Young, 2012). Biodegradation of hexachlorocyclohexane (HCH) isomers by several white-rot fungi after 10 d in liquid medium are as follows: *Phanerochaete chrysosporium*, *Phanerochaete sordid*, *Bjerkandera adusta*, *Poliporus ciliates*, *Phlebia radiate*, *Lentinus tigrinus*,

Stereum hirsutum, Pleurotus eryngii, Irpex lacteus and *Trametes hirsutus* (Quintero et al., 2008). Several white-rot fungi like *Pleurotus* sp., *Trametes versicolor, Panus tigrinus, Phanerochaete chrysosporium, Trametes villosa, Poliporus pinsitus, Cladosporium* sp., are used in the treatment of wastes containing phenols, chlorophenols, oligomeric polyphenols coming from oil, coking and olive mill wastewaters (Tišma et al., 2010). Tables 14.8–14.11 prove the unique power of white-rot fungus for bioremediation of the most stubborn and diverse xenobiotics (Bogan et al., 1996; Pickard et al., 1999; Raghukumar, 2000; Mori et al., 2003; Potin et al., 2004; Varasaritha et al., 2010; Norton, 2012; Singh, 2013).

TABLE 14.8   White-Rot Fungi Used in Decolorization of Industrial Effluents.

White-Rot fungi	Industry
Coriolus versicolor	Textile
Funalia trogii	
Pleurotus ostreatus	
Phanerochaete chrysosporium	Pulp and paper
Funalia trogii	
Pleurotus ostreatus	
Phanerochaete crasa	
Fomitopsis sp. IMER2	
Trametes hirsute	
Flavodon flavus	
Marasmius quercophilius	
Polysporus ostreatus	
Pycnoporus cinnabarinus	
Pycnoporus coccineus	Food
Trametes pubescens MB 89	
Trametes versicolor	
Pleurotus sajor	
Panus tigrinus	
Pleurotus ostreatus	
Phanerochaete chrysosporium	Coke

TABLE 14.9 White-Rot Fungi Used in Treatment of Polycyclic Aromatic Hydrocarbons (PAHs).

White-Rot fungi	Waste Treated
Pleurotus ostreatus	PAHs aged creosote contaminated soil
Bjerkandera adusta, Irpex lacteus, Lentinus tigrinus	PAHs in forest and salt marsh soils
Bjerkandera adusta	Hexachlorocyclohexane (HCH) isomers present in a spiked soil
Phanerochaete chrysosporium, Pleurotus pulmonarius	Aromatic hydrocarbons in an aged contaminated soil containing high concentrations of heavy metals
Phanerochaete laevis HHB-1625, *Phlebia lindtneri, Coriolopsis gallica* UAMH-8260, *Stropharia rugoso-annulata, Cladosporium sphaerospermum*	PAH

TABLE 14.10 White-Rot Fungi Used in Treatment of Dyes.

White-Rot fungi	Dye(s)
Phanerochaete chrysosporium, Trametes versicolor, Bjerkandera sp., *Clitocybula dusenii, Pleurotus eryngii*	Anthraquinone dyes, Azo dyes, indigo carmine, Direct Red-80
Phanerochaete chrysosporium, Coriolus versicolor, Pleurotus ostreatus, Pleurotus sajorcaju	Reactive Blue 4, Reactive Red 2
Phanerochaete chrysosporium, Phanerochaete ostreatus	Direct Blue 71, Direct Red 80, Direct Yellow 106, Reactive Blue 222, Reactive Red 195, Reactive Yellow 145, Reactive Black 5, Acid Blue 62, Acid Yellow 49, Acid Red 266
Coriolus versicolor	Poly S119
Pleurotus ostreatus, Coriolus versicolor, Funalia trogii	Remazol Brilliant Blue Royal, Drimaren Blue
Pleurotus ostreatus, Cladosporium cladosporioides	Acid Orange 7, Acid Orange 8, Mordant Violet 5, Acid Blue 193, Acid Black 210, Crystal Violet, Reactive Black B(S), Reactive Black BL/LPR
Dichomitus squalens, Daedalea flavida, Irpex flavus, Polyporus sanguineus	Coracryl dyes (black, pink, violet, red), Reactive dyes (yellow, orange, red), Rathiodal dyes (scarlet)
Geotrichum candidum, Coriolus versicolor, Phanerochaete chrysosporium, M. sterile	De-colorization of digested molasses spent wash (DMSW)

TABLE 14.11 Other Applications of Bioremediation by White-Rot Fungi.

White-Rot fungi	Xenobiotics	References
Polysporus sp. S133, *Pleurotus tuberregium* (Fr.) Sing.	Crude oil	Isikhuemhen et al. (2003), Kristanti et al. (2011)
Phanerochaete velutina, Stropharia rugosoannulata, Gymnophilus luteofolius	Nitro-aromatic compounds	Tuomela et al. (2013)
Trametes trogii UAMH 8156, *Trametes hirsuta* UAMH 8165, *Trametes versicolor* IFO 30340, *Phanerochaete chrysosporium* ATCC 24725, *Tyromyces palustris* IFO 30339	Dibenzyl sulphide	Van Hamme et al. (2003)
Pleurotus ostreatus	Alkaline batteries	Calamur (2012)
Cladosporium resinae	Polyurethane	Russell et al. (2011)
Cladosporium cladosporioides	Rubber	Linos and Steinbüchel (2001)
Cladosporium sp., *Pleurotus ostreatus, Phanerochaete chrysosporium*	Ochratoxin A	Abrunhosa et al. (2002, 2010)
Cladosporium sp.	Hydrocarbon	Adekunle and Oluyode (2005)
Stropharia coronilla	Benzo[α]pyrene	Steffen et al. (2003)
Phanerochaete chrysosporium	RDX	Harms et al. (2011)

14.5.5.1 PRACTICAL IMPLEMENTATION OF MYCOREMEDIATION USING WHITE-ROT FUNGI

In order to use white-rot fungi successfully for bioremediation, knowledge of fungal physiology, biochemistry, enzymology, ecology, genetics, molecular biology, and engineering, among other cognate subjects is required. Substrates such as wood chips, wheat straw, peat, corn cobs, sawdust, a nutrient-fortified mixture of grain and sawdust, bark, rice, annual plant stems and wood, fish oil, alfalfa, spent mushroom compost, sugarcane bagasse, coffee pulp, sugar beet pulp, okra, canola meal, cyclodextrins, and surfactants can be used as co-substrates in inoculum production either off-site or on-site, or as mixed with contaminated soils to improve the processes of degradation. It is critical to attain the correct nitrogen/carbon ratio in the substrates used, so to avoid any impeding effect on the efficiency of the fungi in the bioremediation process. Fungal inocula 'encapsulated' with alginate, gelatin, agarose, carrageenan, chitosan, in the form of pellets, may offer a better outcome than with inocula produced using bulk substrates. They preserve the viability of the inoculum and contribute nutrients to maximally support the degradation of pollutants. This, furthermore, increases the survival and effectiveness of the introduced species (Rhodes, 2014).

14.5.6 MYCOREMEDIATION BY ENDOPHYTES

Endophytes are hyper-diverse microorganisms that live within the inner tissues of plants, without causing any overt disease symptoms. These organisms enter their hosts by penetrating exterior surfaces, and some play a key role in plant decomposition following host tissue death. These endophytic fungi contribute to decomposition of lignocellulose polymers and are major contributors to the carbon cycle. Table 14.12 gives the list of these microorganisms which have the ability to degrade a polymer as complex as lignocellulose suggesting that these organisms offer promise for their ability to degrade other complex polymers, such as those present in plastics, PAH, leachate, mine-tailings. The unique biological niche of endophytes as endosymbionts of tissues rich in complex carbon polymers justifies the investigation of their wider metabolic capabilities. Each of the more than 300,000 land plant species on Earth potentially hosts multiple endophyte species. Only a small sampling of plants has been examined for their endophytic associations, yet many of these organisms can be readily cultured. Individual trees can harbor hundreds of endophytic species, some of which are known but many of which are yet new to science (Russell et al., 2011).

TABLE 14.12   Mycoremediation of Xenobiotics by Endophytes.

Endophytic Fungi	Xenobiotics	References
Ceratobasidium stevensi	Polyaromatic hydrocarbon	Dai et al. (2010)
Colletotrichum gloeosporiodes	Leachate	Rashmi et al. (2014)
Alternaria sp., *Cladosporium* sp., *Paecilomyces* sp., *Penicillium* sp., *Aspergillus* sp.	Mine tailings	Garza et al. (2013)
Lasiodiplodia sp., *Bionectria* sp., *Alternaria dauci*, *Nectria* sp., *Pestalotiopsis microspora*, *Edenia gomezpompae*, *Pleosporales* sp., *Phaeosphaeria* sp., *Plectosphaerella* sp., *Neonectria* sp.	Synthetic polymer	Russell et al. (2011)
Phomopsis sp.	Phenanthrene	Tian et al. (2007)
Paecilomyces lilacinus	Rubber	Linos and Steinbüchel (2001)
Paecilomyces	Hydrocarbon	Adekunle and Oluyode (2005)

14.6   CONCLUSION AND FUTURE DIRECTION

There is always a great demand for sustainable, cheap and tailor made technologies for 'bioremediation'. There is always a high impetus to translate powerful ecosystem of fungi into ecology based technologies. The degree of mechanical intervention in natural attenuation of soil is quite low, which probably favors the establishment of filamentous fungi. Many fungi produce exudates those might serve as auxiliary carbon source for pollutant degrading bacteria. Therefore, it is crystal clear that fungi have important biochemical roles in the biosphere. Fungi being ubiquitous member of sub-aerial and sub-soil environments, they become dominant group in metal polluted habitats. Their oligotrophic growth and explorative filamentous growth habit, flexible growth strategies, and resistance to extreme environmental factors including metal toxicity, irradiation, and desiccation make 'fungi' successful colonizers of earth surface. Among the numerous fungal species the 'White Rot Fungi' deserve a special mention as a highly promising and efficient bioremediation agent. They have a capacity to produce a specialized group of peroxidases called 'The Ligninolytic System' which is capable of biodegradation of anthropogenic chemicals. An important aspect of passive bioremediation technologies involving fungi, besides their low cost is the common acceptance of risk-based clean-up standards such as those currently present in the legislation of two countries, viz. USA and UK. Thus, there are important financial, ecological, and legal reasons for a better understanding of fungal activities and their implementation in environmental technology leading to 'fungal bioremediation'.

KEYWORDS

- **Biodegradation**
- **Bioindicator**
- **Bioprospecting**
- **Bioremediation**
- **Biosorption**
- **Biotransformation**
- **Co-metabolism**

- **Endophytes**
- **Mycoremediation**
- **Natural attenuation**
- **Recalcitrant compounds**
- **Unique lignolytic system**
- **White-rot fungi**
- **Xenobiotics**

REFERENCES

Abrunhosa, L.; Serra, R.; Venâncio, A. Biodegradation of Ochratoxin A by fungi isolated from grapes. *J Agric Food Chem* 2002, 50, 7493–7496.

Abrunhosa, L; Paterson, R. R.; Venâncio, A. Biodegradation of Ochratoxin A for food and feed decontamination. *Toxins (Basel)* 2010, 2(5), 1078–1099.

Acharya, C.; Sukla, L. B.; Misra, V. N. Biological elimination of sulphur from high sulphur coal by Aspergillus-like fungi. *Fuel* 2005, 84, 1597–1600.

Adekunle, A. A.; Oluyode, T. F. Biodegradation of crude petroleum and petroleum products by fungi isolated from two oil seeds (melon and soybean). *J Environ Biol* 2005, 26(1), 37–42.

Bacic, M. K.; Newell, S. Y.; Yoch, D. C. Release of dimethylsulfide from dimethylsulfoniopropionate by plant-associated salt marsh fungi. *Appl Environ Microbiol* 1998, 64, 1484–1489.

Becker, J. G.; Shaw, C. G. *Fungi in domestic sewage-treatment plants.* State College of Washington, Pullman, 1954, pp 173–180.

Bisht, J.; Harsh, N. S. K.; Palni, I. M. S.; Pande, V. Effect of repeated application of endosulfan on fungal population of pine forest soil. *Biotechnol Int* 2014, 7(1), 11–20.

Bogan, B. W.; Lamar, R. T. Polycyclic aromatic hydrocarbon degrading capabilities of *Phanerochaete laevis* HHB-1625 and its extracellular ligninolytic enzymes. *Appl Environ Microbiol* 1996, 62, 1597–1603.

Calamur, S. Mycoremediation of household hazardous waste through *Pleurotus ostreatus*, 2012, https://www.clu-in.org/download/studentpapers/Fungal-Bioremediation-of-Household-Hazardous-Waste.pdf

Dai, C. C.; Tian, L. S.; Zhao, Y. T.; Chen, Y.; Xie, H. Degradation of phenanthrene by the endophytic fungus *Ceratobasidium stevensii* found in *Bischofia polycarpa*. *Biodegredation* 2010, 21, 245–255.

Fetzner, S. Biodegradation of xenobiotics. In *Biotechnology, Encyclopedia of Life Support Systems*, Doelle, H. W.; Da Silva, E. J., Eds.; UNESCO, Eolss Publishers: Oxford, UK, 2002. http://www.eolss.net/Eolss-sampleAllChapter.aspx

Fragoeiro, S. I. D. S. *Use of fungi in bioremediation of pesticides. Cranfield University*: Bedford, UK, 2005, pp 1–241.

Gadd, G. M. Geomycology: Biogeochemical transformations of rocks, minerals, metals and radionuclides by fungi, bioweathering and bioremediation. *Mycol Res* 2007, 111(1), 3–49.

Garza, J. R.; Allieri, M. A. A.; Murrieta, M. S. V.; Hernandez, M. O. F.; Tovar, A. V. R.; Hu, E. T. W. Study of endophytic and soil fungi from tailing mines metallopytes plants: Diversity and potential in bioremediation. *XV Congreso Nacional de Biotecnología y Bioingeniería*, 2013.

Gautam, S. P.; Bundela, P. S.; Pandey, A. K.; Awasthi, M. K.; Sarsaiya, S. Isolation, identification and cultural optimization of indigenous fungal isolates as a potential bioconversion agent of municipal solid waste. *Ann Environ Sci* 2011, 5, 23–34.

Hansen, J. Mycoremediation, 2012, http://rydberg.biology.colostate.edu/phytoremediation/2012/Mycoremediation%20by%20James%20Hansen.pdf

Harms, H.; Schlosser, D.; Wick, L. Y. Untapped potential: Exploiting fungi in bioremediation of hazardous chemicals. *Nat Rev Microbiol* 2011, 9, 177–192.

Hofrichter, M.; Scheibener, K.; Schneegab, I.; Fritsche, W. Enzymatic combustion of aromatic and aliphatic compounds by manganese peroxidase from *Nematoloma frowardii*. *Appl Environ Microbiol* 1998, 64, 399–404.

Isikhuemhen, O. S.; Anoliefo, G. O.; Oghale, O. I. Bioremediation of crude oil polluted soil by the white rot fungus, *Pleurotus tuberregium* (Fr.) Sing. *Environ Sci Pollut Res Int* 2003, 10(2), 108–112.

Kim, Y. H.; Lee, J.; Ahn, J. Y.; Gu, M. B.; Moon, S. H. Enhanced degradation of an endocrine-disrupting chemical, butyl benzyl phthalate by *Fusarium oxysporum* f. sp. *pisi* cutinase. *Appl Environ Microbiol* 2002, 68(9), 4684–4688.

Kristanti, R. A.; Hadibarata, T.; Toyama, T.; Tanaka, Y.; Mori, K. Bioremediation of crude oil by white rot fungi *Polyporus* sp. S133. *J Microbiol Biotechnol* 2011, 21(9), 995–1000.

Linos, A.; Steinbüchel, A. Biodegradation of natural and synthetic rubbers. In *Biopolymers, Vol. 2*, Koyama, T., Steinbüchel, A., Eds.; Wiley-VCH: Sendai, Japan, 2001, pp 321–334.

Magan, N.; Fragoeiro, S.; Bastos, C. Environmental factors and bioremediation of xenobiotics using white rot fungi. *Mycobiology* 2010, 38(4), 238–248.

Mali, A.; Pandit, V.; Majumder, D. R. Biosorption and desorption of zinc and nickel from wastewater by using dead fungal biomass of *Aspergillus flavus*. *Int J Tech Res Appl* 2014, 2(6), 42–46.

Mali, A.; Pandit, V.; Majumder, D. R. Biosorption of heavy metals by dead fungal biomass. *IJCSEIERD* 2014, 4(3), 11–20.

Martínková, L.; Vejvoda, V.; Kaplan, O.; Kubáč, D.; Malandra, A.; Cantarella, M.; Bezouška, K.; Křen, V. Fungal nitrilases as biocatalysts: Recent developments. *Biotechnol Adv* 2009, 27(6), 661–670.

Mori, T.; Kitano, S.; Kondo, R. Biodegradation of chloronaphthalenes and polycyclic aromatic hydrocarbons by the white-rot fungus *Phlebia lindtneri*. *Appl Microbiol Biotechnol* 2003, 61, 380–383.

Nakagawa, A.; Osawa, S.; Hirata, T.; Yamagishi, Y.; Hosoda, J.; Horikoshi, T. 2,4-dichlorophenol degradation by the soil fungus *Mortierella* sp. *Biosci Biotechnol Biochem* 2006, 70(2), 525–527.

Norton, J. M. *Fungi for bioremediation of hydrocarbon pollutants*. University of Hawaii at Hilo: Hawaii, USA, 2012, pp 18–21.

Pickard, M. A.; Roman, R.; Tinoco, R.; Vazquez-Duhalt, R. Polycyclic aromatic hydrocarbon metabolism by white rot fungi and oxidation by *Coriolopsis gallica* UAMH 8260 laccase. *Appl Environ Microbiol* 1999, 65, 3805–3809.

Potin, O.; Veignie, E.; Rafin, C. Biodegradation of polycyclic aromatic hydrocarbons (PAHs) by *Cladosporium sphaerospermum* isolated from an aged PAH contaminated soil. *FEMS Microbiol Ecol* 2004, 51, 71–78.

Quintero, J. C.; Moreira, M. T.; Feijoo, G.; Lema, J. M. Screening of white rot fungal species for their capacity to degrade lindane and other isomers of hexachlorocyclohexane (HCH). *Cien Inv Agr* 2008, 35(2), 159–167.

Raghukumar, C. Fungi from marine habitats: An application in bioremediation. *Mycol Res* 2000, 104(10), 1222–1226.

Rashmi, P. A.; Joseph, D.; Joy, J. T.; Mathew, L. Bioactivities of *Colletotrichum gloeosporioides* - an endophyte of *Justicia adhatoda*. *Res J Pharm Biol Chem Sci* 2014, 5, 552–558.

Rauch, M. E.; Graef, H. W.; Rozenzhak, S. M.; Jones, S. E.; Bleckmann, C. A.; Kruger, R. L.; Naik, R. R.; Stone, M. O. Characterization of microbial contamination in United States Air Force aviation fuel tanks. *J Ind Microbiol Biotechnol* 2006, 33(1), 29–36.

Rhodes, C. J. Mycoremediation (bioremediation with fungi) - growing mushrooms to clean the earth. *Chem Spec Bioavailab* 2014, 26(3), 196–198.

Russell, J. R.; Huang, J.; Anand, P.; Kucera, K.; Sandoval, A. G.; Dantzler, K. W.; Hickman, D.; Jee, J.; Kimovec, F. M.; Koppstein, D.; Marks, D. H.; Mittermiller, P. A.; Núñez, S. J.; Santiago, M.; Townes, M. A.; Vishnevetsky, M.; Williams, N. E.; Vargas, M. P. N.; Boulanger, L. A.; Bascom-Slack, C.; Strobel, S. A. Biodegradation of polyester polyurethane by endophytic fungi. *Appl Environ Microbiol* 2011, 77(17), 6076–6084.

Sierra-Alvarez, R. Removal of copper, chromium and arsenic from preservative-treated wood by chemical extraction-fungal bioleaching. *Waste Manage* 2009, 29(6), 1885–1891.

Singh, S. S. *Waste treatment by white rot fungi*. Bioinfo Publications: Mumbai, India, 2013.

Singh, S. B.; Lal, S. P.; Pant, S.; Kulshrestha, G. Degradation of atrazine by an acclimatized soil fungal isolate. *J Environ Sci Health B* 2008, 43(1), 27–33.

Spain, J. C. Biodegradation of nitro-aromatic compounds. *Ann Rev Microbiol* 1995, 49, 523–525.

Steffen, K. T.; Hatakka, A.; Hofrichter, M. Degradation of benzo[a]pyrene by the litter-decomposing basidiomycete *Stropharia coronilla*: Role of manganese peroxidase. *Appl Environ Microbiol* 2003, 69, 3957–3964.

Takey, M.; Shaikh, T.; Mane, N.; Majumder, D. R. Bioremediation of xenobiotics: Use of dead fungal biomass as biosorbent. *Int J Res Eng Technol* 2014, 3(1), 565–570.

Tian, L. S.; Dai, C. C.; Zhao, Y. T.; Zhao, M.; Yong, Y. H.; Wang, X. X. The degradation of phenanthrene by endophytic fungi *Phomopsis* sp. single and co-cultured with rice. *Chin Environ Sci* 2007, 27(6), 757–762.

Tišma, M.; Zelić, B.; Vasić-Rački, D. White-rot fungi in phenols, dyes and other xenobiotics treatment - a brief review. *Croat J Food Sci Technol* 2010, 2(2), 34–47.

Tuomela, M.; Winquist, E.; Anasonye, F.; Räsänen, M.; Steffen, K. Bioremediation of TNT contaminated soil with fungi. *European Conference of Defence and the Environment: Conference Proceedings*, 2013, 141, http://www.defmin.fi/files/2512/15_Steffen_Kari.pdf

Van Hamme, J. D.; Wong, E. T.; Dettman, H.; Gray, M. R.; Pickard, M. A. Dibenzyl sulfide metabolism by white rot fungi. *Appl Environ Microbiol* 2003, 69(2), 1320–1324.

Varasaritha, Y.; Avasn, M.; Mukkanti, K. Potential fungi for bioremedation of industrial effluents. *Bioresources* 2010, 5(1), 8–22.

Young, D. *Bioremediation with white-rot fungi at Fisherville Mill: Analyses of gene expression and number 6 fuel oil degradation*. Mosakowski Institute for Public Enterprise, Clark University: Worcester, Massachusetts, 2012, pp 1–82.

INDEX

Printed and bound by CPI Group (UK) Ltd, Croydon, CR0 4YY

23/10/2024

01777696-0016